AF847088

Jie Zhang, Wei Qin, Lihui Wu, Junliang Wang, Youlong Lv, Xiaoxi Wang
Wafer Fabrication

Also of interest

Productivity and Organizational Management
C. Machado, J. P. Davim (Eds.), 2017
ISBN 978-3-11-035545-1, e-ISBN (PDF) 978-3-11-035579-6,
e-ISBN (EPUB) 978-3-11-038661-5

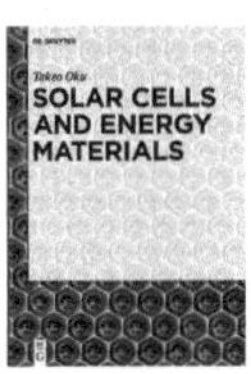

Solar Cells and Energy Materials
T. Oku, 2016
ISBN 978-3-11-029848-2, e-ISBN (PDF) 978-3-11-029850-5,
e-ISBN (EPUB) 978-3-11-038106-1

Innovation Management
C. Machado, J. P. Davim (Eds.), 2015
ISBN 978-3-11-035872-8, e-ISBN (PDF) 978-3-11-035875-9,
e-ISBN (EPUB) 978-3-11-038667-7

Industrial Software Applications
R. Geisler, 2015
ISBN 978-3-11-037098-0, e-ISBN (PDF) 978-3-11-037099-7,
e-ISBN (EPUB) 978-3-11-039678-2

Jie Zhang, Wei Qin, Lihui Wu, Junliang Wang,
Youlong Lv, Xiaoxi Wang

Wafer Fabrication

Automated Material Handling Systems

DE GRUYTER

華中科技大學出版社
http://www.hustp.com
中国 · 武汉

Authors
Prof. Jie Zhang
College of Mechanical Engineering
Donghua University
Shanghai
China
mezhangjie@dhu.edu.cn

Wei Qin
Institute of Intelligent Manufacturing
and Information Engineering
Shanghai Jiao Tong University
Shanghai
China
wqin@sjtu.edu.cn

Lihui Wu
School of Mechanical engineering
Henan University of Technology
Zhengzhou
China
wulihui@haut.edu.cn

Junliang Wang
Institute of Intelligent Manufacturing
and Information Engineering
Shanghai Jiao Tong University
Shanghai
China
starsmall@sjtu.edu.cn

Youlong Lv
College of Mechanical Engineering
Donghua University
Shanghai
China
lvyoulong@dhu.edu.cn

Xiaoxi Wang
emilywang0210@gmail.com

ISBN 978-3-11-048690-2
e-ISBN (PDF) 978-3-11-048747-3
e-ISBN (EPUB) 978-3-11-048723-7

Library of Congress Cataloging-in-Publication Data
Names: Zhang, Jie, 1963 September 21- author.
Title: Wafer fabrication : automatic material handling system / Jie Zhang,
Wei Qin, Lihui Wu, Junliang Wang, Youlong Lv, Xiaoxi Wang.
Description: Berlin ; Boston : Dr Gruyter, [2018] | Includes bibliographical references.
Identifiers: LCCN 2018007853| ISBN 9783110486902 (hardcover) | ISBN 9783110487473 (pdf) |
ISBN 9783110487237 (epub)
Subjects: LCSH: Semiconductor wafers–Automatic control. | Manufacturing
processes–Automatic control.
Classification: LCC TK7871.85 .Z43 2018 | DDC 621.39/5–dc23 LC record available at
https://lccn.loc.gov/2018007853

Bibliographic information published by the Deutsche Nationalbibliothek
The Deutsche Nationalbibliothek lists this publication in the Deutsche Nationalbibliografie;
detailed bibliographic data are available on the Internet at http://dnb.dnb.de.

© 2018 Walter de Gruyter GmbH, Berlin/Boston
Typesetting: Integra Software Services Pvt. Ltd.
Printing and binding: CPI books GmbH, Leck
Cover image: Patrick landmann/science photo library

www.degruyter.com

Preface

At present, the semiconductor manufacturing industry has become the high-tech key industry that has been promoted and developed in China. With the rapid development of the semiconductor manufacturing technology, wafer sizes have developed from 6 inch, 8 inch, 12 inch, to 18 inch. For a typical 300 mm wafer fabrication system, there are usually 200–600 processing operations for each wafer. As many as hundreds of vehicles are used to fulfil the material handling work. The total delivery distance of each wafer reaches 8–10 miles. Therefore, the automated material handling system with high operating efficiency has been widely adopted in order to improve the wafer processing machine utilization and shorten the chip delivery due date to ensure that wafer manufacturing enterprises have high competitiveness in the market.

In the 1990s, well-known experts in the field of wafer fabrication, such as Kumar and Leachman in the United States, had begun to study scheduling problems in wafer fabrication systems. Their results have led to the recent interest in the scheduling problems in wafer fabrication systems. Although there are a lot of methods reported in the literature, the scheduling optimization problems in automated material handling systems have been rarely studied. These scheduling problems are NP-hard problems due to their large-scale, dynamic and real-time features.

We have worked on investigating theories and techniques of modeling and scheduling in automated material handling systems in semiconductor wafer fabrication systems. In particular, with the support of National Natural Science Foundation Programs and National High Technology Research and Development Programs, we have published a large number of papers in the field of scheduling in automated material handling systems. This book is a systematic summary of these research results. This book presents methods and technologies of scheduling problems in automated material handling systems comprehensively and systematically.

We are grateful to Yinbin Sun and Cong Pan for their assistances in the preparation of this book. Meanwhile, Jungang Yang, Qiong Zhu, Peng Zhang, Xiaolong Yang, Tengda Li and Yaping Zhou completed many auxiliary works, and we thank all of them. We wish to acknowledge a large number of references in the completion of the manuscript. We are sorry for any errors if there is negligence.

In addition, we very much appreciate the comments by Prof. Zhengcheng Duan at Huazhong University of Science and Technology, Prof. Yiming Rong at Tsinghua University, and Prof. Guoquan Huang at the University of Hong Kong.

The writing of this book has been supported by the National Nature Science Foundation of China (Grant Nos 51435009, U1537110, and 51275307), and by the National Science and Technology Academic Works Publishing Fund.

The theories, methods and applications of scheduling for automated material handling systems in semiconductor wafer fabrication systems are rapidly developing and have become a hot topic in the field of scheduling. If you have any questions about shortcomings and mistakes of this book, please do not hesitate to contact us.

https://doi.org/10.1515/9783110487473-201

Contents

1 Semiconductor wafer fabrication system

1.1 Semiconductor Manufacturing Industry

In the twenty-first century, modern technology experienced rapid development, especially the extensive application of electronic technology, which marked the human society entering the information age. In this "new economy" era, the information industry has become one of the largest and fastest growing industries in the world's economy. Information and Internet technologies have a revolutionary impact on human economic and social life, while the semiconductor industry is its foundation. Integrated circuits (IC), microwave devices and optoelectronic devices made by semiconductor materials are the core products of the electronics industry and the information industry. They have been widely applied in people's daily lives and exist all over the world. They have infiltrated into digital entertainment, mobile communications, e-commerce, automotive industry, medical equipment, aerospace and so on. In fact, every step of human society's development nowadays is dependent on semiconductor chip technology. It can be said that the country that owns the semiconductor manufacturing industry will embrace the future of the world [1].

1.1.1 Development and Current Status of the Semiconductor Manufacturing Industry

The semiconductor manufacturing industry is developed on the basis of vacuum tube electronics, radio communications and solid physical technology that emerged in the first half of the twentieth century. After World War II, scientists at Bell Telephone Laboratory began to study the solid-state silicon and germanium semiconductor devices. In 1947, they invented the first solid-state transistor, which led to the solid materials and technology-based modern semiconductor industry. Robert Nois in Fairchild Semiconductor and Jack Kilby in Texas Instruments independently invented the IC in 1959. The concept of an IC is intended to interconnect different components on planar silicon. This concept motivates engineers to design more sophisticated electronic circuits to meet customers' new demands [2].

With the development of science and integrated technology, the carrier of ICs moved from the era of ordinary silicon material into the wafer era. Semiconductor chips are manufactured through a number of processes in the wafer to create a large number of complementary metal oxide semiconductor components that are connected with a metal wire to become a chip with logic function or storage function.

https://doi.org/10.1515/9783110487473-001

Manufacturing processes are becoming increasingly complex; meanwhile, the semiconductor manufacturing machine costs and semiconductor manufacturing plant construction costs are increasing. More European and American companies have moved out of the semiconductor manufacturing field and focused on design. Semiconductor manufacturing industry enterprises can be classified into three major classes: semiconductor equipment suppliers, semiconductor design company (fabless mode) and semiconductor manufacturing company (foundry mode). Of course, there are some companies, such as Intel Corporation, which set up their own manufacturing plants to produce advanced chips in order to protect the key technology of semiconductor chips. For this kind of manufacturers, its manufacturing process is at least one generation ahead of the manufacturing process of a professional pure semiconductor factory.

Wafer manufacturing is the core and foundation of the semiconductor IC industry, which has shown a positive growth trend for a very long time. The Global Semiconductor Consortium (GSA) and the market research firm IC Insights jointly conducted a survey to study the growth trend in the industry. The survey report indicated that the percentage of ICs manufactured by wafer foundries in the entire chip market increased from 21% in 2004 to 24% in 2009 and jumped to 37% in 2014. This suggests that the semiconductor industry is transitioning from Integrated Device Manufacturers (IDM) to fab-site or fabless plant-based models, which is currently in the steep part of the industrial life cycle S-curve. GSA and IC Insights expect sales of ICs manufactured by foundries to

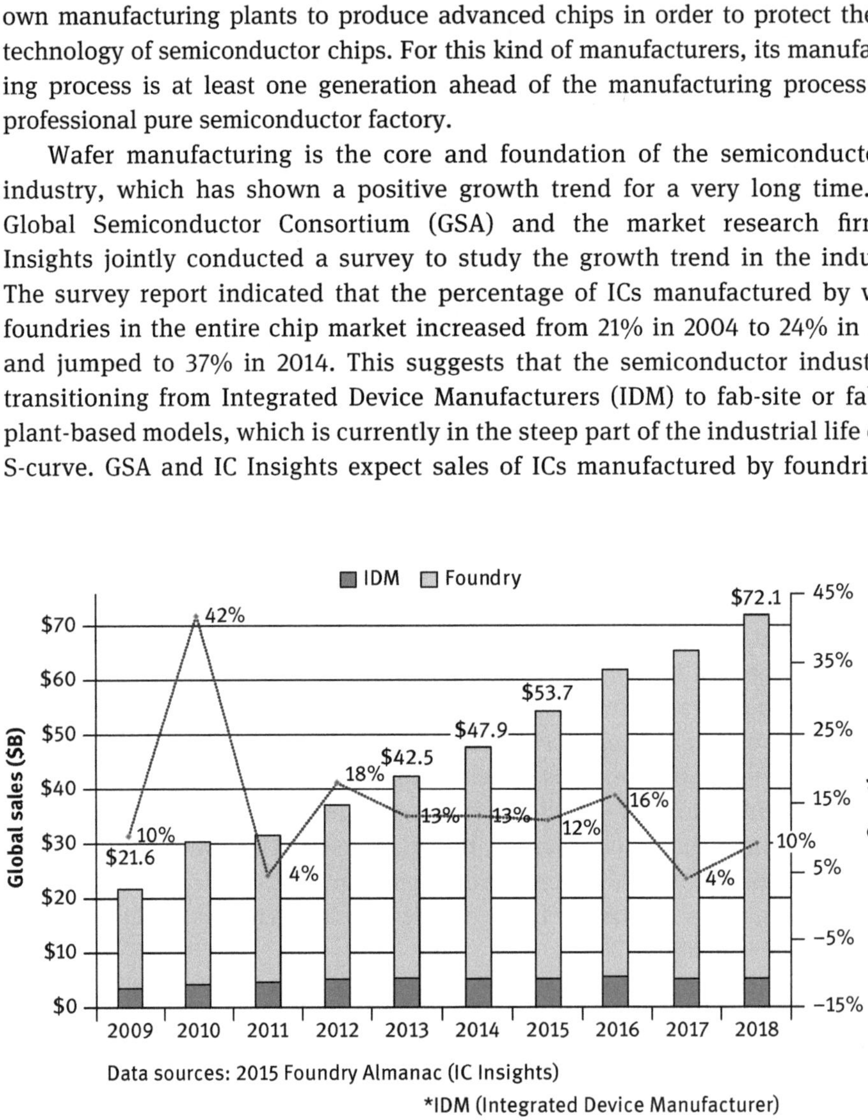

Figure 1.1: Forecast of sales and growth rates for global foundries (2015–2018).

approach 46% of total industry sales in 2018. Figure 1.1 shows the forecast of sales and growth rates for the global foundries in 2015–2018. Encouraged by demand, foundry companies are also gradually expanding their capacities, and actively developing to become super-factory (MegaFab) models. According to industry consensus, the so-called super factory refers to a 12-inch wafer factory with the monthly production capacity of 100,000–15 million and the amount of $70–$80 billion investment.

Semiconductor ICs are applied in almost all electronic equipment. Because of its strong promoting role and large multiplier effect, it is of great significance to computers, household appliances, digital electronics, automation, communications, aerospace and other industries. According to the China Semiconductor Industry Association statistics in 2010–2014, semiconductor IC industry sales in China maintained a double-digit, high growth rate. The proportion of GDP is increasing, as shown in Figure 1.2. Sales revenue data of 2015 must be available now. At present, the output value of the information industry in developed countries occupies 40–60% of the total output value of their national economy, and 65% of the GDP growth is related to the IC industry. Therefore, we can seize the initiative of the national economic development by further developing the IC industry. The semiconductor IC industry in China is also rapidly developing. China has developed a number of highly competitive enterprises including SMIC and Huahong Grace, which have entered the global top ten foundry business (shown in Table 1.1).

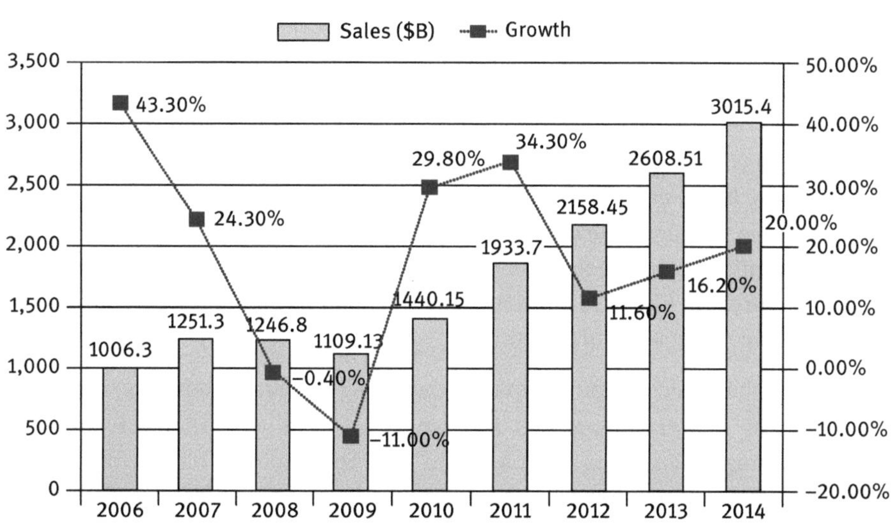

Figure 1.2: IC industry sales in China (2006–2014).

Table 1.1: The global foundry business ranking (2013).

Ranking	Company	Country/region	Sales (100,000,000$)	Year-on-year growth (%)
1	TSMC	Taiwan, China	198.5	17
2	GlobalFoundries	Milpitas, CA, USA	42.61	6
3	UMC	Taiwan, China	39.59	6
4	Samsung	Korea	39.50	15
5	SMIC	Mainland, China	19.73	28
6	Powerchip	Taiwan, China	11.75	88
7	Vanguard	Taiwan, China	7.13	23
8	Huahong Grace	Shangai, China	7.10	5
9	Dongbu	Korea	5.70	6
10	TowerJazz	Israel	5.09	−20

1.1.2 Future Challenges in the Semiconductor Manufacturing Industry

In the past half century, the semiconductor industry has undergone tremendous changes. In the coming two decades, the semiconductor industry will continue to face dramatic changes and challenges. Three aspects are presented as follows [3]:

1. Technological change is accelerating
   The semiconductor chip industry began to develop rapidly in 1960, and the smallest feature size of the IC has been drastically reduced in less than 60 years. The size of the chip reduced from 50 μm in 1960 to 0.18 μm in the early twenty-first century. Up to now, the 40 nm IC has been very stable in the market. The 28 nm IC has also officially entered mass production and now it seems to have evolved to 22 nm, 16 nm and even more fine scale. At present, the 22 nm technology in Intel (the world's most advanced technology company) has entered mass production (leading industry one–two years); 14 nm process is also mature. Rapid technological change leads to the emergence of new products and shortens products' life cycles. Therefore, the primary challenge of semiconductor manufacturers is to achieve the maximum profit in the increasingly shortened product life cycle by shortening the manufacturing cycle and reducing production costs at a molar speed.

2. The size of the semiconductor chip manufacturing industry needs to be expanded
   As the cost of developing and implementing next-generation processes is increasing rapidly, leading companies in the industry must be the forefront of semiconductor process development, and only large companies can afford these costs. Based on this, the acquisition of wafer industry has become the focus of the industry since 2008. Tower Semiconductor Ltd acquired Jazz Semiconductor Inc.

in 2008. Global Foundries acquired Chartered Semiconductor in 2009, and became the second largest wafer foundry. NEC and Manulife Semiconductor (leading chipmakers in Japan) have also merged in 2010. This wave of mergers and acquisitions has also affected China. Therefore, the important development trend of the future semiconductor manufacturing industry is to enhance its scale.

3. Diversification of market demand
The dependence on and demand for digital electronic products in modern society will continue to rise rapidly in terms of quantity, quality and diversity. Technological innovation shortens the life cycles of semiconductor products, and fierce market competition reduces the profit in the semiconductor chip unit. In the shortest possible time, enterprises have to meet the dynamic demands of market diversification to produce a large number of new high-quality products to meet customer requirements and to quickly seize the market in order to earn the maximum profits. Therefore, one of the major challenges of semiconductor manufacturing enterprises is to effectively integrate internal and external resources in order to use the cost of less varictics and mass production to meet the demand of multi-species and mass production.
The development of the semiconductor industry is of great strategic significance in China; however, it is still in its early stages and faces challenges in technical expertise and has deficiencies in management. The most effective way to grow in the field would be to explore the development path by combining technology and management, which would actively promote the development of the semiconductor industry in China.

1.2 Processing of Semiconductor Chips

The entire semiconductor chip manufacturing process consists of five phases: wafer preparation, wafer fabrication, wafer sort, chip packaging and chip final measurement. Among them, wafer fabrication and wafer sort are incorporated as "front-end operations"; chip packaging and chip final measurement are referred to as "back-end operations" (shown in Figure 1.3) [4].

In a semiconductor wafer fabrication system (SWFS), dozens or hundreds of specific IC chips are fabricated on a wafer with silicon as a substrate. The operational risks involved include a large number of workloads, large quantities of expensive processing machines and production steps, and a high degree of re-entrant flows.

1.2.1 Wafer Preparation

Silicon is the main semiconductor material used to make IC chips and the most important material in the semiconductor industry. High-purity silicon used to make

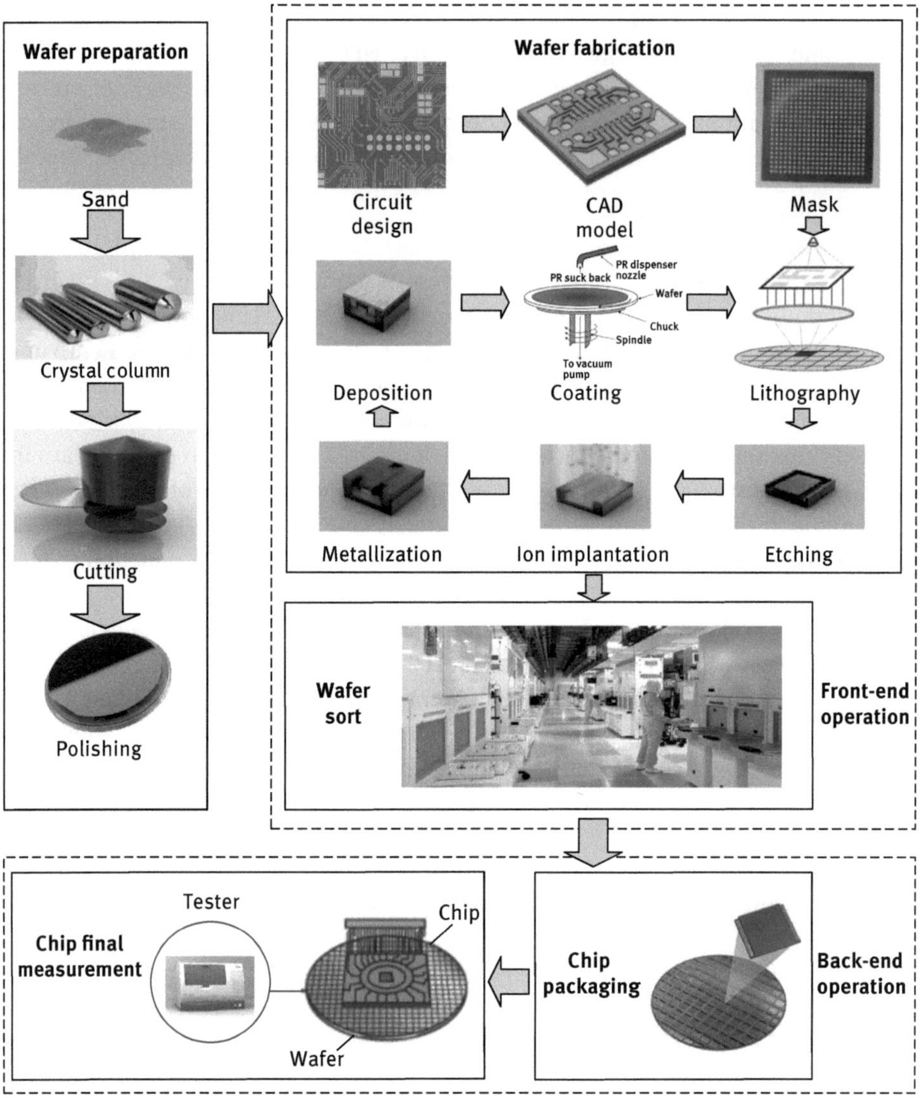

Figure 1.3: The main steps in semiconductor chip manufacturing process.

the chip is called semiconductor-grade silicon; sometimes, it is referred to as electronic-grade silicon [4]. The basis of wafer preparation is to convert a polycrystalline semiconductor-grade silicon block to a large single-crystal silicon, called a silicon ingot. The head and tail of the monocrystalline silicon rod are cut off and then the rod is mechanically trimmed to the appropriate diameter in order to obtain a "silicon rod" with a suitable diameter and a certain length. Since the silicon is very hard, a diamond saw is used to cut the silicon rod into wafer thin slices of equal thickness. The wafer is then ground in order to reduce the saw marks and damage to the front and back surfaces of the wafer and to thin the wafer. This process is called milling, after which the wafer is cleaned using chemicals, (i.e. sodium hydroxide, acetic acid and nitric acid to clean the surface of the wafer) in order to reduce damage and cracks generated during grinding. Thereafter, in the chamfering process the edges of the wafer are rounded to reduce the possibility of damage during future circuit fabrication. Finally, the wafer surfaces are polished to reduce surface roughness, thereby further reducing the possibility of breakage. The wafer is formed after these steps.

1.2.2 Wafer Fabrication

Wafer fabrication is the core of semiconductor chip manufacturing. Wafer manufacturing process is similar to building a house. The wafer is built up layer by layer, and these layers are distinguished by the lithography process. Each layer includes deposition, lithography, etching, ion implantation, electroplating and other processes. The number of process steps varies with the product. In general, the production process of a product consists of at least hundreds of processes [5].

In the wafer fabrication process, the most important feature is its re-entrant characteristic [6]. In 1993, Kumar proposed Re-entrant Lines (unlike Job shop and Flow shop) according to the typical characteristics of the regular reentries of the machining paths in the semiconductor manufacturing system. A product in a re-entrant manufacturing system repeatedly revisits the same machine or the same machine group at different processing stages of a process route.

1. Deposition

   The first step in fabricating a wafer is to deposit a layer of non-conductive silicon dioxide film on the wafer. In the subsequent manufacturing process, the silicon dioxide layer is grown; the deposition process will be carried out many times. The silicon dioxide film forming technique includes physical vapor deposition (PVD) and chemical vapor deposition (CVD). The PVD technique is to apply thermal or kinetic energy to the source of the material to be deposited to decompose it into an aggregate of atoms or atoms, to impinge on the wafer surface, to bond or agglomerate to form a film. The PVD technique consists of three methods: resistance heating deposition method, electron gun evaporation method and sputtering method. The CVD technique is to introduce the reaction gas into the high temperature furnace to produce a chemical effect on the wafer surface with the gaseous chemical raw material, and to deposit a thin film on the wafer surface.

2. Lithography

   Lithography is the basis to build up a semiconductor Metal-Oxide-Semiconductor Field-Effect Transistor (MOS) tube and a circuit on a flat wafer, which contains many steps and processes. First, the silicon is coated with a layer of corrosion-resistant photoresist, and then light is let through a hollow mask engraved with a circuit pattern to irradiate the wafer. The parts to be irradiated (such as the source and drain regions) are deteriorated, and the gate area is not irradiated and its photoresist remains sticky. The next step is to clean the wafer with a corrosive liquid, and the deteriorated photoresist will be removed to expose the underlying wafers. The gate area is not affected due to protection by the photoresist.

3. Etching
   Etching is a technology to remove the material by using chemical reactions or physical impact. Etching techniques contain wet etching and dry etching. Wet etching is to use chemical solutions to etch by chemical reactions, while dry etching is generally plasma etching. Plasma etching may be the physical effect on the surface of the chip by high-energy particles in the plasma, it may be a complex reaction between the active radicals in the plasma and the atoms on the chip surface, or it may be a combination. After the etching process is completed, the photoresist on the wafer surface is removed with ionized oxygen. Then the wafer is cleaned with a chemical reagent.

4. Ion implantation
   Ion implantation (ion doping) is a technique by which impurities are injected into a specific region of the semiconductor assembly in an ion form to obtain accurate electronic properties. These ions must first be accelerated to a sufficiently high energy and velocity to penetrate (inject) the film to reach a predetermined implant depth. The ion implantation process can precisely control the impurity concentration in the implantation area. Basically, the impurity concentration (dose) is controlled by the ion beam (the total number of ions in the ion beam) and the scanning rate (the number of times the wafer passes through the ion beam). And the depth of ion implantation is determined by the size of the ion beam energy.

5. Electroplating
   The function of the chip is achieved by the connection of the transistors. The conductive metal (usually aluminum) is deposited on the wafer surface by a metallization process. Unwanted metals are removed by using photolithography and etching processes to leave circuit metal between connected transistors. Complex wafers need a lot of insulators. A chip must be connected to millions of conductive lines, which include horizontal connections in the same layer and the vertical connections between layers.

1.2.3 Wafer Sort

The wafer test results are determined by wafer production and quality verification, also known as die sort or wafer sort. During testing, the electrical capacity and circuit function of each chip are detected. The wafer test includes the following three objectives: first, before the wafer is sent to the packaging plant, the qualified chips are identified. Second, the electrical parameters of the device/circuit are assessed, and engineers need to monitor the distribution of parameters to obtain the quality of the process. Third, the chip yield is detected. The yield test provides full performance feedback to wafer producers. Qualified chips and defective products on the wafer are recorded on the computer in the form of a wafer chart.

In the test, the wafer is fixed on a chuck with vacuum suction. The hair-like probe made of gold wire mounted on the test head is kept in contact with the pad on the grain to test its electrical characteristics. Unqualified grains are marked. When the wafer is cut in units of grain, the marked unqualified grain will be eliminated and will no longer enter the next process so as not to increase manufacturing costs.

As the size of the IC continues to increase and the line width shrinks, the circuit density continues to increase. The cost of wafer testing is high and continues to rise because longer test times, more sophisticated power supplies, mechanical devices and computer systems are required to perform test work and monitor test results. The visual inspection system is also more sophisticated and expensive as the chip size expands. The chip designer is asked to introduce the test pattern into the storage array. Test designers are exploring how to make the testing process more streamlined and effective. For example, simplified test procedures are adopted after the chip parameters are evaluated. It is also possible to interlace the chips on the wafer or to test multiple chips at the same time.

1.2.4 Chip Packaging

The purpose of chip packaging is to protect the chip from the external environment and to provide good working conditions, which causes the IC to have a stable and normal function. Specifically, the chip and other elements that are arranged in the frame or substrate are fixed and connected by using membrane technology and micro-processing technology in order to constitute the overall three-dimensional structure of the process.

The package provides protection for the chip. Non-encapsulated IC chips are generally not directly available. Chip packaging features include:

1. Transmission power, which mainly refers to the power supply voltage distribution and conduction.
2. The transmission of the circuit signal is mainly to minimize the delay of the electrical signal. When wiring, the path between the signal line and the chip and the path through the encapsulated I/O interface should be as short as possible.

3. To provide cooling channels, in a variety of chip packaging, distribution of the heat gathered by long-term works of components should be taken into consideration.
4. Structural protection and support, chip packaging for the chip and other connecting parts can provide a solid and reliable mechanical support that can adapt to a variety of working environments and changing conditions.

1.2.5 Chip Final Measurement

After the packaging process of the semiconductor chip is completed, the chip package testing is carried out. That is, the structures and electrical functions of the semiconductor components are confirmed so as to ensure that the semiconductor components meet the requirements of the system, which is also known as the final measurement. Package testing includes both quality and reliability testing. Quality inspection is mainly to detect the availability of the chip after packaging, i.e., the chip performance. Reliability testing is to test encapsulation reliability related parameters. The general reliability test project has six parts: pretreatment test, temperature cycling, thermal shock test, high-temperature storage test, temperature and humidity test, and high-temperature boiling test.

1.3 Constitution of the Semiconductor Wafer Fabrication System

In the entire semiconductor chip manufacturing process, wafer fabrication is the core stage. Its role is to print multi-layer circuits on silicon or gallium-arsenic compound wafers. In order to achieve this goal, in accordance with a certain pattern, a variety of interrelates involved in the wafer fabrication process are integrated with the interaction of the relevant elements by the wafer fabrication system.

According to the physical composition, a wafer fabrication system consists of machine layer, unit layer and system layer (shown in Figure 1.4). The machine layer includes processing machines, buffers, material handling equipment and so on. In general, semiconductor processing machines can be further classified as SPMs (single lot processing machines) and BPMs (batch processing machines). In order to avoid contamination of the wafers during the production process, the wafers are delivered with a standard container in a manufacturing system. In a semiconductor manufacturing system, quantities of wafers are grouped into a standard container, which is called a lot. A lot usually contains 25 wafers. In a wafer fabrication system, a part (e.g., a wafer) may revisit the same machine pool multiple times throughout the manufacturing process.

The unit layer of the wafer fabrication system includes an SPM group, a BPM group and a material handling equipment group. The SPM group consists of several

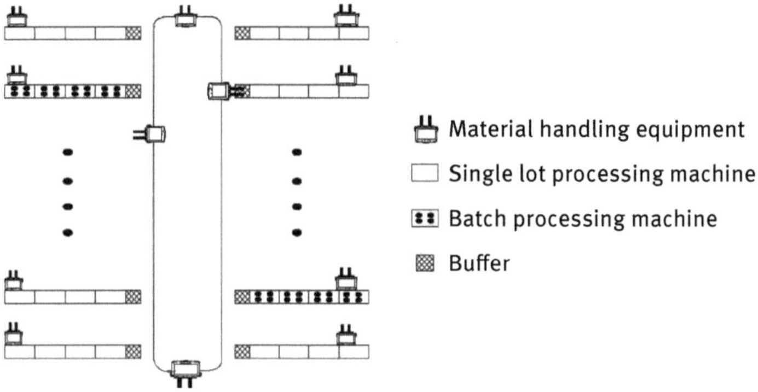

Figure 1.4: The physical layout of a wafer fabrication system.

SPMs with the same processing function. The BPM group consists of several BPMs with the same machining function. The material handling equipment group consists of buffer, material handling equipment and other materials handling equipment.

The wafer fabrication system is composed of two parts: a wafer processing system and a material handling system.

1.3.1 Wafer Processing System

The wafer processing system comprises several main work areas, such as lithography and etching areas. Each area is responsible for completing a certain number of processing operations. Each work area consists of several workstations, equipment group or bay. Each workstation contains several processing machines, work in process (WIP) buffers and transfer mechanisms. In the non-automated wafer processing system, there will be corresponding machine operators. The wafer processing machines in the workstation have substantially the same function, or they can be combined to complete a particular processing step. The entire wafer processing system can be seen as a four-layer architecture consisting of a wafer fab, a machining area, several workstations and equipment, as shown in Figure 1.5. The fab contains four areas: lithography processing zone, etching processing zone, ion implantation processing zone and deposition processing zone.

In general, semiconductor processing machines can be further classified as SPMs and BPMs.

In a semiconductor manufacturing system, quantities of wafers are grouped into a standard container, which is called a lot. The wafers in a lot are all the same type of products and travel as a unit between machines. Each SPM is capable of performing manufacturing operations on many wafers at a time and processing wafers one by one. After all the wafers in a lot have been completed, this lot of wafers will leave the SPM together. SPMs can be further classified into three classes: single lot – single

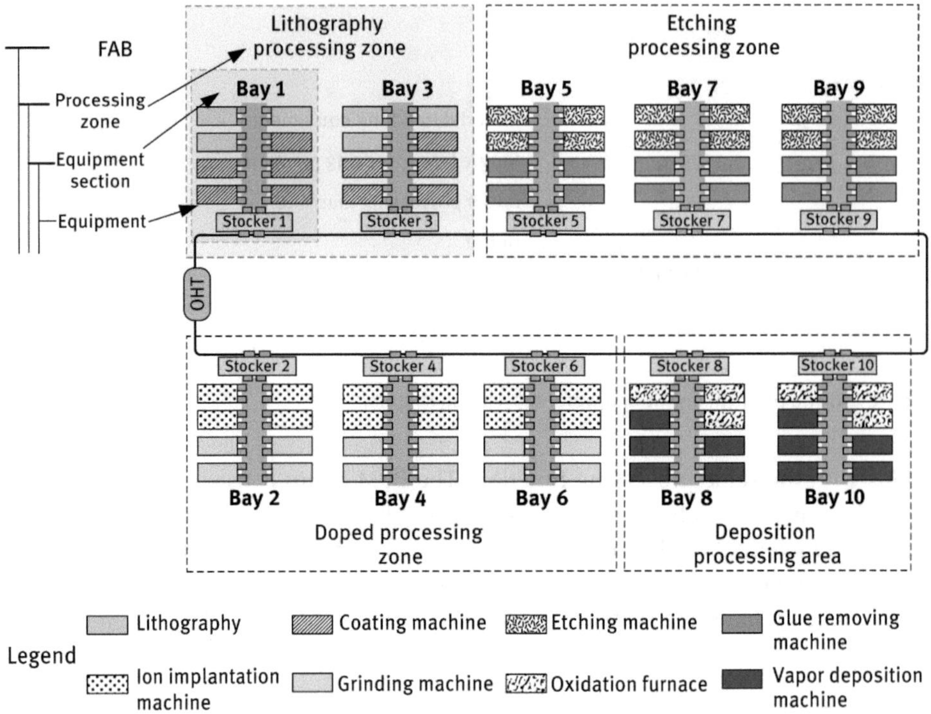

Figure 1.5: The four-layer architecture diagram of the wafer processing system.

piece processing machines (Figure 1.6), single lot–multiple pieces parallel processing machines (Figure 1.7) and MPM (single lot–multiple pieces multi-chamber processing machines). MPM can be subclassified as MPM_SC (multi-chamber processing machine with same chambers) and MPM_DC (multi-chamber processing machine with different chambers). For MPM_SC, wafers can be processed in any available chamber, one by one. For MPM_DC, many wafers are loaded, and should be processed in several chambers successively, according to its processing sequence. A typical SPM includes RTP machine, ion implanter, CVD machine, coating and exposure device in lithography process, ion etcher and so on.

Each BPM is capable of simultaneously performing manufacturing operations on multiple lots of wafers with the same processing sequence. All the wafers in multiple lots will be processed together, and they will leave the BPM simultaneously. BPMs can be further classified into two classes. The first one is BPPM (batch parallel processing machines) (Figure 1.8), in which multiple lots of wafers are processed in parallel machines. The other one is WPM (wet-bench processing machine) (Figure 1.9), in which multiple lots of wafers are processed in serial machines. Typical BPM includes horizontal and vertical oxidation furnaces, metal-etching machines, wet etching machine and so on.

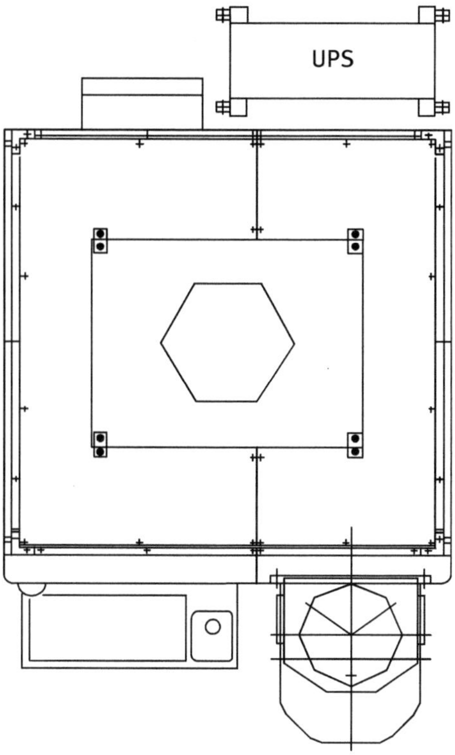

Figure 1.6: Single lot – single piece processing machine.

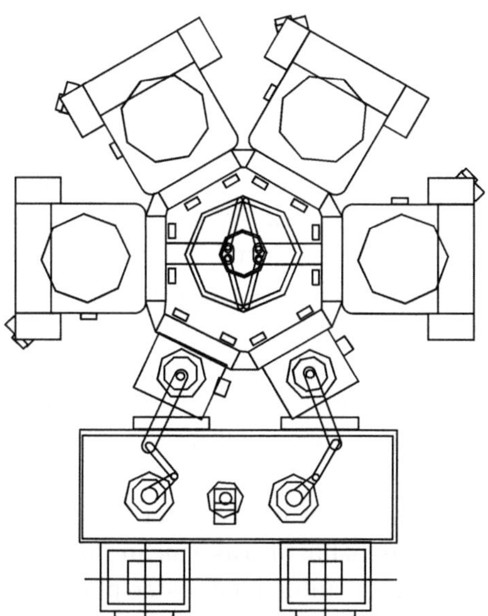

Figure 1.7: Single lot – multiple pieces multi-chamber processing machine.

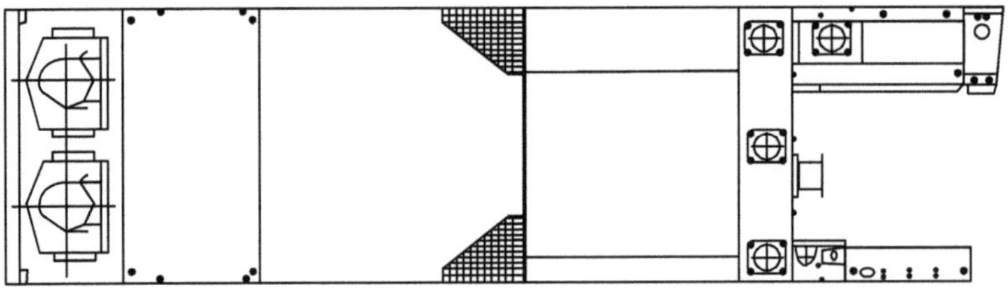

Figure 1.8: Batch parallel processing equipment.

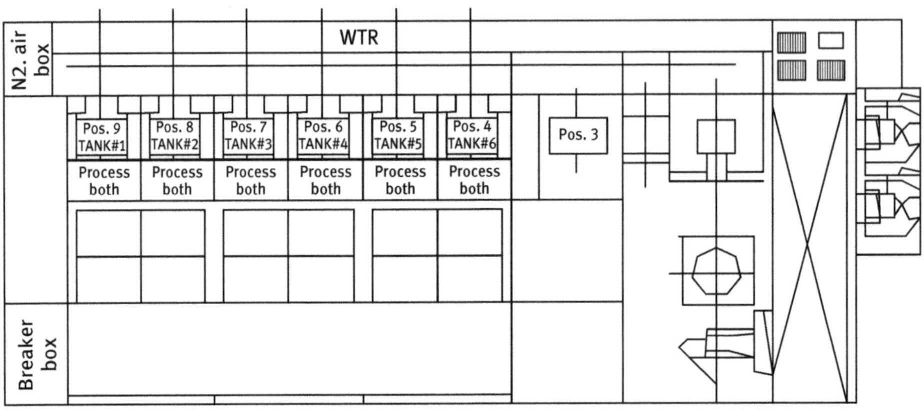

Figure 1.9: Wet-bench processing machine.

1.3.2 Material Handling System

The material handling system mainly includes buffer and material handling equipment. The buffer is a temporary storage system of various types. The most common storage system is the automatic shelf, which can be further subclassified as the Cartesian type and the carousel type. Another storage system is UTS (under track storage), which is able to improve the efficiency of AMHS (automated material handling system). The material handling equipment in semiconductor manufacturing systems consists of five types of equipment: vertical lifters, overhead tracks with transport vehicles, RGV (rail-guided vehicle), AGV (automated guided vehicle) and conveyors. The material handling system to transmit wafers in a production unit is called Intrabay AMHS, while the material handling system to transmit wafers among production units is called Interbay AMHS. AGV is widely applied as flexible and intelligent material handling equipment.

The AMHS is adopted in wafer fabs because it can be used not only as an automated warehouse for temporary storage of wafers but also as a means for rapid delivering among different work areas. In a highly automated wafer plant, the AMHS connects the relatively independent production units throughout the plant through the transport tracks so as to complete the automatic material transmission of the entire factory. The benefits include:

1. reduction of personnel

   With the rapid development of wafer processing technology and rapid increase in production, the traditional way requires a large number of material handling personnel. For example, the output of the wafer production line is 40,000 pieces per month, and 200 people are required to carry the material for each shift. The use of the AMHS can greatly reduce the number of staff who are responsible for material handling and improve the accuracy of material transmission, which has become the driving factor for the development of the AMHS.

2. reduction of labor intensity of operators

   The wafer container quality of a 200 mm wafer is about 4.5 kg and the wafer container quality of a 300 mm wafer is about 9 kg. Although a person can easily carry such a wafer container once or twice, it will cause repetitive stress damage. Therefore, when converting to a 300 mm wafer processing system, the use of the AMHS can greatly reduce the labor intensity of operators on the shop floor.

3. wafer damage risk reduction

   During wafer processing, the total value of wafers in a wafer container is between \$10,000 and \$1 million. In the traditional manual handling system there is possibility of wafer containers falling, thus causing damage to the wafers. The use of an AMHS can effectively improve the safety of wafer handling.

4. increase in machine utilization

   Traditional handling methods in a wafer factory can likely cause a loss of 10–20% because the machines are inactive while they wait for material to be delivered. In comparison, an AMHS can effectively shorten the material handling time, thereby improving machine utilization. According to statistics, the use of an AMHS to improve machine utilization can save several billion dollars during the life of the plant.

5. shortening the production cycle

   The AMHS can shorten the wafer handling time between machines and provide a set of automatic predictable work flow planning systems, thereby shortening the processing cycle. The processing cycle is a very important indicator for wafer fabs because faster product launches or shorter lead times can significantly affect the value of a product.

   The efficient performance of an AMHS can effectively improve the wafer factory's production capacity and improve the production machine utilization. The AMHS itself has a high handling accuracy; it can greatly reduce the production losses due to operator error. The AMHS has particular advantage for 300 mm wafer and above and is more widely used in large-size wafer fabrication plants.

1.4 Production Scheduling in the Semiconductor Wafer Fabrication System

With rapid changes in market demand and intensified international competition, the complexity of production scheduling in semiconductor manufacturing systems lies in the following aspects [7, 8]:

1. Large-scale production

   In a typical wafer fab, there are dozens of significant processes. Each process consists of 300–900 processing steps on more than 100 hundred machines. The average production cycle time in the production line is approximately 30–60 days by repeatedly implementing manufacturing processes including oxidation, deposition, metallization, lithography, etching, ion implantation, photo-resist strip, cleaning, inspection and metrology. In addition, various types of products are processed in a semiconductor production line at the same time, which are different in thousands of operation sequences; even if similar products have different versions. This large-scale production increases the complexity of semiconductor manufacturing. Furthermore, orders from different customers containing dozens of product types are processed in the semiconductor production line at the same time. Therefore, production scheduling problems become a great challenge in semiconductor manufacturing systems.

2. Re-entrant flow

   The nature of semiconductor manufacturing systems is re-entrant. In a traditional manufacturing system such as job shop and flow shop, the re-entrant flow occurs either in individual processes or in a few rework processes, which belongs to the local phenomenon. However, in a re-entrant semiconductor manufacturing system, a part (e.g., a wafer) may re-visit the same machine pool multiple times throughout the manufacturing process. At different stages in the re-entrant flow, a wafer has to compete with others to be processed in the same machine pool. In particular, two reasons make the production scheduling in re-entrant manufacturing systems different from that in traditional manufacturing systems. First, the hierarchical structure of semiconductor products is so strongly related that each layer is processed in similar operation sequences with different materials or precision. Hence, the re-entrant flow is adapted to process the next layer of the same product. Second, since machines required in wafer fabrication are very expensive, there is need to introduce re-entrant flow to improve machine utilization. Because of the re-entrant flow, wafers visit the same machine pool many times throughout the manufacturing process. This results in significant increase in the number of wafers to be scheduled in the same machine pool, which enlarges the solution space and increases the complexity of the production scheduling process.

3. Mixed processing mode
   In general, in semiconductor manufacturing systems, 25 or 50 wafers are grouped into a standard container, called a lot. Lots of wafers are processed in SPM and BPM. One lot of wafers can be processed in an SPM at a time, while lots of wafers with the same serial number or the same name of the manufacturing process can be processed in a BPM at the same time. The maximum batch size is regarded as an important parameter of BPMs, which indicates the maximum processing capacity. As a major feature in semiconductor manufacturing systems, batch processing is an important factor in meeting customers' demands for delivery due dates, and it affects the total production cycle of the semiconductor production line. Considering differences between SPM and BPM, a different scheduling method should be proposed to properly arrange operation sequences in order to improve the productive efficiency of the total semiconductor production line.

Production scheduling in wafer fabrication systems is one of the most complex production scheduling problems [9]. According to the different scheduling objects, production scheduling problems can be classified into three categories: scheduling problems of processing systems, scheduling problems of AMHSs and integrated scheduling problems of processing systems and material handling systems. Production scheduling problems of AMHSs can be further classified as: Intrabay scheduling problem, Interbay scheduling problem and Intrabay and Interbay integrated scheduling problem according to the scheduling scope. The authors have investigated these scheduling problems for over ten years and published some papers.

At present, with rapid changes in market demand for semiconductor ICs and the intensification of global economic competition, wafer manufacturing enterprises face enormous pressure to improve production efficiency and reduce production costs. The scheduling methods and techniques of AMHSs with excellent performance are one of the effective and fundamental ways to improve the overall efficiency of AMHSs and wafer processing systems. This book will introduce solutions and techniques of scheduling problems in AMHSs comprehensively and systematically.

References

[1] Huanzi company. China's integrated circuit industry analysis report for 2011. Beijing, 2011.
[2] Wang, Y.Y. Adapt to the need of the industry of the development of microelectronics technology in the 21st century. Physics, 2004, 33(6): 407–413.
[3] Jiang, Z.H. The development of integrated circuit industry in China. Proceedings of the 2005 Asia and South Pacific Design Automation Conference, Shanghai, China, 2005, pp. 7–8.
[4] Zant, P. Chip Manufacturing, Semiconductor Technology Process and Practical Tutorial (Fourth Edition). Beijing: Electronic Industry Press, 2004.
[5] Quirk, M. Semiconductor Manufacturing Technology. Beijing: Electronic Industry Press, 2004.

[6] Narahari, Y., Khan, L.M. Performance analysis of scheduling polices in re-entrant manufacturing system. Computer Operations Research, 1996, 23(1): 37–51.

[7] Uzsoy, R., Lee, C.Y., Martin-Vega, L.A. A review of production planning and scheduling models in the semiconductor industry. Part I: system characteristics, performance evaluation and production planning. IIE Transactions, 1992, 24(4): 47–60.

[8] Uzsoy, R., Lee, C.Y., Martin-Vega, L.A. A review of production planning and scheduling models in the semiconductor industry. Part II: shop-floor control. IIE Transactions, 1994, 26(5): 44–55.

[9] Kumar, P.R. Scheduling semiconductor manufacturing plants. IEEE Control Systems Magazine, 1994, 14(6): 33–40.

2 Automated material handling systems in SWFSs

2.1 Development of Material Handling Systems in SWFSs

In the initial stage of 150 mm and 200 mm semiconductor wafer fabrication systems (SWFSs), material in the workshop is transported and stored manually. Therefore, its transportation process might affect its proper functioning, and can easily lead to incorrect operations, low material handling efficiency, uneven material distribution, idle equipment and long product cycle. With the development of manufacturing technologies and rapid changes of market demands, the traditional manual material handling mode is gradually disappearing from the historical stage. Wafer manufacturing material handling systems have evolved from the stage of semi-automation to automation and are now rapidly moving toward more intelligent ones.

2.1.1 Semi-automated Material Handling Systems

After the 200 mm SWFS factory underwent short-term manual operation to fulfill the transportation constraints, storage and distribution of work-in-process (WIP), it then got transferred quickly to semi-automated material handling phase. The material handling in the processing regions (Bay) is developed on the basis of manual operations in combination with the Interbay automated transportation and storage system. However, WIP transportation between machines is still not automated. In order to improve production efficiency, the automated material handling system (AMHS) (shown in Figure 2.1) is adopted in many 200 mm SWFS factories. It is a semi-automated material handling system, i.e., the first generation of an AMHS [1].

In the first generation of an AMHS, WIP is intuitively controlled and manual transportations without any added value are substituted. The storage capacity and transfer speed of an AMHS must meet production demands and ensure that operators obtain the required material on time. In a 200 mm SWFS factory, the AMHS has played a very important role, which helps implement mass production in semiconductor factories and improve production efficiency and shorten products' cycle times.

2.1.2 Automated Material Handling Systems

With the emergence of 300 mm semiconductor factories, the automated material system (i.e., the second generation of an AMHS) came into being. It has been greatly improved on the basis of the original semi-automated material handling system.

https://doi.org/10.1515/9783110487473-002

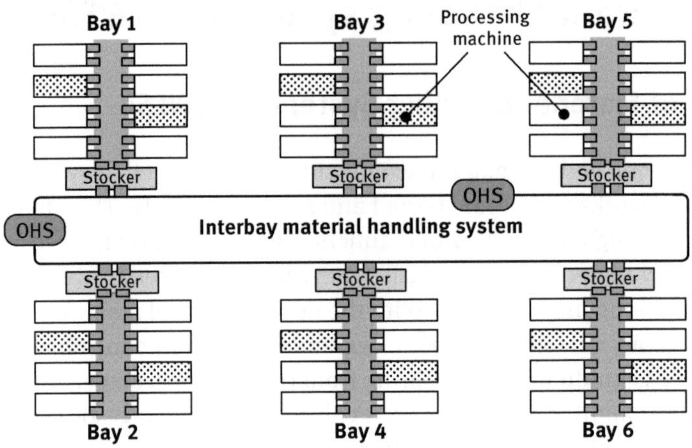

Figure 2.1: An AMHS in 200 m semiconductor manufacturing factory.

First, the Intrabay delivery system is added to the existing Interbay system. Accordingly, additional features such as transfer interface between Interbay and Intrabay systems are added to the stockers (shown in Figure 2.2). In addition, the stocker capacity is increased for the Intrabay WIP storage. In the second generation of an AMHS, a new vehicle – overhead hoist transport (OHT) – is utilized to deliver cassettes directly to the processing machines. The cassette is first moved from the Load Port of the processing machine to the OHT and then transported to the destination stocker in another bay. It is then moved to the Load Port of the Interbay by the robot of the destination stocker and then loaded to the overhead shuttle (OHS). Then, the cassette is moved to the stocker which is the nearest machine for the next operation. When this cassette is to be processed, the OHT in Intrabay automatically loads and transports this cassette to the destination processing machine and places it at the Load Port of the processing machine. The entire delivering sequence is shown in Figure 2.2.

The second generation of an AMHS is a highly mechanized and segmented material handling system. The entire delivery process consists of two types of transportation to stockers and OHT only works in the specified Intrabay. Each segmented part (Intrabay, stocker, Interbay) is united by the central control system. Transport management is confined to the area related to transport operations and the transfer speed of the entire system is restricted. This AMHS design is established on the basis of the early process flow of the semiconductor factories, in which the system should assign OHTs to each Intrabay according to its production rate. The flexibility of an AMHS is limited by the segmented layout, which is unable to respond quickly to changed process or other abnormal events.

In order to achieve the goal of the entire manufacturing system, segmented 300 mm AMHS can respond according to the demands of the manufacturing

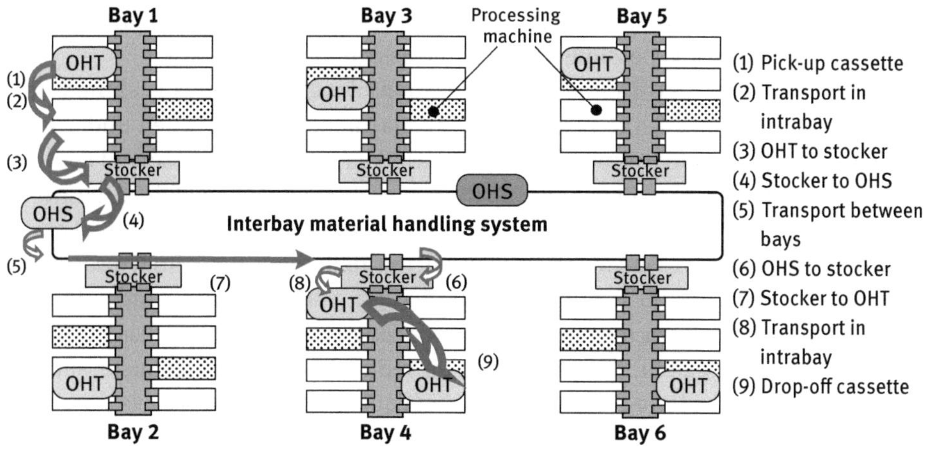

Figure 2.2: The segmented AMHS in a 300 mm semiconductor factory.

execution system (MES). However, the number of OHSs in Interbay and the number of OHTs in Intrabay are fixed. Reconfiguration of the vehicles should be accomplished manually, which restricts the AMHS's response speed to the changes of the plant. In this design, the stocker not only plays a role in storing wafers, but also plays a more important role of acting as a bridge in delivering wafers. Therefore, the stability of stockers, to a large extent, will affect the production capacity of all machines in this bay. For example, when there is a breakdown of the stocker of a bay (i.e., it does not work properly), the wafers to be processed in this bay cannot be sent to the specified processing machines and the processed wafers cannot be delivered to next machine on time.

2.1.3 Intelligent Material Handling Systems

On the basis of the 300 mm automated material handling system, equipment suppliers and solution providers of an AMHS apply various advanced technologies and try to further improve the agility and flexibility of AMHSs toward the intelligent direction.

In the embryonic intelligent AMHS, an OHT system is extended to the Interbay area, in which OHTs are able to roam anywhere in the factory. Thus, the wafer cassette will not be temporarily stored in the stocker when it is delivered between different bays. Instead, the wafer cassette is directly delivered to the processing machine, which takes a shorter delivering time. The entire material handling system can achieve dynamic load balancing and "predict" required production rate and allocate proper number of vehicles in each bay. Tasks are dynamically allocated to the nearest vehicle: when a vehicle completes a task, the system will rationally assign it another task based on its location. Even if this task is assigned to a farther away vehicle at this time, it is possible to adjust the assignment [2].

In the intelligent AMHS, the storage location of wafer cassettes is shifted from stockers to a place nearest to next processing machine, which is called under-track storage (UTS) or zero-footprint storage (ZFS). UTS uses part of the reserved space in the clean room and thus will not take up the space of processing machines (shown in Figure 2.3). UTS has the following significant advantages: (1) it places wafer lot closer to processing machines in order to shorten the delivering time from processing machines to stockers; (2) it minimizes the footprint of stockers in order to save space for the clean room to place more processing machines; (3) it has cut down the cost to a large extent. UTS not only can reduce the fixed investment such as stockers, but also can improve the reliability of equipment in order to save the cost of regular maintenance.

To support the integrated AMHS, dynamic routing selection and UTS storage, the transportation control system and the material control system of an AMHS should have intelligent functions such as correct allocation of tasks, efficient utilization of resources, dynamic balance of fab load, continuous detection of hundreds of parameters associated with WIP and timely responses. Through the unity and integration of Interbay and Intrabay systems, the control system can visually monitor and coordinate each delivery step. By dynamically controlling the position of OHTs, the resources within the material handling system are fully optimized and utilized. The dynamic decision function enables the AMHS to transfer from the design concept which shortens cycle time by rapid mechanical motion in the early stage to the new design concept which controls the vehicle's location intelligently and allocates tasks properly.

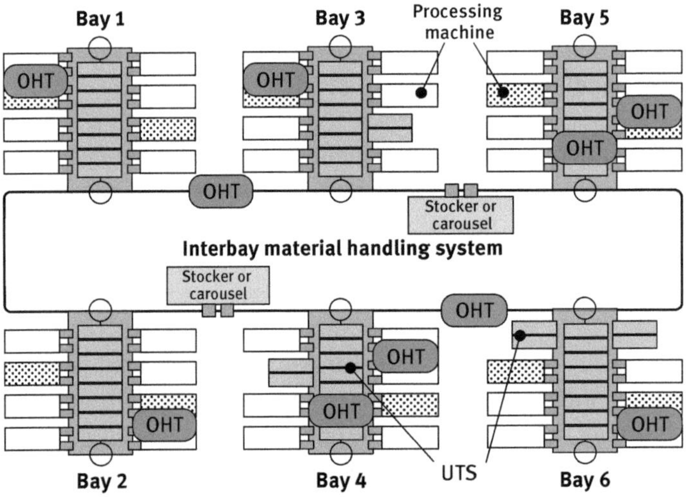

Figure 2.3: An AMHS with UTS in the 300 mm semiconductor factory.

In addition, different information systems and automated handling systems can be integrated together by a fab management software package. Additional values and benefits are obtained when these systems are pre-integrated and their advantages are fully realized. By this kind of integration, the demands of material can be expected by an AMHS, and can be dynamically adjusted according to changes in the manufacturing system. The current 300 mm AMHS can achieve the function which cannot be achieved by traditional fab and the early AMHS, which brings more added values to the semiconductor wafer fab.

2.2 Components of the Automated Material Handling System

The AMHS in the semiconductor wafer fabrication system consists of five key important components: carriers, transport system, storage system, tracking system and control system.

2.2.1 Carriers

The wafer carrier is a wafer cassette in the transport process, which is also known as wafer lot. In the automated material handling system, the wafer carrier as a unit is transported by the vehicle, and usually its capacity is limited to 25 wafers. One vehicle can transport only one to two wafer carriers. In both of the early 200 mm AMHS and the current 300 mm AMHS, the design of wafer carriers is related to whether it can automatically transport wafers, which is also the basis of the hardware of an AMHS and a key step to optimize the AMHS. As there is development in wafer fabrication technologies and as the wafer size becomes larger, the size and structure of wafer carriers are also changing, which mainly has the following types.

1. Standard enclosed wafer carrier
   In the 200 mm wafer factory, the standard mechanical interface (SMIF)-enclosed wafer carrier is commonly adopted. In each processing machine, there is a standard port to open the wafer carrier. In addition, in the factory adopting SMIF, each machine has a micro environment, which provides a clean environment when a wafer is moved out of the wafer carrier. The SMIF wafer carrier has a universal automation interface which prevents wafers from the potential contamination in the factory; therefore, the SMIF wafer carrier is ideal for the AMHS.

2. Opening cassette
   The wafer in the opening cassette is exposed to the surrounding environment. Thus, with respect to the factory adopting SMIF carriers, the factory using opening

cassettes should be kept cleaner. Opening cassettes could be transported separately and also be stored in a sealed box before transportation. For opening cassette and storage boxes, one major drawback is that the automation is not taken into consideration. Therefore, the interface and the robot must be customized additionally in order to move the opening cassette into or out of the automated material handling system.

3. Front opening unified pod
   In the transition to 300 mm wafer, the focus of the worldwide factories is on the standardization of the wafer carrier. Eventually, an integrated 1wafer carrier with a friendly microenvironment and automation ready, which is called the front opening unified pod (FOUP), is selected as the industry standard, as shown in Figure 2.4.
   Compared with 200 mm SMIF wafer carrier, 300 mm FOUP seals wafers within a controllable environment. Moreover, the robot handling SMIF wafer carriers can also handle FOUPs. There are two differences between FOUP and SMIF: (1) FOUP opens a door at the front, while SMIF wafer carriers open a door at the bottom; (2) FOUP has an embedded wafer holder inside, while SMIF wafer carriers has a removable wafer holder inside.

2.2.2 Transport System

Material movement within the AMHS is usually fulfilled by the transport system which consists of various transport vehicles, rails and control system. Transport vehicles generally include: OHT, OHS, AGV, RGV and conveyors.

1. Overhead hoist transport
   As shown in Figure 2.5, the OHT runs along the track mounted on the ceiling and these tracks are connected to the pick-up/drop-off ports of processing machines. The OHT can lift wafer carriers to these pick-up/drop-off ports. The

Figure 2.4: Front opening unified pod.

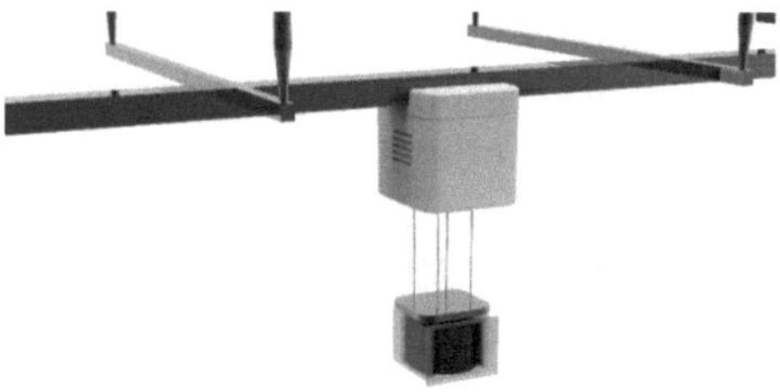

Figure 2.5: Overhead hoist transport (OHT).

Figure 2.6: Overhead shuttle (OHS).

OHT system is used in both the 300 mm and 200 mm SMIF factories. One major advantage of OHTs is that it does not require any floor space in the factory. Another important advantage is that it can adapt flexibly to changes in the plant layout.

2. Overhead shuttle
Similar to OHT, the OHS also runs along the track mounted on the ceiling, as shown in Figure 2.6. The OHS is responsible for delivering wafer carriers between stockers. Like the transport vehicles that are used between bays, the OHS is also applied in both 200 mm and 300 mm SMIF factories. An OHS can transport one or two wafer carriers and often completes transport tasks together with OHT, RGV and AGV. There are two kinds of OHS: one uses its own robot to move the wafer carrier from the stockers and send to the OHS, and the other uses the robot of the stocker to load wafer carriers.

3. Rail-guided vehicle
The RGV is a transport vehicle used on the ground and can be used to handle material within and between workshops. The RGV is one of the fastest

material handling vehicles and thus is commonly used in workshops that demand the highest production rate. For security reasons, the RGV cannot coexist with factory personnel, which becomes one of its limitations. Another limitation is that RGV cannot support the flexibility of changes in the factory layout. Although RGV has been used in the 200 mm SMIF factory, it is gradually being replaced by OHT in the 300 mm factory.

4. Automated guided vehicle
   The AGV is also a transport vehicle that operates on the ground and can reach anywhere in the factory by program control. The AGV can be used for material handling within and between workshops. The AGV is slower than other material handling vehicles but it adapts flexibly to changes in the factory layout. Under certain conditions, the AGV can temporarily coexist with workers in the same workshop. The AGV is mostly used in 200 mm SMIF factories and is rarely used in 300 mm SMIF factories.

5. Conveyor
   The conveyor is placed in the ceiling and can send 200 mm and 300 mm wafer carriers. Although the conveyor system is slower than other material handling systems, it provides high material throughput because the movement of wafer carriers will not be influenced by the number of vehicles. The conveyor is usually used in the point-to-point transport, such as transporting wafer carriers between segmented factory buildings. The conveyors in combination with OHTs are taken into consideration to be adapted in high material throughput areas.

2.2.3 Storage System

In the process of wafer fabrication, specified storage equipment and its corresponding management system are designed to store wafers. Currently, there are two types of storage system: stocker and under-track storage (UTS) system.

The stocker is a storage system that is used to store large number of wafer carriers. A stocker consists of one or more input/output ports, shelves for placing wafer carriers and a robot for handling wafer carriers between shelves. The shape of stockers varies and its capacity ranges from 50 to thousands of wafer carriers. In the factory, there are two kinds of stockers: descartes type and carousel type. There is only one robot in descartes stocker, which can run in three dimensions inside the system and move the wafer carriers from the input/output port to the vertical shelves. Descartes stockers are popular in 200 mm and 300 mm wafer factories. The wafer carriers in carousel stockers are placed on horizontal or vertical shelves and the robot moves in one direction inside the system.

Therefore, the carousel stockers are placed more densely and are more reliable than descartes stockers. In the 300 mm factory, carousel stockers and OHTs are combined for automation applications.

UTS is also a common method for wafer storage, which can be combined with OHTs. UTS contains a shelf under the track of OHTs, which can store wafer carriers, as shown in Figure 2.7 UTS has three outstanding advantages: (1) it does not occupy floor space; (2) it is highly reliable and (3) it improves the transport efficiency of an AMHS.

Because the wafer carrier can be stored beside the next processing machine, the time to store wafer carriers in stockers is reduced. In addition, UTS provides more input/output ports instead of two ports in stockers, which alleviates the congestion of OHTs and improves the efficiency of an AMHS.

2.2.4 Tracking System

In modern 300 mm wafer fabs, the number of wafer WIPs is up to 30,000 to 50,000 pieces and each wafer typically has 200–600 machining processes. The transportation distance to complete all the processes is up to 8–10 miles. Faced with such a large-scale material handling, it is necessary to adopt tracking systems to trace the wafer carriers. The tracking system consists of hardware-level identification (such as RFID, two-dimensional code, etc.), reading devices and software-level management systems.

In recent years, RFID technology has been developing rapidly. Because the tags can be read without sight, RFID technology has become the mainstream for

Figure 2.7: Under track storage system.

tracking wafers. In IBM's semiconductor manufacturing factory located in New York, RFID technology is used to track thousands of orders of chips. The factory covers an area of 140,000 square feet and has equipment worth billions of dollars. It is the first fully automated semiconductor manufacturing factory and operates since mid-2002. In its tracking system, each wafer carrier is customized according to the 300 mm wafer standard, is made of polycarbonate and is implanted a passive RFID tag with unique identification code which can be read by a RFID reader spread over the processing machines. Each processing machine contains a RFID reader, a lithographic exposure tool, metal deposition equipment, chemical mechanical polishing, and an oven, and involves processes such as chemical etching and ion implantation. The reader can also be overall planned and distributed in the device of material handling systems. The reading range of thousands of RFID readers is 3–6 inches. The tracking system in the factory can trace every wafer, including the specific locations where they are being processed. By integrating RFID technology in the production tools, the manufacturing system can determine which product to process next, how to process and when to process.

2.2.5 Control System

The control system in an AMHS consists of stocker controller (STC), transport controller (TSC) and auxiliary control software. At different levels, the control system is divided into equipment control system and material control system. The former is used to manage vehicles, stockers and other hardware devices and responsible for their proper operations. The latter is used to guarantee the on-time delivery of wafers and the transport planning of vehicles, including material control software, scheduling and dispatching software, and so on.

1. Material control system (MCS)
   The MCS is a real-time central control system and it is mainly responsible for the management and control of products, masks, and other transport and storage material. The MCS translates the instructions of operators and other software to the commands that material handling device can understand. In addition, the MCS has a database to identify and track material in the factory. In order to optimize the factory, the MCS must include algorithms dealing with advanced tracking, merging, space splitting, prevention of material shortage and pollution. In order to adapt to continuous changes in the factory, it should have the abilities of being dynamically expanded and repeatedly arranged.
2. Scheduling and dispatching software
   The scheduling and dispatching system will analyze the status of the wafer fabrication system and trigger material movement in the MCS system, which

can improve on-time delivery rate and reduce WIP. These software packages review the current status of the factory and continuously decide when to start and terminate products and how to balance demands of customers' shipments to minimize the inventory holding cost. Many rules are applied in scheduling and dispatching software of wafer material handling in order to optimize the efficiency of material handling system and improve on-time delivery rate, and also reduce inventory levels, production cycle times and its fluctuations.

2.3 Layout of the Automated Material Handling System

The layout design is an important part in the design phase of the automated material handling system, which aims to allocate all kinds of production processes and related resources to reasonable positions. Under this circumstance, all the WIP can flow with the highest efficiency in the factory, which reduces capital and product costs and shortens the cycle times. In terms of the topological structure, layouts commonly adopted in current AMHSs can be classified as follows [3–8].

2.3.1 Single-spine Configuration

The single-spine configuration is the general layout for 300 mm fabs. In this layout, a central aisle is connected with several machines installed in several bays. As shown in Figure 2.8, stockers are located at the intersection of the main aisles of

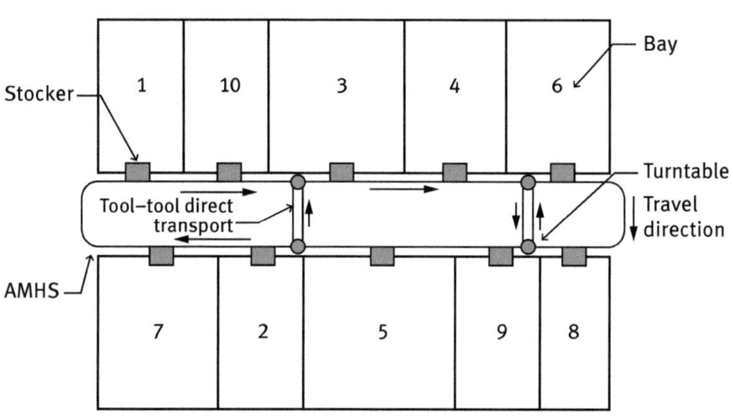

Figure 2.8: Spine configuration for fab layouts.

each bay. An AMHS based on a spine configuration consists of one directed flow loop and several crossover turntables to change the travel directions.

2.3.2 Double-spine Configuration

In some overhead material handling systems, a double-spine configuration is also adopted, as shown in Figure 2.9.

2.3.3 Integrated Configuration

The traditional segmented AMHS is highly automated, but wafer delivery among different production regions must be completed via stockers. This way not only affects the transportation efficiency, but also brings disastrous effects to the critical processing machines when stockers break down. Therefore, an integrated material handling system comes into being, in which inter-regional transport for key processing machines can be completed via OHT, i.e., Tool–Tool direct transport. The feature of this design is that wafer transport between key processing machines does not rely on transit of stockers in order to improve the efficiency of an AMHS.

As shown in Figure 2.10, in the factory without Tool–Tool direct transport, the transport route from Tool A to Tool B is A→Stocker1→Stocker4→Tool B. But in the factory with Tool–Tool direct transport (shown in Figure 2.11), the transport route from Tool A to Tool B is Tool A→Tool B, which is achieved through inter-regional transport by OHT.

2.3.4 Perimeter Configuration

The perimeter configuration is another widely applied configuration. As shown in Figure 2.12, there are two physically separated loops in this layout. Several crossover turntables are adopted to switch the direction loops.

Feindel and Kaempf suggested connecting the Interbay with Intrabay material transportation facilities by using a conveyor-type continuous flow transport (CFT)

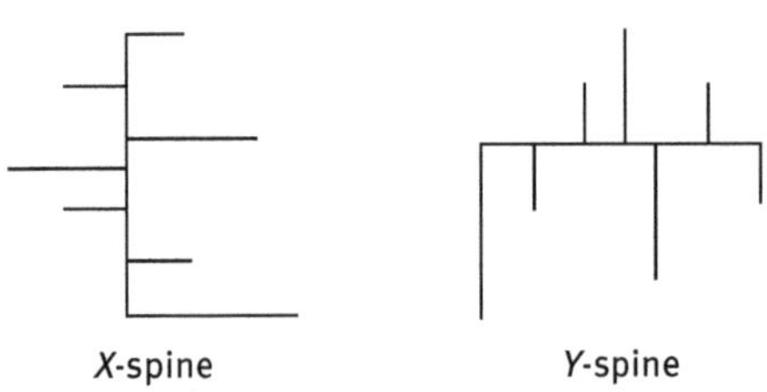

Figure 2.9: Double-spine layouts.

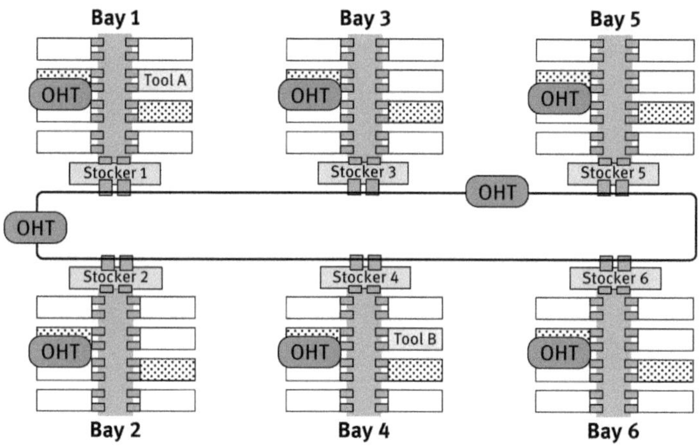

Figure 2.10: Segmented AMHS without Tool-Tool handling ability.

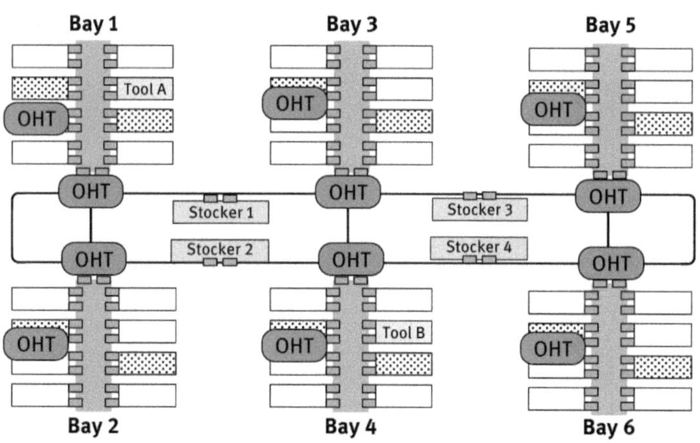

Figure 2.11: Integrated AMHS with Tool-Tool handling ability.

facility in order to reduce the MH time. The requirements of large end-of-bay stockers were reduced by using local buffering facilities. An overhead or floor-mounted conveyor was proposed for high-capacity Interbay with long-distance and point-to-point pod transport. It is suitable for small zero-footprint buffers close to processing tools. The standard deviation caused by irregular MH equipment availability was eliminated, and it was possible to obtain more predictable delivery times. As shown in Figure 2.13, a buffer loop was used to provide space for WIP to wait until a tool became available. A robotic hoist or elevator was proposed to transfer pods between the CFT buffer and the tool load port.

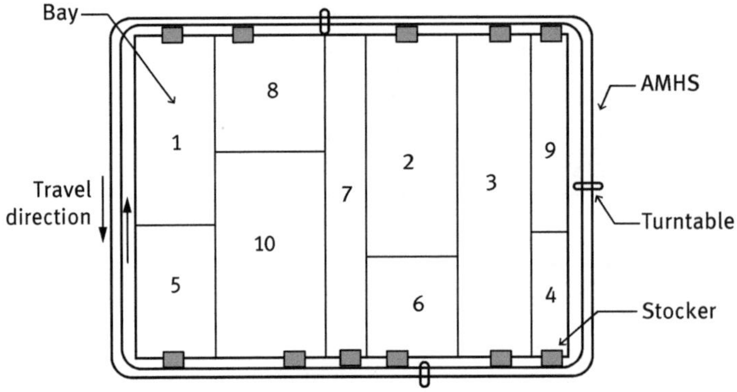

Figure 2.12: Perimeter configuration for fab layouts.

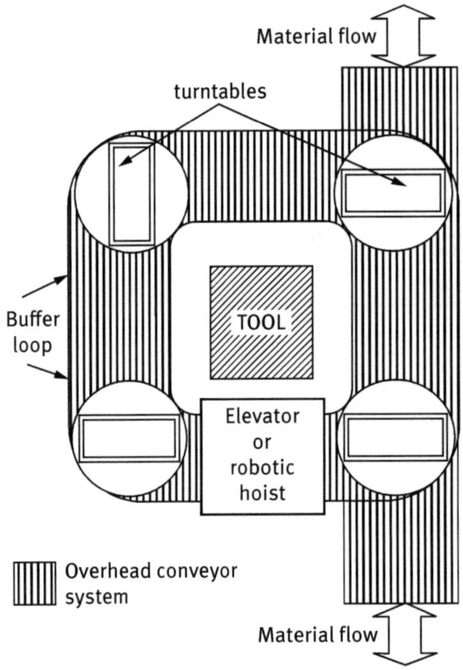

Figure 2.13: Plan view of zero footprint CFT buffer loop.

2.3.5 Mixed Configuration

Some scholars and engineers proposed a configuration that is composed of two Interbay track loops, one is the spine type and the other is the perimeter type, as shown in Figure 2.14.

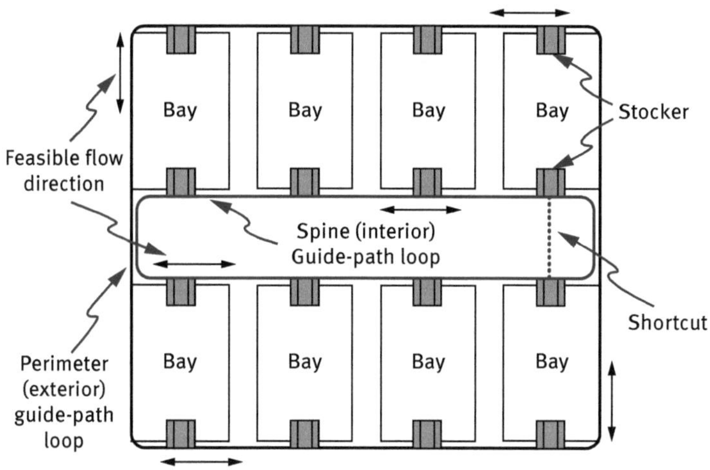

Figure 2.14: An example of a two-row dual-loop bay layout.

2.4 Features of the Automated Material Handling System

The characteristics of an automated material handling system are summarized as follows:

1. Large-scale. In a typical 300 mm wafer fabrication system, there are usually hundreds of processing machines. Each cassette has 200–600 processing operations. The total delivery distance of each cassette reaches 8–10 miles.
2. Stochastic and dynamic. In the Interbay material handling system, a number of vehicles are assigned to deliver different cassettes in a same period. Therefore, the moving requests normally arrive at the system in a random way and unexpected blockage of vehicles often arises. The uncertainty of the vehicle location and status significantly increases the complexity of the scheduling problem.
3. Multiple-objective. Normally there are two types of indicators used to measure the performance of an Interbay material handling system: the measurements for transportation and manufacturing performance. In practical industrial manufacturing, the production scheduling process of Interbay material handling is required not only to increase the transport efficiency but also to meet the production strategy. The balance of various objectives needs to be considered in scheduling Interbay material handling.

The measurements for transportation performance mainly refer to: (a) mean no-load arrival time of vehicles; (b) mean transport time of vehicles; (c) mean waiting time of cassettes; (d) mean delivery time of cassettes and (e) mean utilization of vehicles. The measurements for manufacturing performance include: (a) throughput of the

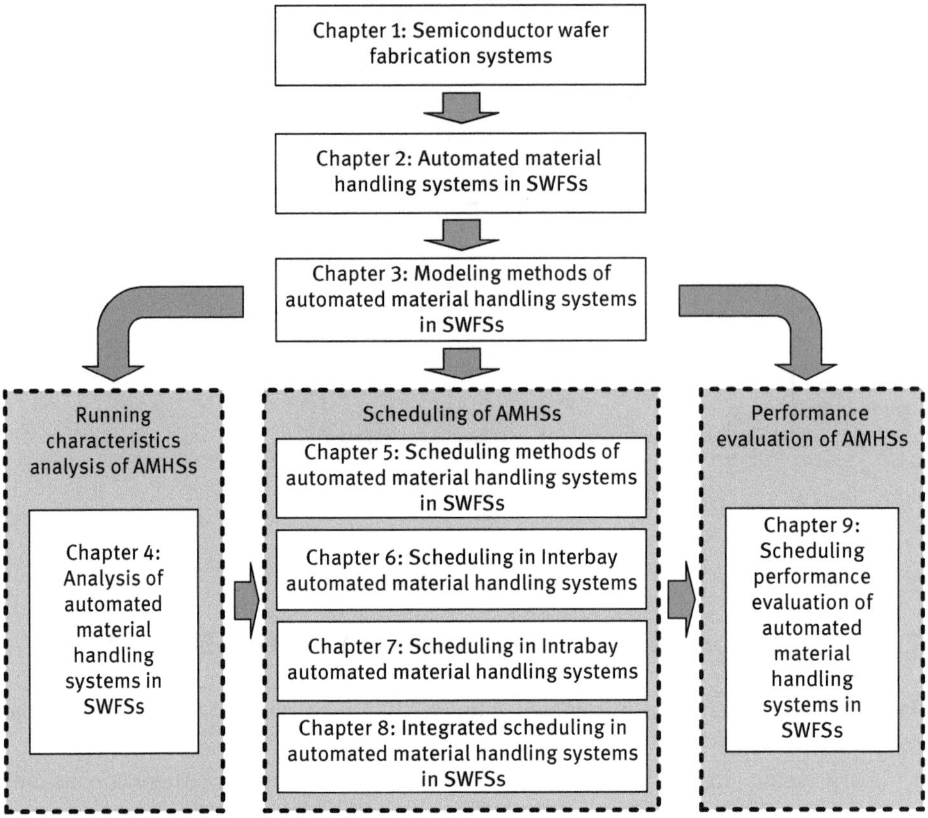

Figure 2.15: Organizations of this book.

manufacturing system; (b) mean production cycle time; (c) utilization of key equipment and (d) average due date satisfaction rate.

Due to these features, it has become a huge challenge to model, analyze and schedule automated material handling systems in semiconductor wafer fabrication systems. In this book, on the basis of Chapter 1, this chapter introduces SWFSs and AMHSs, and the following chapters will address modeling, running characteristics analysis and performance evaluation methods of AMHSs. The organization of each chapter is shown in Figure 2.15.

References

[1] Van Antwerp, K. Automation in a semiconductor fab. Semiconductor International, December, 2004 [Online] Available http://www.semiconductor.net/article/CA483805.html.

[2] Kondo, H., Harada, M. Study for realizing effective direct tool-to-tool delivery semiconductor manufacturing. ISSM 2005, IEEE International Symposium. 2005(13): 21–24.

[3] Wiethoff, T., Swearingen C. AMHS software solutions to increase manufacturing system performance. Advanced Semiconductor Manufacturing Conference. 2006(17): 306–311.
[4] Zhao, W., Mackulak, G.T. Reducing model creation cycle time by automated conversion of a CAD AHMS layout design. Simulation Conference Proceedings in 1999 Winter. 1999(1): 779–783.
[5] Kaempf, U. Automated wafer transport in the wafer fab. Proceedings of the IEEE/SEMI Advanced Semiconductor Manufacturing Conference. 1997: 356–361.
[6] Lin, J.T., Wang, F.K., Wu, C.K. Connecting transport AMHS in a wafer fab. International Journal of Production Research, 2003, 41(3): 529–544.
[7] Montoya-Torres, J.R. A literature survey on the design approaches and operational issues of automated wafer-transport systems for wafer fabs. Production Planning & Control, 2006, 17(7): 648–663.
[8] The International Technology Roadmap for Semiconductors (ITRS), 2004 edition, SIA Semiconductor Industry Association. 2006, http://www.itrs.net.

3 Modeling methods of automated material handling systems in SWFSs

The characteristics of an automated material handling system (AMHS) in semiconductor wafer fabrication systems (SWFSs) are large-scale, real-time, stochastic and multi-objective. In order to design an effective scheduling method to optimize the wafer material handling and production process, it is necessary to establish a reasonable system model that is an abstract description of the system. Modeling is an important means and prerequisite to analyze and make decisions for the system [1]. Specifically, we hope to use the established model to reveal the process and mechanism of inherent operational change in an AMHS to determine a variety of factors (e.g., the number of vehicles, the number of shortcuts, rail layout, vehicle speed, vehicle jam probability, wafer card waiting time distribution) on operation performances. At the same time, with the model of the AMHS, the performance of a scheduling method can be evaluated quickly and effectively in order to verify the effectiveness of the scheduling method.

At present, modeling methods of wafer manufacturing include network flow model, queuing theory model, mathematical programming model, Markov model, simulation model and Petri net model.

3.1 Modeling Methods Based on the Network Flow

The network flow theory is a kind of theory and method in the graph theory. It primarily studies a class of optimization problems on networks. In 1955, T.E. Harris first proposed the problem of seeking maximum traffic volume between two points on a given network when studying the maximum railway throughput. L.R. Ford and D.R. Fulkerson et al. proposed an algorithm to solve these problems, and established the network flow theory [2].

3.1.1 Basic Theory of the Network Flow Model

The graph G is an ordered pair (V, E), where V is regarded as the vertice set, E is the edge set, and E and V do not intersect. They can also be written as $V(G)$ and $E(G)$.

The directed graph is used to represent the ordered binary, and the direction of each arc in the directed graph is given. In general, an arc can be represented as an ordered pair (i, j), which points to node j from node i. Arrows are used to specify the orientation, as shown in Figure 3.1.

One common representation is the node-arc incidence matrix of a graph. The structure is constructed by listing the nodes vertically and arcs horizontally. Then

https://doi.org/10.1515/9783110487473-003

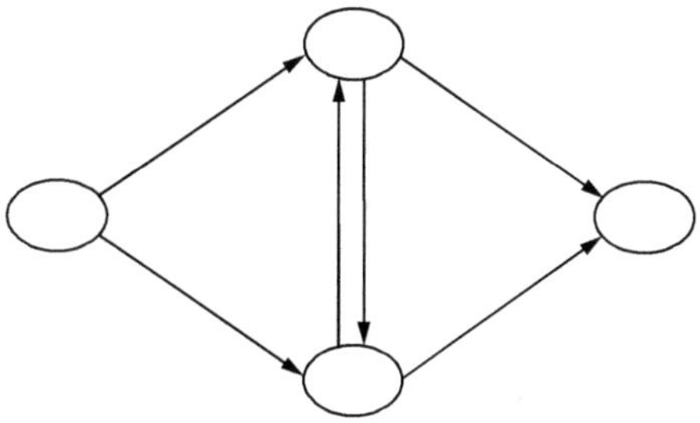

Figure 3.1: Directed graph.

below the arc in the column, a "1" is at the position of the corresponding node i, and a "–1" is at the position corresponding to node j. The incidence of Figure 3.1 is represented as follows:

$$
\begin{array}{c} \\ 1 \\ 2 \\ 3 \\ 4 \end{array}
\begin{array}{c}
\begin{array}{cccccc} (1,2) & (1,4) & (2,3) & (2,4) & (4,2) & (4,3) \end{array} \\
\left(\begin{array}{cccccc}
1 & 1 & & & & \\
-1 & & 1 & 1 & -1 & \\
 & & -1 & & & 1 \\
 & -1 & & -1 & 1 & -1
\end{array}\right)
\end{array}
$$

Suppose that $V=\{|v_1, \ldots, v_m\}$ and $E=\{e_1, \ldots, e_n\}$. Introduce an $m \times n$ node-arc incidence matrix A, and its elements a_{ij} are represented as follows:

$$
a_{ij} = \begin{cases} 1, & \text{node } v_i \text{ is the starting point of the arc } e_j; \\ -1, & \text{node } v_i \text{ is the ending point of the arc } e_j; \\ 0, & \text{otherwise} \end{cases} .
$$

There may be a stream along the arc, and a stream in a directed arc (i, j) is denoted as $x_{ij} \geq 0$. The flow must meet the preservation requirements at each node in the network. That is, the traffic volume of node i is

$$
\sum_{j=1}^{n} x_{ij} - \sum_{k=1}^{n} x_{ki} = b_i
$$

The first term on the left is the sum of the flows flowing out of node i, and the second term is the sum of the flows entering node i; b_i is the storage requirement of node i. Of course, if the arc from node i to node j does not exist, x_{ij} does not exist either.

In particular, the traffic volume can neither be created nor be lost at the node unless the node is a starting point or an ending point, i.e., the total traffic volume entering a node must be equal to the total traffic volume flowing out of the node. Therefore,

$$
\sum_{j=1}^{n} x_{ij} - \sum_{k=1}^{n} x_{ki} = 0
$$

The first term on the left is the sum of the flows flowing out of node i, and the second term is the sum of the flows entering node i.

Many practical network flow problems such as transportation problems, distribution problems, shortest path problems and maximum flow problems can be transformed into such network flow planning problem.

3.1.2 Network Flow Modeling Process for AMHSs

The method based on the network flow model usually simplifies the material handling problem in AMHSs to a deterministic directed graph model. The material handling time is simplified to a definite time. The long-term behavior of the directed graph model is analyzed statistically, and then the network flow planning problem is solved.

In general, an available wafer fabrication workshop is divided into equal-sized unit-length squares (workstations) by using a space filling curve (SFC) method, where each machining area consists of several workstations. In Figure 3.2, the width of each workstation is the height of the rectangle. Each machining area is assigned a unique value as its address. The address numbering sequence is the same as the direction of the material flow in the AMHS, and the box from the upper-left corner serves as the starting point.

The position of the loading/unloading point (P/D) of the machining area is indicated by the boundary distance from the starting workstation of the machining area. Therefore, the P/D position of the upper row of the machining area is represented by the distance between the P/D point and the left side of the starting rectangle. And the P/D position of the lower row of the machining area is represented by the distance between the P/D point and the right side of the starting rectangle. Obviously, the P/D point is close to the material handling system. Based on the SFC method, it is first necessary to develop an initial flow sequence to construct a layout, as shown in Figure 3.3. Once the spatial fill curve and the initial flow order are determined, the design of the material handling system becomes a matter of determining the number of shortcuts required in the layout and each shortcut location. The goal of this model is to find the optimal solution, which minimizes the increase in the shortcut costs while minimizing the reduction in handling costs.

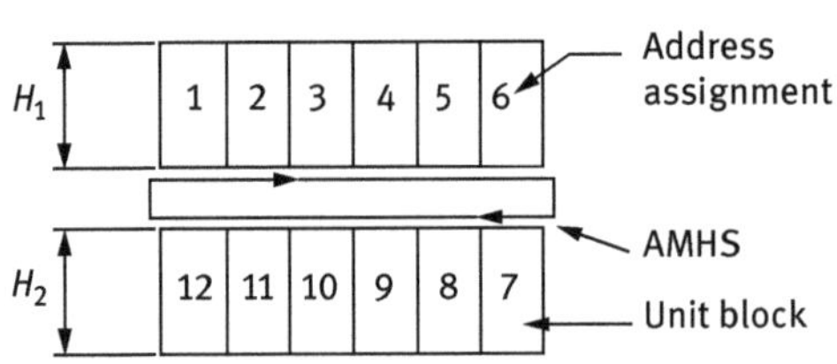

Figure 3.2: Address assignments for spine configuration.

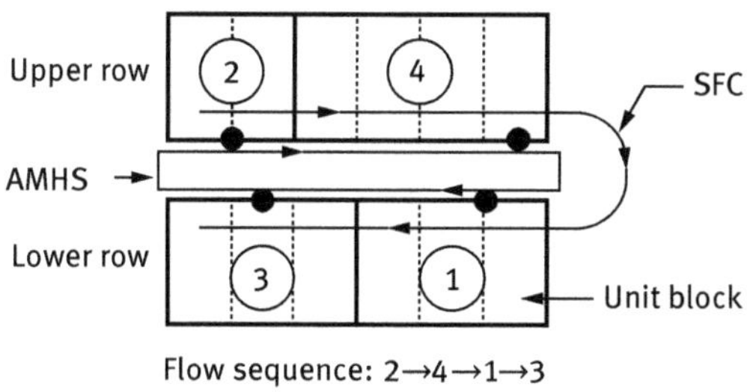

Figure 3.3: SFC example for spine configuration.

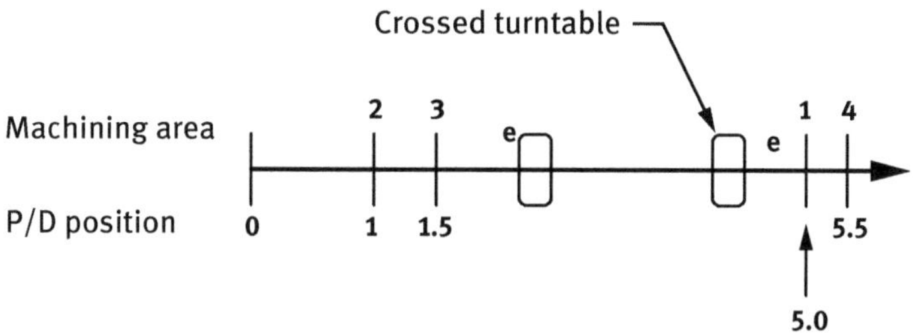

Figure 3.4: P/D point sorting example.

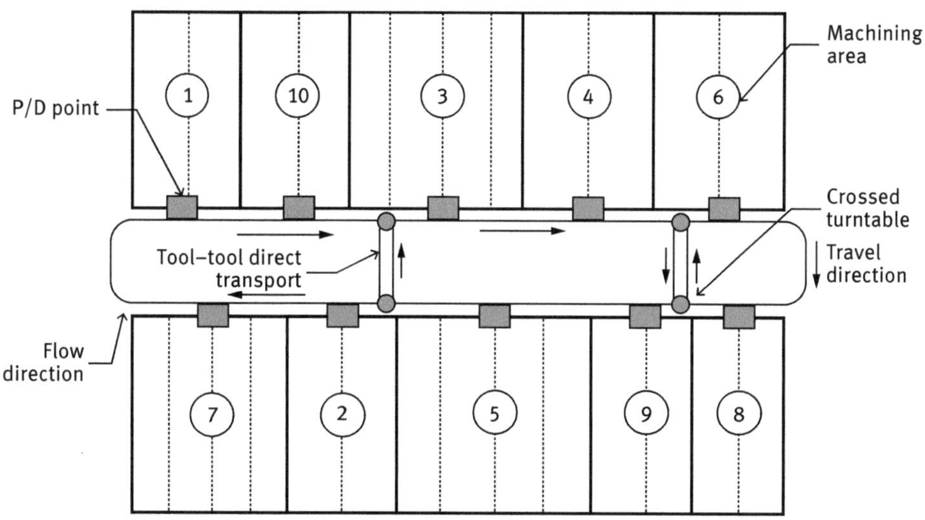

Figure 3.5: Spine configuration integrated design.

The position of the crossed turntable is determined. First, from the left border, the P/D point is placed onto the horizontal axis. The two-dimensional coordinates of the P/D point are represented by a single dimension, as shown in Figures 3.4 and 3.5. In this example, there are three optional positions available for setting the crossed turntable: one between machining area 2 and machining area 3, one between machining area 3 and machining area 1 and one between the machining area 1 and machining area 4.

If the vertex set (V) represents the P/D point of the machining area and the crossed turntable, and the edge combination (E) represents the directional flow curve between the vertices, then the standard network representation $G = (V, E)$ indicates the problem.

Figure 3.5 shows the spine configuration integrated design. A feasible layout of the AMHS and a candidate shortcut scheme are considered. Each arc has the same length. δ_{ij} is denoted as the directional flow distance of arc(i, j). The directional flow density between the P/D points of the machining area is k; then the set of directional flow density is defined as K, $k \in K$. u^k is the distance cost of transporting unit k, and c_{ij}^k is the cost of transporting k on arc(i, j), and $c_{ij}^k = u^k \delta_{ij}$.

Let F be the cost for installing a crossed turntable per period. The following notations are used:

- A_i area of the machining zone;
- H_1 height of the upper row;
- H_2 height of the lower row;
- L length of the AMHS loop;
- a fixed distance across the center spine;
- x_{ij}^k flow of commodity k on the directed arc;
- y_{ij} crossed turntable on arc (i, j);
- $D(k)$ drop-off point of commodity k;
- $P(k)$ pick-up point of commodity k;
- $\bar{S}$ set of directed shortcut arcs, $\bar{S} \subset E$;
- S set of undirected shortcut arcs, $S \subset \bar{S}$.

The width of the unit rectangle in the upper row is assumed to be one. Then, the width of the unit rectangle in the lower row is $b = H_1/H_2$. Note that L, A_i, a and b are used to calculate the arc length δ_{ij}. The network flow model of the integrated shortcut design problem can be expressed as

Minimize

$$Z = \sum_{k \in K} \sum_{(i,j) \in E} c_{ij}^k x_{ij}^k + \sum_{(i,j) \in S} F y_{ij} \tag{3.1}$$

Subjected to

$$\sum_{j \in V} x_{ij}^k - \sum_{j \in V} x_{ij}^k = \begin{cases} -1, & \text{if } i = D(k) \\ 1, & \text{if } i = P(k),\ \forall i \in V \text{ and } \forall k \in K \\ 0, & \text{if otherwise} \end{cases} \tag{3.2}$$

$$x_{ij}^k,\ x_{ji}^k \le y_{ij},\ \forall (i,j) \in S \ \text{ and } \ \forall k \in K \tag{3.3}$$

$$x_{ij}^k \geq 0,\ \forall (i,j) \in E \ \text{and}\ \forall k \in K \tag{3.4}$$

$$y_{ij} \in \{0,1\},\ \forall (i,j) \in S \tag{3.5}$$

Assume that the flow sequence is given. Equation (3.1) represents the optimization objective function, which minimizes the reduction in handling costs while minimizing the increase in the shortcut costs. Equation (3.2) shows the flow conservation constraint for each commodity k. If (i, j) is not in the design, then the flow of each commodity k on the arc is zero in both directions. This is a multi-commodity network flow model with fixed expenditure variables on some arcs.

Finally, this integrated design problem is solved by using iterative descent heuristic pairwise exchange heuristics, as shown in Figure 3.6, where N is the total number of workstations covered in the layout. The experimental results show that the integrated design method is superior to the traditional improved quadratic ensemble method.

The modeling method of a wafer manufacturing system based on the network flow is often developed on the basis of some assumptions such as steady state and

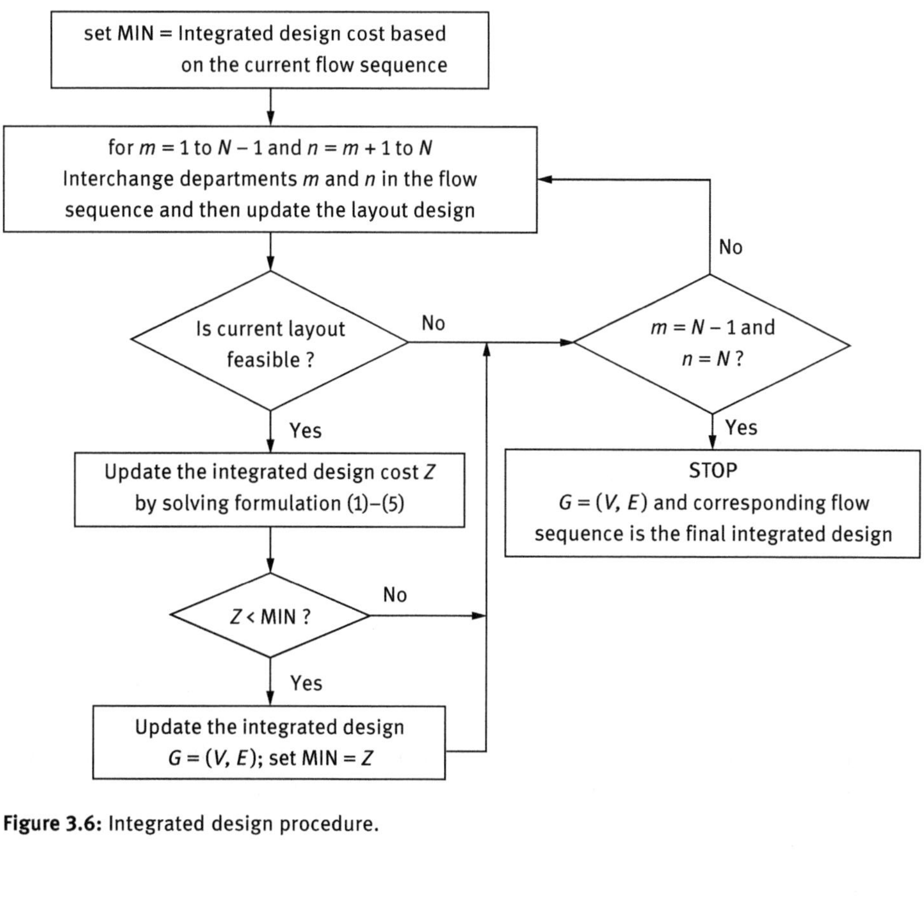

Figure 3.6: Integrated design procedure.

independence. The stochastic characteristics of the material handling process are simplified during modeling. And these conditions in reality may not be satisfied, and thus cannot guarantee the accuracy of its analysis results. Moreover, this method only analyzes the dynamic process of the system, and it is difficult to analyze the operation of the material handling system in the stochastic dynamic environment. The above limitation leads to the application limitation of this method.

3.2 Modeling Methods Based on the Queuing Theory

The method based on the queuing network model usually describes the material handling system as a queuing network. The method is only suitable for the operation analysis of the small-scale material handling system. For large-scale, random and complex wafer manufacturing problems, the difficulty of solving problems will increase exponentially.

3.2.1 Basic Theory of the Queuing Theory Model

Queuing theory [3–6] is also known as stochastic service system theory. First, the arrival time and the service time of the service object are statistically studied, and some statistical indexes (such as waiting time, queue length, busy period and so on) are obtained. Then, according to these indexes, the structure of the service system is improved or the service objects are reorganized. Hence, the service system can meet the needs of clients and optimize the cost or certain indicators. It is a branch of mathematical operations research disciplines. It also studies the randomization of queuing phenomena in service systems. It is widely applied in the random service systems with sharing resources such as computer networks, production and transportation, inventory and so on. Three aspects such as statistical inference, system behavior and system optimization are mainly investigated in the queuing theory. Its purpose is to correctly design and effectively run various service systems.

Although the contents of the queue problems are different, they have the following common characteristics:

1. A person or thing that requests a service (e.g., a waiting patient, an aircraft requesting a landing, etc.) is regarded as a "customer."
2. The person or thing that serves the customer (e.g., doctors, aircraft runways, etc.) is called "waiter." The service system is formed by customers and waiters.
3. The customer randomly comes to the service system one by one (or a batch). The service time of a customer is not necessarily determinate. This randomness of the service process will result in a long queue of customers, and sometimes the waiter is free.

In order to describe a given queuing system, the following components of the system must be defined:

1. The input process is the probability distribution that the customer comes to the service desk. On the basis of the original data, the experience distribution of the queuing problem is developed according to the rules of the customer arrivals. And then its theoretical distribution is determined by using the statistical method (such as chi-square test), and its parameter values are estimated. The probability distribution of the customer arriving at the service desk is a Poisson distribution, and the arrival of the customer is an independent smooth input process. "Smooth" means that the expected value of the distribution and variance parameters are not affected by time.
2. The queuing rules are those that define how customers queue and wait. The queuing rules generally consist of real-time systems and waiting systems. The rules of the real-time system are the rules that the customers leave when the service desk is occupied; while the rules of the waiting system are rules that the customers queue to wait for the service when the service desk is occupied. Waiting service rules include first-come first-served (FCFS) rule, random service rule, priority service rule and so on. The FCFS rule is investigated in this book.
3. The service agency consists of no waiter, one waiter or more waiters. The service agency provides service for an individual customer or for a group of customers. In the input process, most service times are random. It is always assumed that the distribution of service time is smooth. If ξn is the time required to service the nth customer, the probability distribution of the sequence of the service time $\{\xi n, n = 1, 2, \ldots\}$ expresses the service mechanism of the queuing system. The successive service times $\xi 1, \xi 2, \ldots$ are independent, and the time interval sequence $\{Tn\}$ for the arrival of any two customers is also independent.

The purpose of studying the queuing problem is to study the operating efficiency of the queuing system, to evaluate the service quality, to determine the optimal value of the system parameters, to determine whether the system structure is reasonable and to design improvement measures. Therefore, it is necessary to determine basic quantitative indicators for the system.

1. Length: it is the number of customers (including waiting customers and served customers) in the system, and its expected value is Ls. The queuing length is the number of waiting customers in the queue, and its expected value is Lq. The number of customers in the system (length) = the number of customers waiting for service + the number of customers being served. Hence, the greater the Lq or Ls, the lower the service efficiency.
2. Waiting time: it is the period from the arrival time of the customer to the time to accept the service, and its expected value is Wq. The staying time is the period from the arrival time of the customer to the time to complete the

service, which is the total time spent by the customer in the system, and its expected value is Ws.
staying time = waiting time + service time.
3. Busy period: it refers to the continuous busy time of the service desk, which is the time from the arrival of the customer to the idle service desk until the service desk becomes idle again. This is the most important indicator; it is directly related to the waiter's work intensity. The idle period refers to the length of time for which the desk is kept idle. Obviously, the busy period and the idle period are alternately present in the queuing system.

In addition, the service desk utilization (that is the percentage of the busy time of the waiter in the total time) is an important indicator in queuing theory.

The standard form of the queuing model is presented by $X/Y/Z/A/B/C$,
where
X is the distribution type of customer arrival time;
Y is the distribution type of service time;
Z is the number of waiters;
A is the system capacity;
B is the number of customers;
C is the service rule.

The waiting queue model of the FCFS rule is mainly composed of three parameters: $X/Y/Z$.

For example, "$M/M/1/k/\infty$/FCFS" means that the customer arrival time and the service time obey the negative exponential distribution.

"$M/M/c$" means that the customer arrival time and the service time obey the negative exponential distribution, and it consists of c service stations.

"$M/G/1$" means that the customer arrival time obeys the negative exponential distribution and the service time obeys the general random distribution; it has a single service station.

The input process of the system is subject to Poisson distribution, the service time obeys the negative exponential distribution and the queuing system of a single service station has the following three situations:
1. standard type: $M/M/1$ ($M/M/1/\infty/\infty$);
2. limited system capacity type: $M/M/1/N/\infty$;
3. finite customer source type: $M/M/1/\infty/m$.

For a queuing system of a single service station, parallel C of the service desk, with the single-service desk similar to the multi-service queue system, has the following three situations:
1. standard type: $M/M/C$ ($M/M/C/\infty/\infty$);

2. limited system capacity type: $M/M/C/N/\infty$;
3. limited customer source type: $M/M/C/\infty/m$.

For scheduling problems in AMHSs, there are waiting queues in the wafer carts and transport vehicles; therefore, many scholars investigate queuing problems.

3.2.2 Queuing Theory Modeling Process for AMHSs

From the point of view of the queuing system, the Automated Guided Vehicle (AGV) system can be regarded as a resource, and a transport request can be regarded as an arriving customer. The required number of servers (the size of the fleet) depends on the parameters related to the customer arrival (transport request) and the service time (AGV handling time).

In Figure 3.7, the allocation wait time is the delay time before the transport request for the workpiece is assigned to the cart. In terms of the queuing theory, the mean and variance of vehicle handling time are used to estimate the assignment waiting time. The idle times of the vehicle under different assignment rules are estimated by the analysis model proposed by Koo and Jang (2002) [14]. Once the assignment waiting time and the idle time have been calculated, the expected value of the waiting time of a part can also be estimated, and this parameter will be directly used to estimate the size of the fleet.

The assumptions are presented as follows:
1. The handling time between the loading station and the unloading station is unique and deterministic.
2. The average delivery request rate for different sites is known. However, the interval time between delivery requests is subject to a certain probability distribution.

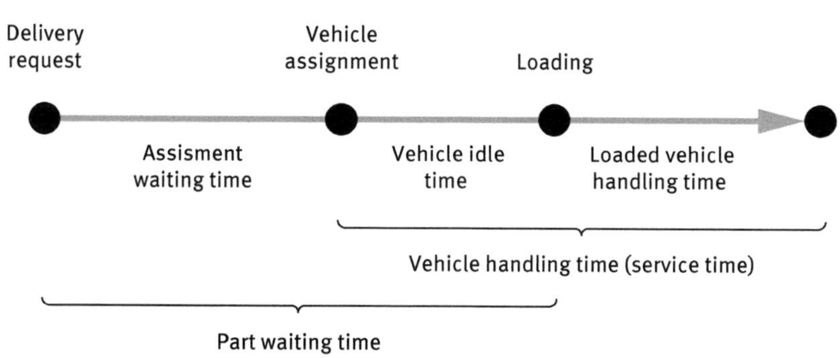

Figure 3.7: Part waiting time in a material handling system.

3. If there is no delivery request waiting for the vehicle, then the vehicle stops at the current site to wait for the next request.
4. A vehicle can only respond to one delivery request at a time.
5. If more than one idle vehicle responds to a delivery request, the vehicle is selected according to the assignment rules. If multiple delivery requests wait for a free vehicle, the vehicle will respond to the delivery request according to the FCFS rule.

In view of the above assumptions, the queuing network theory is used to investigate the fleet size problem in the system. First, the mean and variance of the idle time and the loaded handling time of the vehicle are estimated, and then the expected waiting time of the part during the handling process is estimated. The model parameters are defined as follows:

n the number of pick-up/drop-off locations;
f_{ij} the delivery request rate from location i to location j;
t_{ij} the vehicle travel time from location i to location j;
m the number of vehicles;
ρ vehicle utilization;
$S(k,i)$ the set of all locations that are closer to location i than location k;
lu the sum of the loading time and the unloading time;
F the delivery request rate between all locations $\left(F=\sum_{i=1}^{n}\sum_{j=1}^{n}f_{ij}\right)$;
$q(k,\ i)$ the probability that an idle vehicle at location k is selected when a delivery request is issued at location i;
f_{s_i} the proportion of delivery requirements from location i;
f_{d_k} the proportion of the delivery requirements to location k.

The loaded vehicle handling time includes the loading time and the unloading time; its mean and variance are calculated as follows:

$$E(t_l)=\sum_{i=1}^{n}\left[\sum_{j=1}^{n}\{(f_{ij}/F)(t_{ij}+lu)\}\right] \tag{3.6}$$

$$V(t_l)=\sum_{i=1}^{n}\left[\sum_{j=1}^{n}\left\{(f_{ij}/F)(t_{ij}+lu)^2\right\}\right]-E^2(t_l) \tag{3.7}$$

Its expected value is a weighted average of all possible values. Formulas (3.6) and (3.7) are independent to the selection rule of the idle vehicle.

1. **The idle vehicle handling time**
Suppose that the idle time of the vehicle is independent. Therefore, the number of idle vehicles Z at the moment of issuing the delivery request follows a quadratic distribution:

$$b(m,\,1-\rho),\ i.e.,\ P(z)=pr(Z=z)=C_z^m(1-\rho)^z\rho^{m-z},\ where\ C_z^m=\frac{m!}{(m-z)!z!}$$

Case 1: If there is no free vehicle response to the delivery request, the delivery request is added to the waiting queue. The probability of this situation is ρ^m or $P(0)$. In this case, when the vehicle is idle, the delivery request is responded in accordance with the FCFS rule.
Case 2: If at least one vehicle responses when a delivery request arrives, the free vehicle is arranged according to the assignment rule. The probability of this situation is $(1-\rho^m)$ or $\sum_{z=1}^{m}P(z)$.
Then the idle handling time is estimated by using different dispatching rules.
(1) Random vehicle selection

$$E(t_e)=\sum_{i=1}^{n}\left[fs_i\sum_{j=1}^{n}(fd_k t_{ki})\right] \tag{3.8}$$

$$V(t_e)=\sum_{i=1}^{n}\left[fs_i\sum_{j=1}^{n}fd_k t_{ki}^2\right]-E^2(t_e) \tag{3.9}$$

where $fd_k=\frac{1}{F}\sum_{i=1}^{n}f_{ik},\ fs_i=\frac{1}{F}\sum_{j=1}^{n}f_{ij}$.

(2) Longest idle vehicle selection and least utilized vehicle selection are the same as eqs (3.8) and (3.9).
(3) Nearest vehicle selection

$$E(t_e)=\sum_{z=1}^{n}\left[P(z)\sum_{i=1}^{m}\left[fs_i\sum_{i=1}^{n}\{q(k,i)t_{ki}\}\right]\right]+P(0)\sum_{i=1}^{n}\left[fs_i\sum_{i=1}^{n}(fd_k t_{ki})\right] \tag{3.10}$$

$$\begin{aligned}V(t_e)=&\sum_{z=1}^{n}\left[P(z)\sum_{i=1}^{m}\left[fs_i\sum_{i=1}^{n}\{q(k,i)t_{ki}^2\}\right]\right]\\&+P(0)\sum_{i=1}^{n}\left[fs_i\sum_{i=1}^{n}(fd_k t_{ki}^2)\right]-E^2(t_e)\end{aligned} \tag{3.11}$$

$$q(k,i) = \left[1 - \sum_{r \in S(k,i)} q(r,i)\right] A(k,i)$$

where

$$A(k,i) = 1 - \left(1 - \frac{fd_k}{\sum_{r \notin S(k,\ i)} fd_r}\right)^z, i,\ k = 1,\ 2, \dots, n$$

(4) Farthest vehicle selection is the same as eqs (3.10) and (3.11).
Assume that the handling time and the idle time of a loaded vehicle are independent; then

$$E(t_v) = E(t_l) + E(t_e) \tag{3.12}$$

$$V(t_v) = V(t_l) + V(t_e) \tag{3.13}$$

The queuing theory is used to estimate the waiting time of the assignment process. According to the $G/G/m$ queuing system proposed by Kimura (1991) [15], the assignment waiting time is calculated by

$$W_q = W_o\left(c_a^2 + c_v^2\right) gw/2 \tag{3.14}$$

where

$$g = \begin{cases} Exp\left[\dfrac{-2(1-\rho)(1-c_a^2)^2}{3\rho(c_a^2 + c_s^2)}\right], & if\ c_a 1; \\ 1, & if\ c_a \geq 1; \end{cases}$$

$$\omega = \begin{cases} \left[c_a^2 + c_v^2 - 1 + \left(1 - c_a^2\right)/(1-4\gamma) + \left(1 - c_v^2\right)/(1+\gamma)\right]^{-1} \\ if\ c_a < 1; \\ [2(\left[c_a^2 + c_v^2 - 1 + \left(1 - c_a^2\right)(1-4\gamma) + \left(1 - c_v^2\right)(1+\gamma)\right])\left(c_a^2 + c_v^2\right), \\ if\ c_a \geq 1. \end{cases}$$

$$\gamma = \min\left[(1-\rho)(m-1)\left(\sqrt{4+5m} - 2\right)/(16m\rho), 0.25\left(1 - 10^{-6}\right)\right].$$

2. Determination of the number of vehicles

The initial number of vehicles is set to be $(T_l/T_a)^+$, where T_l is the loaded vehicle handling time. T_a is the valid time available for the vehicle, and the integer $(x)^+$ is greater than or equal to x. In terms of the initial number of vehicles, the

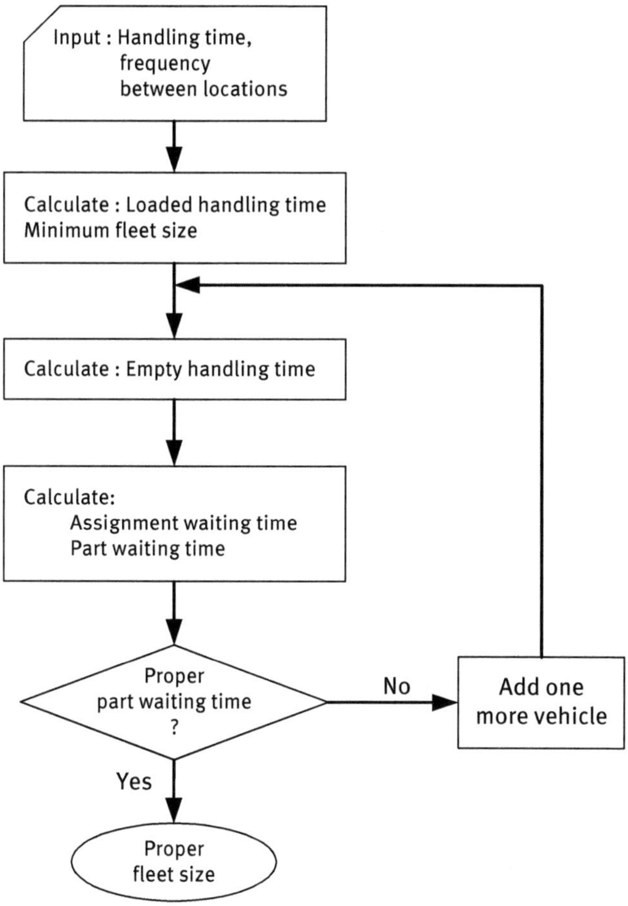

Figure 3.8: Fleet sizing procedure.

expected value of the part waiting time is equal to the sum of the assignment waiting time and the idle time. The program shown in Figure 3.8 is executed until the part waiting time is less than the desired value.

The experimental results show that the algorithm can estimate the minimum number of vehicles, which can guarantee the preset material waiting time limit. And the result can be adjusted according to different vehicle scheduling rules. However, the effectiveness of this method is reduced when the randomness of vehicle running time between sites increases.

The queuing theory-based approach usually simplifies the material handling system to a process in which the material is queued for carriage service. In the process of modeling, it is necessary to simplify the problem greatly, and to neglect the factors such as the jam of the vehicle and the idle time of the handling vehicle, which influence the accuracy of the operation analysis result of an AMHS.

3.3 Modeling Methods Based on Mathematical Programming

Mathematical programming is an important branch of applied mathematics, which does not refer to a particular mathematical problem-oriented computer programming technology. The term appeared in the late 1940s and was first used by Robert Dorfman in Harvard University in the United States. Mathematical planning includes many research branches such as linear programming, nonlinear programming, multi-objective programming, dynamic programming, parameter programming, combinatorial optimization, integer programming, stochastic programming, fuzzy programming, non-smooth optimization, global optimization, variation inequalities and complementary problems, and it is also widely used in the wafer manufacturing process in AMHSs.

3.3.1 The Mathematical Programming Model

The mathematical programming theory is an important part of the operation research theory; it is also a theoretical tool widely used in mathematical modeling. Mathematical programming is to seek a maximal or minimal optimization model of an objective function. According to the constraints of decision-making variables, they can be further classified as constraint mathematical programming and unconstrained mathematical planning.

The general form of the mathematical programming is

$$\begin{aligned} &\max f(x) \quad (\text{or} \min f(x)) \\ &\text{s.t.} \quad h_i(x) = 0, \ i = 1, 2, \ldots, l \end{aligned}$$

$$g_j(x) \geq 0, \quad j = 1, 2, \ldots, m$$

If the objective function and the constraint function are linear functions, it is called linear programming – otherwise it is called nonlinear programming; if the variable is limited to an integer value, it is called integer linear programming; if the decision variable limit is only 0 or 1, it is called 0–1 planning; if only some of the variables in the decision variables are integer, it is called mixed integer programming. If several objective functions are considered in a certain sense, it is regarded as a multi-objective programming problem.

First, we should determine point x that satisfies all the constraints, which is regarded as a feasible solution. A feasible set of points is called a feasible domain. By using the basic properties of the elements in the feasible domain, we can change the value form of the objective function so as to judge the value of the objective function. If x^* makes $f(x^*)$ reaches the maximum or minimum in S, then x^* is the optimal solution and $f(x^*)$ is the optimal value.

The basic form of linear programming is presented as follows:

$$\min z = \sum_{j=1}^{n} c_j x_j$$

$$\text{s.t.} \sum_{j=1}^{n} a_{ij} x_j = b_i, i = 1, 2, \ldots, m$$
$$x_j \geq 0, j = 1, 2, \ldots, n$$

Other forms of linear programming problems can be transformed as the above form. The basic solution to this problem is the simplex method.

The basic form of unconstrained nonlinear programming is $\min_{x \in R^n} f(x)$.

1. The necessary conditions for the local minimal point are as follows: on the local minimal point $x^{(0)}$, the gradient of function $f(x)$ is zero, i.e., $\nabla f(x^{(0)}) = 0$.
2. Numerical solution of unconstrained extreme value: To find the next point, which is closer to the optimal solution than the previous one, from a determined initial point $x^{(0)}$. One of the commonly used methods is to find the falling direction of the function value, to find the minimal function value in this direction, and then find out the optimal solution, realize an iteration, so it can get close to the minimal solution.
3. Commonly used numerical solutions are steepest descent method, conjugate gradient method for quadratic function, Newton method, DFP algorithm in variable scale algorithm, trust region method and so on.

From the point of view of the basic principle, the general form of constrained nonlinear programming is

$$\min_{x \in R^n} f(x)$$

$$\text{s.t. } g_i(x)0, \quad i = 1, 2, \ldots, m$$

$$h_j(x) = 0, j = 1, 2, \ldots, l$$

The basic solutions are as follows:

1. Optimal conditions

   According to the Lagrange multiplier method of the multivariate function, the optimal solution of the above problem x^* should satisfy the *Kuhn – Tucher* condition:

$$\nabla_x L(x^*, \lambda^*, \mu^*) = \nabla f(x^*) - \sum_{i=1}^{m} \lambda_i^* \nabla g_i(x^*) - \sum_{j=1}^{l} \mu_j^* \nabla h_j(x^*) = 0$$

where $\lambda_i^* g_i(x^*) = 0, i = 1, 2, ..., m$

$$\lambda_i^* \geq 0 ....i = 1, 2, ..., m.$$

2. Feasible direction method (*Zoutendijk*)
3. Penalty function method: the outside point method and the interior point method.

The general form of multi-objective programming problem is

$$(vp)\begin{cases} \min\,(f_1(x), f_2(x)..., f_p(x))^T, p > 1, x \in E^p \\ \text{s.t.}...g_i(x) \geq 0, i = 1, 2, ..., m \\ h_j(x) = 0, j = 1, 2, ..., l \end{cases}$$

The concept of PARETO optimal solution is introduced by using the optimization problem of vector set: effective point (minimal vector) and weak effective point (weak minimal vector), which can better compare the values of multi-objective problems.

The mathematical programming method usually simplifies the material handling problem in AMHSs, and it is equivalent to the mathematical programming model. Then, it analyzes the operating characteristics of the material handling system and determines the operating parameters of the optimal system.

3.3.2 Mathematical Programming Modeling Process for AMHSs

Spinal layouts are most common in AMHSs and can be regarded as a straight-line Steiner tree (the least costly distribution tree) with a minimal total handling distance. Since the single-spine layout has only one spindle to contact all the loading/unloading points, it limits the linear Steiner tree. The single-spine layout is shown in Figure 3.9, with its spindle perpendicular to the x-axis (x-spine), or perpendicular to the y-axis (y-spine). Each loading/unloading point is connected to the main track by a linear track.

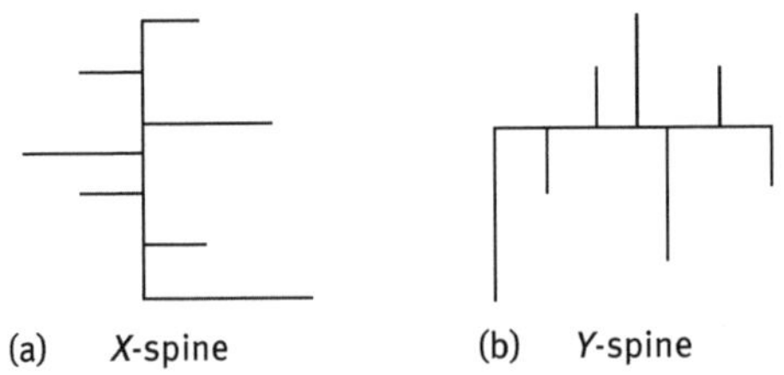

Figure 3.9: Two possible orientations of the single-spine layout: (a) perpendicular to the x-axis and (b) perpendicular to the y-axis.

The optimal single-spine layout of x-spine is considered as an example. Layout problems are mainly concentrated on the loading/unloading point location and the material flow matrix. The variables used in the model are defined as follows:

x_0 the location of x-spine;
n the total number of loading/unloading points;
$P_i(a_i, b_i)$ the location of the ith loading/unloading point;
d_{ij} the distance between P_i and P_j;
$d_{ij}^1 = |a_i - x_0| + |b_i - b_j| + |x_0 - a_j|$, if $b_i \neq b_j$;
$d_{ij}^2 = |a_i - a_j|$, if $b_i = b_j$.

The flow set F is partitioned into two sets:

$$F^1 = \{f_{ij} : b_i \neq b_j\}$$

$$F^2 = \{f_{ij} : b_i = b_j\}$$

Minimize total flow × distance

$$\begin{aligned} &= \sum_{i=1}^{n}\sum_{j=1}^{n} f_{ij}d_{ij} = \sum_{f\in F^1} f_{ij}d_{ij}^1 + \sum_{f\in F^2} f_{ij}d_{ij}^2 \\ &= \sum_{f\in F^1} f_{ij}\left(|a_i - x_0| + |b_i - b_j| + |x_0 - a_j|\right) + \sum_{f\in F^2} f_{ij}|a_i - a_j| \\ &= \sum_{f\in F^1} f_{ij}\left(|a_i - x_0| + |x_0 - a_j|\right) + \sum_{f\in F^1} f_{ij}|b_i - b_j| + \sum_{f\in F^2} f_{ij}|a_i - a_j| \end{aligned}$$

Since all a_i and b_i are known, and $\sum_{f\in F^1} f_{ij}|b_i - b_j| + \sum_{f\in F^2} f_{ij}|a_i - a_j|$ is a constant, the objective can be transferred to

$$\begin{aligned} \sum_{f\in F^1} f_{ij}\left(|a_i - x_0| + |x_0 - a_j|\right) &= \sum_{f\in F^1} f_{ij}|a_i - x_0| + \sum_{f\in F^1} f_{ij}|x_0 - a_j| \\ &= \sum_{f\in F^1} (f_{ij} + f_{ji})|a_i - x_0| \\ &= \sum_{i=1}^{n} w_i|a_i - x_0| \end{aligned}$$

This problem is equivalent to a straight-line distance single facility layout problem, where $w_i = \sum_{f\in F^1} (f_{ij} + f_{ji})|a_i - x_0|$, i.e., the total flow into and out of the unloading point i in the flow set F1. The optimal solution can be obtained by using most theories or the semi-sum method. Most theories refer to a weight that is at least equal to the sum of the other weights, and the optimal spine lies at the loading/unloading point with the most weight. In the semi-sum method, the optimal x-coordinate satisfies no more than half of the weight on the left side of x, and no more than half of the weight on the right side of x. Thus, the optimal spine is located at least through a loading/ unloading point.

The experimental results show that this method can obtain the optimal solution in the linear time. At the same time, numerical analysis further shows that the total length of a single-spine layout track is shorter relative to a two-way spine layout, but it leads to an increase in the material handing flow.

3.4 Modeling Methods Based on the Markov Model

Markov was the representative of the Petersburg School of Mathematics. He proposed and studied a general scheme for the study of natural processes by mathematical analysis from 1906 to 1912 called the Markov chain. At the same time, a stochastic process – Markov process – is developed. Markov found that the state of the nth experiment was often determined by the results of the previous ((n−1)th) experiment by observing the trials many times. Markov pointed out that: for a system, there was a transition probability from one state to another state. This transition probability can be deduced based on its previous state, and it was independent of the original state of the system and the Markov process prior to this transition. The Markov model is a statistical model describing the Markov process and is used to describe the stochastic process of discrete events with Markov properties in mathematics. In the process, in the case of a given current knowledge or information, the past (i.e., the current historical state) is irrelevant for predicting the future (i.e., the present future state)

3.4.1 Basic Theory of the Markov Model

Both time and state of a Markov process are discrete; this kind of process is regarded as a Markov chain. The Markov chain is a sequence of random variables X_1, X_2, X_3 . . . The range of these variables, that is, the set of all their possible values, is called the "state space" and the value of X_n is the state at time n. If the conditional probability distribution of $X_n + 1$ for the past state is only a function of X_n, then

$$P(X_{n+1} = x|X_0, X_1, X_2, ..., X_n) = P(X_{n+1} = x|X_n)$$

where x is a state in the process. The above identity can regarded as the expression of a Markovian property.

Markov first made such a process in 1906. This generalization to the countable infinite state space was given by Colmogolov in 1936.

The Markov chain is a stochastic process that satisfies the following two assumptions:

1. At $t+1$, the probability distribution of a system state is only related to the state at time t, and is independent of the state before time t;
2. The state transition from time t to time $t+1$ is independent of t. A Markov chain model can be expressed as (S, P, Q), where the meanings of the elements are as follows:
   (1) S is a non-empty state set composed of all possible states of the system, and sometimes it is referred to as the state space of the system. It can be a finite set, a set of columns or any non-empty set. It is assumed that S is a countable set (i.e., finite or rankable). The status is represented by lowercase i, j (or S_i, S_j).
   (2) $P = [P_{ij}]_{n \times n}$ is the system state transition probability matrix, where P_{ij} denotes the probability that the system is in state i at time t, the state j at the next time $t+1$, and N is the number of all possible states of the system. For any $i \in s$, $\sum_{j=1}^{N} P_{ij} = 1$.
   (3) $Q = [q_1, q_2, ..., q_n]$ is the initial probability distribution of the system; q_i is the probability that the system is in the state i at the initial time $\sum_{j=1}^{N} q_i = 1$.

The Markov chain is represented by a conditional distribution. $P(X_{n+1}|X_n)$ is referred to as a "transition probability" in a stochastic process. This is sometimes referred to as a "step transition probability." The transition probabilities of two, three and more steps can be derived from the one-step transition probability and the Markov properties:

$$P(X_{n+2}|X_n) = \int P(X_{n+2}, X_{n+1}|X_n)dX_{n+1} = \int P(X_{n+2}|X_{n+1})dX_{n+1}$$

$$P(X_{n+3}|X_n) = \int P(X_{n+3}|X_{n+2}) \int P(X_{n+2}|X_{n+1})P(X_{n+1}|X_n)dX_{n+1}dX_{n+2}$$

These equations can be generalized to any future time $n+k$ by multiplying the transition probabilities.

$P(X_n)$ is the distribution of the state at time n with an initial distribution $P(X_0)$. The change in the process can be described by the following time step:

$$P(X_{n+1}) = \int P(X_{n+1}|X_n)P(X_n)dX_n$$

This is one version of the Frobenius-Perron equation. In this case, one or more state distributions satisfy:

$$\pi(X) = \int P(X|Y)\pi(Y)dY$$

where Y is a nominal variable to facilitate the integration of variables. This distribution is called a stationary distribution or a steady-state distribution. A stationary distribution is a characteristic equation of the conditional distribution function corresponding to an eigenvalue of 1. Where a stationary distribution exists, and if it is unique, it is determined by the particular nature of the process.

If the state space is finite, the transition probability distribution can be expressed as a matrix with (i, j) elements, which is called the "transition matrix": $P_{ij} = P(X_{n+1} = i|X_n = j)$. For a discrete state space, the integral of the k-step transition probability is the summation, which can be obtained by finding the kth power of the transition matrix.

That is, if P is a one-step transition matrix, P^k is the transition matrix after the k-step transition.

During the AMHS operation, the arrival time of the wafer card in each storage warehouse is similar to the Poisson distribution; hence, the vehicle has a random characteristics at the time of loading/unloading the wafer card at the loading/unloading position of the storage, which causes that the state sequence of the loading process, the delivering process and the unloading process has a random characteristic. At the same time, the state of the vehicle at the loading/unloading position in the storage warehouse is related to its previous state of this moment; that is, the state of the vehicle has non-aftereffect property. According to the Markovian definition, the state sequence with randomness and non-aftereffect property has a Markov feature. Therefore, the state sequence changes in the loading process, the delivering process and the unloading process have a Markov characteristic. The operation model of an AMHS can be established by using the Markov model.

3.4.2 Markov Modeling Process for AMHSs

This section analyzes the impact of single vehicle's speed and loading amount on the performance of the material handling system [7, 8]. Three parameters are used to physically describe the state of the system: 1) the present location of the vehicle, 2) what load(s) the vehicle is carrying and 3) the waiting demands in queues at each pickup/delivery station. Figure 3.10 gives a pictorial diagram of the state space. The figure of the AGV system is represented by using a series of nodes and arcs. The nodes represent pickup/delivery stations while the arcs represent handling times between stations.

In order to model the behavior of the system, it is necessary to consider the following assumptions. First, the arrival rate of the handling request arrives at queue (i) follows the exponential distribution of the mean value λ_i. This assumption indicates that the machining and completion time of a part is random. A second assumption is that the handling time between stations is deterministic. The handling times for the model include not only the time to move between stations, but also the time to pickup or for delivery.

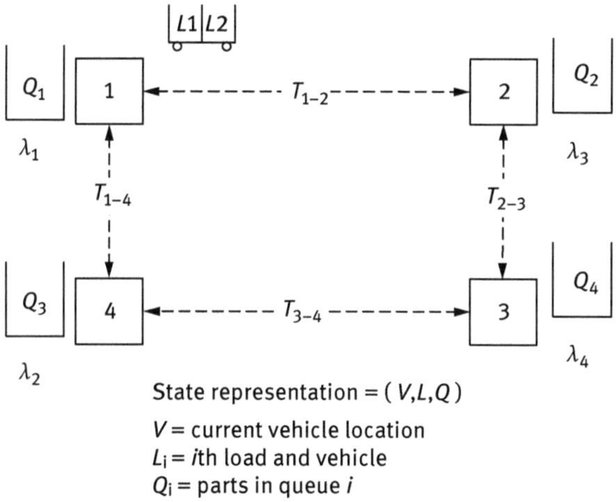

Figure 3.10: State space.

It is easily seen that the process of going from one state to another state is a Markov process. The underlying transition probabilities can be calculated as a function of the arrival rate of the handling request (λ_i), and a discrete probability distribution determines the type of arrival. At each state of the system, the decision to move the vehicle will also affect the outcome of the state transition.

Since even a simple AGV system will have many state numbers, the following constraints are considered in order to easily deal with the problem. 1) The queue size should be limited to one. Increasing the queue by one would increase the number of states by two orders of magnitude. 2) Only up to four stations are considered in the system. 3) The capacity of a vehicle is no more than two parts.

For different layouts, generalized vehicle control strategies are derived from the optimal control strategies. In Figure 3.11, there are four types of layouts that are used to determine the rules of the vehicle.

For a particular AMHS, the optimal control strategy for the orbital layout is dependent on the orbital form. For example, no loading vehicle is at site 1, while there are parts waiting for moving at site 2 and site 3; the optimal control strategy may be "go to site 2." These specific optimal control strategies will be estimated using general rules.

Control logic: (1) If the vehicle is loaded, it is shipped to the destination site. (2) If the vehicle is idle, update the system status. From the site adjacent to the current site, if these sites have a delivery request, then go to the nearest site with the delivery request. If there is no delivery request for the adjacent site, search for other recent sites with delivery requests. Determine the shortest path to the site where the delivery request is located. Select the path to the site with the highest arrival rates. When passing through each site, re-evaluate the state of the system. If there is a part to deliver along the way, then transport the part; otherwise, the vehicle repeats the aforementioned control process.

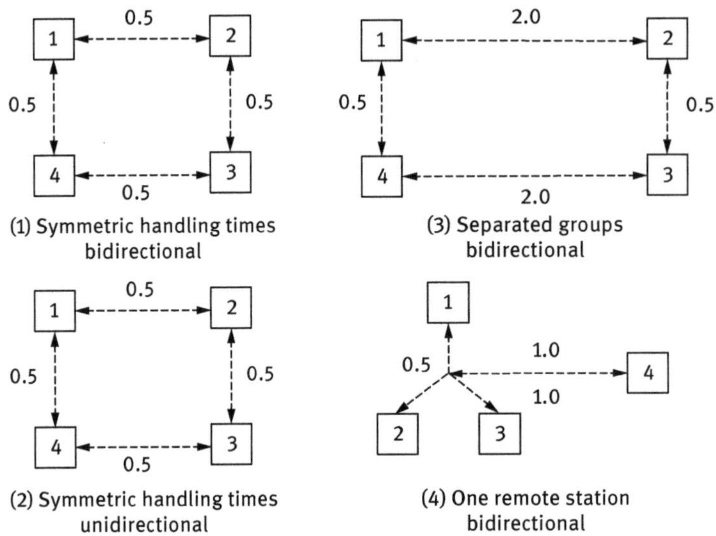

Figure 3.11: Four layouts.

The above control rules try to maximize the number of deliveries each time, rather than considering the waiting time of the part. Finally, the part yield using the control rule is compared with the part yield using the optimized dispatch strategy and the part yield of the control rules used in the current plant. The results show that these rules obtain near optimal scheduling strategies: the maximum part yield and the shortest waiting time of the part.

3.5 Modeling Methods Based on Simulation

A simulation model refers to the decomposition of all concerned phenomena into a series of basic activities and events, and they are combined together according to the logical relationship between activities and events [9]. The simulation model is similar to the object or its structure. It can be a physical model or a mathematical model. However, not all the objects can develop physical models. For example, in order to study the dynamic characteristics of an aircraft, the computer on the ground can only be used to simulate. First of all, we must establish a mathematical model of the object, and then convert it into a form suitable for computer processing. Finally, the simulation test is used to demonstrate the changes of the relevant elements of the system.

3.5.1 Basic Theory of the Simulation Model

A discrete system is a system in which activity and state changes occur only at discrete points. The state of the discrete system is only related to the discrete point,

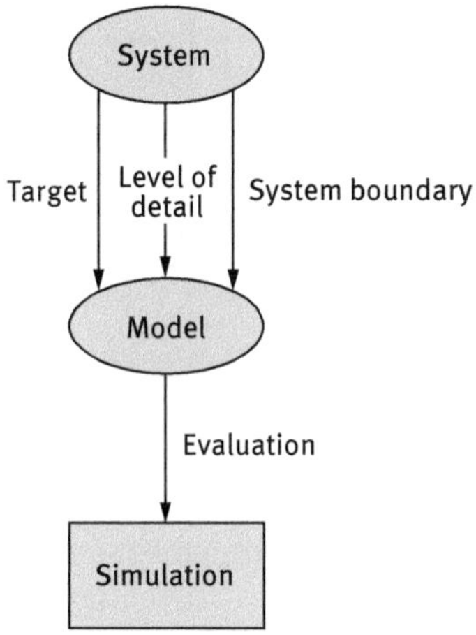

Figure 3.12: The process of the computer simulation.

and the system changes at discrete points. There is no doubt that the manufacturing system simulation problem is a discrete event system simulation problem.

As early as the mid-1960s, American scholars (P.J. Kiviat and M.R. Lackner) have classified the classic simulation strategies; in 1973, G.S. Fishman classified scheduling into three classes: event scheduling, activity scanning and process interaction.

The process of solving problems by computer simulation is divided into three parts: model construction, simulation experiment and data analysis. The process diagram is shown in Figure 3.12.

In the modeling process, we need to consider two issues: one is how to build up a real-world system model; another is how to develop a simulation model in the computer. In the simulation experiment, the researchers set the possible alternatives, the length of running time, the number of repeats and the initial state for the model. At the data analysis phase, three issues should be addressed. First, how to analyze and collect the input data required by the model; second, how to reduce the model output error; third, how to interpret the model output.

Discrete systems involve several concepts of entities, attributes, activities, conditions, states and events. The components of the discrete system interact and relate to each other, and the simulation is carried out. The relationship is shown in Figure 3.13.

Although simulation is widely used in a variety of real-world systems, all discrete-time simulation models have many common elements, and these components have a logical organization that facilitates the programming, debugging and further alteration of the simulation model's computer program. In particular, the following components are found in most discrete-time simulation models using common language programming:

System state: It is a set of state variables that are required to describe the system at a given time;

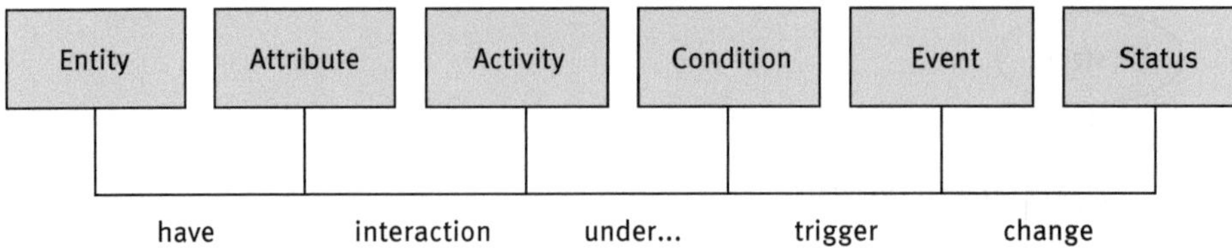

Figure 3.13: The relationship between the various components.

Simulation clock: It gives the current variable of the simulation time;
Event table: It is used to record the next event that will occur;
Statistics counters: These are variables used to store system performance statistic information;
Initialization routine: It is a subroutine that initializes the simulation model at zero time;
A timer routine: It is a subroutine for determining the next event from the event table and advancing the simulation clock to a time at which the next event occurs;
Event routine: It is a subroutine that updates the state of a system when a particular type of event occurs (one for each type of event);
Library routines: It is a set of subroutines used to generate random observations from a defined probability distribution as a part of a simulation model;
Report generator: It calculates the estimated value of the expected performance metric (via a statistical counter) and generates a report at the end of the simulation;
Main routine: It is a subroutine that calls the timing routine to determine the next event, and then transfers control to the appropriate event routine to update the system state as appropriate. The main program can also check the termination condition and call the report generator when the program is stopped.

The logical relationship (control flow) of these components is presented as follows: the simulation process is started at time zero; the main program calls the initialization routine, sets the simulation clock to zero and initializes the system state, the statistical technology and the event table. After control returns to the main routine, it calls the timing routine to determine which type of event occurred first. If type i is the next one to occur, the simulation clock advances to the moment that event type i is about to occur, and control returns to the main routine. Next, the main program invokes event routine i, where three typical activities occur: (1) updating the system state to indicate that an event of type i has occurred; (2) collecting information about system performance by updating the statistics counter and (3) generating the time of occurrence of future events and adding it to the event table. In general, it is necessary to generate random observations from the probability distribution to determine the time of occurrence of these future events; this generated observation is regarded as a random variable. After all procedures are completed, it is checked whether the simulation should be terminated either by event routine i or by the main program (based on a termination condition). If the simulation termination condition is satisfied, the main program calls the report generator to calculate an estimate of the performance expected metric and generate a report. If the simulation has not been terminated, the control returns to the main routine, and the cycle repeats the process (i.e., main routine – timing routine – main routine – event routine – termination check) until the condition of termination is satisfied.

3.5.2 Simulation Modeling Process for AMHSs

The AMHS involves a large number of processing equipment and material handling equipment. Meanwhile, scheduling in AMHSs also has the characteristics of large-scale, real-time, random and multi-objective, and its material handling process includes constraints related to temporary jamming, deadlock, wafer card handling timeliness and in-process balance. The traditional mathematical analysis methods must simplify the scheduling problem in AMHSs, which cannot obtain the desired analysis results. The simulation model is a good way to make up this defect to simulate the operation of an AMHS as much as possible and output comprehensive and detailed statistical data, so as to provide a powerful tool for the researchers. So far, many scholars have applied the simulation model to investigate AMHSs. An example of AMHS simulation modeling based on eM-plant is presented below.

To evaluate the modified Hungarian algorithm (MHA) and the fuzzy-logic control (FLC) method, discrete event simulation models are constructed using the eM-Plant software and VC++.

By analyzing the objects and data of the AMHS in SWFS, the simulation model elements are abstracted and corresponded to the simulation software elements, as shown in Table 3.1.

In addition, there are other elements used in the simulation modeling process, as shown in Table 3.2. Figure 3.14 shows the simulation model.

In order to validate the effectiveness of the MHA and the FLC method, it is compared with several rule-based material handling scheduling methods, including the longest time rule (RLWT) based on a reassignment method, shortest path rule (STD), and the Feed Forward Heuristic Bidding Rule (CLAB) in the case of $N = 3$.

In the simulation model, the moving process of the vehicle is accurately matched with the control process of the vehicle. Therefore, the simulation results can provide a more accurate estimate of the performance of the system. By changing the number of available vehicles, the simulation model analyzes the different scenarios. In addition, four performance parameters will be calculated by simulation.

Table 3.1: Simulation model elements.

Actual objects	Simulation model elements	Icons in simulation software
Track	Track	
Processing machine	Machine	
Import parking area in each processing machine	Port_In	
Export parking area in each processing machine	Port_Out	
Stockers in each intrabay	Stocker	
Wafer lot	lot_a, lot_b, lot_c	
Vehicle	OHT	

Table 3.2: Other elements in the simulation model.

Simulation model elements	Icons in simulation software
Source	
Drain	
Control method	
TableFile	

1. Handling time: The time from when the wafer workpiece is loaded into the car to be moved to the destination storage warehouse;
2. Waiting time: The time the delivery request was sent from the wafer workpiece to the time it was loaded onto the vehicle.
3. Production: During the simulation time, the total number of wafer workpieces completed by the wafer manufacturing system;
4. Vehicle utilization: The percentage of time that a vehicle is actually in use.

Totally, four scenarios were designed for different combinations of two loading ratios (3 cassettes/2.5 h and 3 cassettes/2.75 h) and automated vehicles numbers (8 and 12).

Table 3.3: Performance of different dispatching methods under scenario S1 (3 C/3h, OHT = 12).

Dispatching strategy	Delivery time (s)		Waiting time (s)		Throughput (cassette)	Vehicle utilization
	Mean	Deviation	Mean	Deviation		
MHAFLC	328.9	45,207	128.7	5,220	2,543	0.5247
CLAB	356.8	48,320	141.3	7,903	2,158	0.5267
STD	492.1	49,981	290.8	6,782	2,028	0.5100
RLWT	348.7	45,766	148.0	8,756	2,142	0.5190

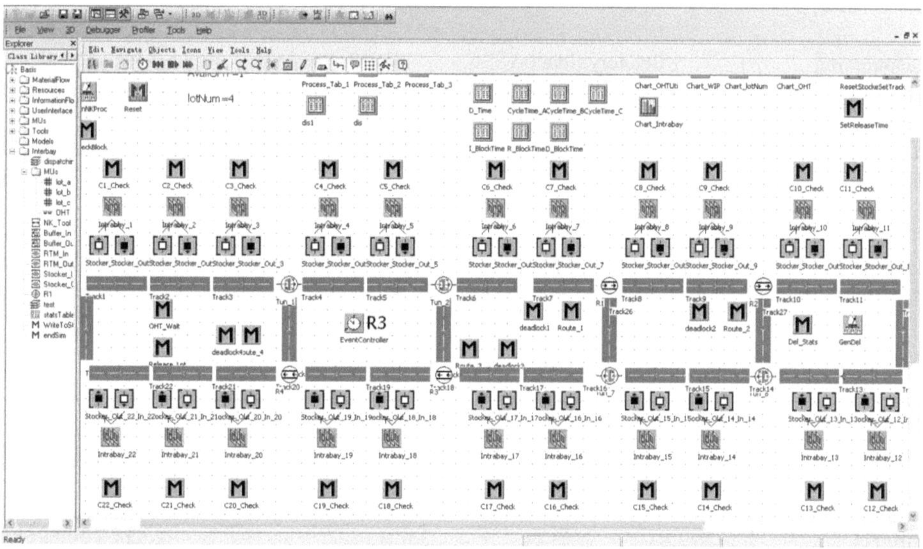

Figure 3.14: The simulation model.

Each scenario was replicated in three simulation runs, and each run simulated a production period of 150 days with an aging time $T_{\max} = 2{,}000$ (s). The total simulation scenarios is $2 \times 2 = 4$, and the experiments repeat three times for each scenario.

The simulation results are shown in Tables 3.3–Table 3.6. In scenarios 1 and 2, the proposed MHAFLC has better performance in terms of mean delivery time, mean waiting time and system throughput. As for vehicle utilization, those dispatching approaches have no significant difference. In scenarios 3 and 4, the proposed MHAFLC has better performance in terms of all measurements except the vehicle utilization. It can be concluded that when the system load is at a high level, the proposed MHAFLC approach may significantly improve the system performance.

For further analysis, an indicator of the comprehensive function D is introduced to compare various dispatching rules. The comprehensive function D is defined as

Table 3.4: Performance of different dispatching methods under scenario S2 (3 C/2.5 h, OHT = 12).

Dispatching strategy	Delivery time (s)		Waiting time (s)		Throughput (cassette)	Vehicle utilization
	Mean	Deviation	Mean	Deviation		
MHAFLC	328.1	45,414	128.5	5,128	2,593	0.6251
CLAB	362.1	48,271	156.5	7,741	2,528	0.5245
STD	494.0	49,181	292.0	6,397	2,369	0.5120
RLWT	348.8	45,966	147.5	8,636	2,505	0.5243

Table 3.5: Performance of different dispatching methods under scenario S2 (3 C/2.5h, OHT = 12).

Dispatching strategy	Delivery time (s)		Waiting time (s)		Throughput (cassette)	Vehicle utilization
	Mean	Deviation	Mean	Deviation		
MHAFLC	521.7	45,414	330.2	5,128	2,589	0.7258
CLAB	641.2	48,271	425.8	7,741	1,261	0.5319
STD	601.2	49,181	491.5	6,397	2,332	0.7215
RLWT	659.5	45,966	379.2	8,636	1,673	0.6167

Table 3.6: Performance of different dispatching methods under scenario S3 (3 C/3h, OHT = 8).

Dispatching strategy	Delivery time (s)		Waiting time (s)		Throughput (cassette)	Vehicle utilization
	Mean	Deviation	Mean	Deviation		
MHAFLC	529.8	45,401	337.8	5,013	2,523	0.7680
CLAB	562.4	48,711	349.2	7,907	1,047	0.5341
STD	590.7	49,317	398.8	7,217	2,026	0.7280
RLWT	673.3	45,106	348.5	8,032	1,344	0.6167

$$D_p(x) = \left\{ \sum\nolimits_{u \in U} w_u \left[\frac{f_u^* - f_u(x)}{f_u^* - f_u^-} \right]^p + \sum\nolimits_{v \in V} w_v \left[\frac{f_v(x) - f_v^-}{f_v^* - f_v^-} \right]^p \right\}^{1/p}$$

where U and V denote the sets of performance measures corresponding to the maximum and minimum objectives, and w_u and w_v are weight coefficients for different measurements. $f_u^* = \max_{x \in X}(f_u(x))$ and $f_u^- = \min_{x \in X}(f_u(x))$ are the permitted maximal and minimal values for index u. f_v^* and f_v^- are defined in the same way. Figure 3.15 shows the D-function values of the proposed dispatching method and other rules ($p = 1$). Table 3.7 reports the ANOVA results of different dispatching methods.

Table 3.7: ANOVA results of different dispatching methods.

Proposed approach	Traditional approaches	Square deviation (SD) × 1,000	Freedom	Mean SD × 1,000	*F*-value
1	MHA	(221.5, 1,112.5)	(1, 23)	(221.5, 48.4)	$4.58 > F_{0.05}(1, 23)$
2	RLWT	(189.1, 914.1)	(1, 23)	(189.1, 39.7)	$4.76 > F_{0.05}(1, 23)$
3	STD	(242.7, 12,01.3)	(1, 23)	(242.7, 52.2)	$4.65 > F_{0.05}(1, 23)$
4	CLAB	(88.9, 612.1)	(1, 23)	(88.9, 26.6)	$3.34 > F_{0.05}(1, 23)$

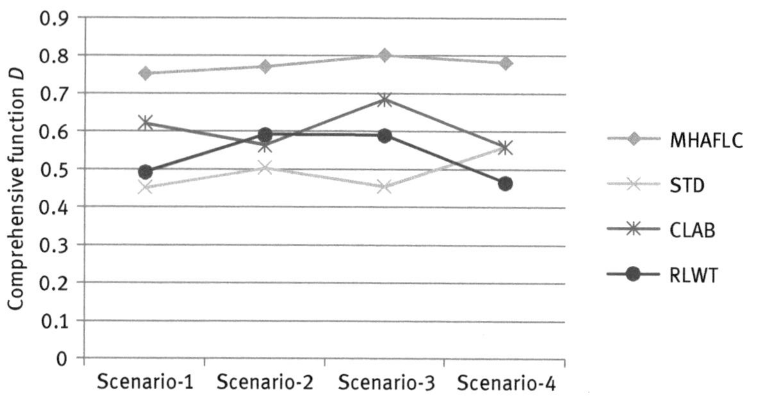

Figure 3.15: Comparison of *D*-function values of proposed dispatching method and other rules.

Furthermore, it is found that the proposed MHAFLC dispatching method has better performance except for vehicle utilization in all scenarios. In terms of *D*-function value, MHA-based dispatching methods have the best performance in comparison with the other rules. However, the fluctuation of its *D*-function value indicates that the robustness needs to be strengthened. The proposed fuzzy-logic-based control method is applied to dynamically adjust the weight and therefore improve the robustness.

Next, how to use the simulation model to evaluate different scheduling rules? In the one-way double closed-loop Interbay material transportation system, the event-driven vehicle assignment strategy has three decision points. 1) When the wafer workpiece needs to be moved to the next storage depot, the transport loop needs to be selected. There are four rules for loop selection: Shortest Distance First (SD), Work in Progress (WN), Work Distance Rule (WD) and Work in Process Rule (WR) on robot transfer mechanism (RTM) 2) Allocate empty vehicles for wafer workpieces. For vehicle assignments, only the recent vehicle priority rule (NV) is available. 3) Arrange the wafer workpiece to be moved to the free vehicle. There are two rules: first encounter first service (FEFS) and the longest waiting time priority rule (LWT). According to the above three decision points and seven rules, you can design eight kinds of rule combinations, as shown in Figure 3.16.

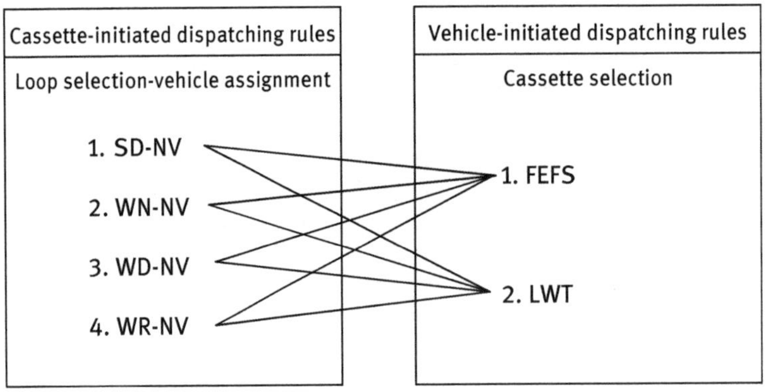

Figure 3.16: Dispatching rules based on casette and vehicle initiation.

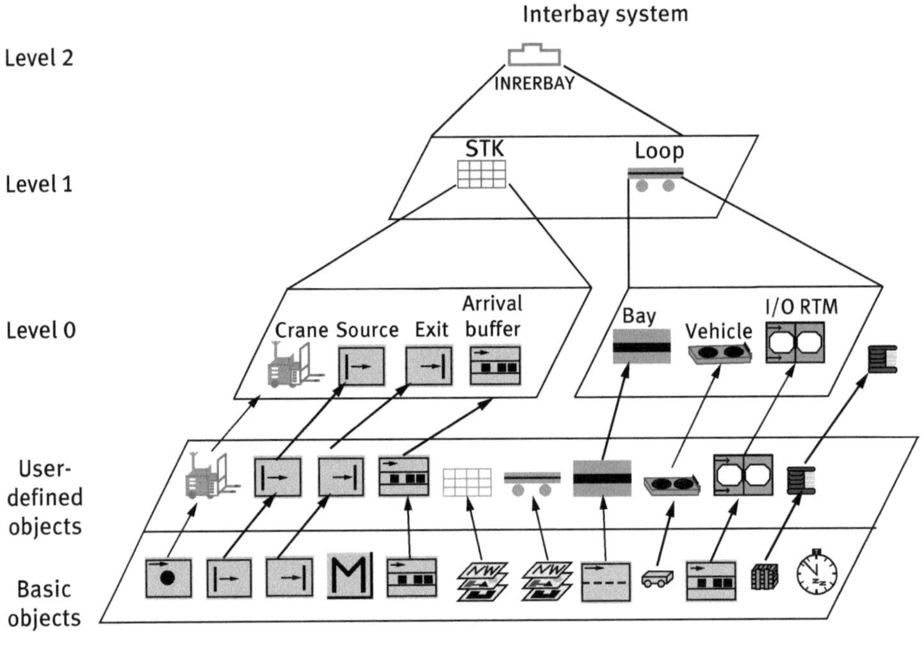

Figure 3.17: Hierarchy of an interbay simulation model.

In order to evaluate the performance of the eight rules as shown in Figure 3.16, a discrete time simulation model of an Interbay material handling system is established based on the simulation software. With the base objects provided by eM-Plant, users can customize more complex objects such as loops that contain buffers, tracks and trolleys. Figure 3.17 shows the objects and their hierarchical relationships of the simulation model for Interbay material handling systems.

In general, productivity in the Interbay system has the highest weight. Two factors affect the performance of the assigned rules: 1) the flow rate of the wafer workpiece and 2) the number of vehicles. The wafer workpiece flow rate can be set to 345,690 or 1,035 cassette h-1. The number of vehicles can be set to (per loop) 8, 12 or 16. Therefore, a total of nine kinds of scenes are possible. Statistical tests using ANOVA show that the number of repetitions of the simulation model does not significantly affect the statistical data of the simulation model. Therefore, each simulation model is repeated only three times. The total number of simulation experiments is 8 (rule) × 9 (scene) × 3 (repeat) = 216 times.

Experimental results show that the assignment rule has significant influence on average transit time, productivity, waiting time and vehicle utilization. Among them, the combination of SD-NV and FEFS rules is obviously superior to other combination of rules. In addition, through the analysis of the corresponding curved surface, the optimal combination of the flow rate and the number of vehicle is obtained.

In general, the scheduling model based on the simulation model can describe the AMHS scheduling process. However, the method also has serious disadvantages such as long modeling period, low modeling efficiency and poor reusability.

3.6 Modeling Methods Based on the Petri Net

The Petri net is a tool developed by C.A. Petri in his doctoral thesis in 1982 as an asynchronous concurrent operation to describe the elements of the system. After several decades of development, the Petri net theory has been widely used in many fields such as computer, electronics, machinery, chemistry and physics, and has become a powerful tool for modeling AMHS [10].

3.6.1 Basic Theory of the Petri Net Model

Petri nets are supported by formalized mathematical methods and can be used to express the static structure and dynamic changes of the discrete event dynamic system (DEDS). It is a structured DEDS description tool that describes the logical relationships of asynchronous, synchronous and parallel systems. These descriptions are used to check and analyze the performance of the system. For example, the Petri net is adopted to describe machine utilization, productivity, reliability and other indicators. A Petri net is a triplet:

$$N = (P, T, F)$$

where $P = \{p_1, p_2, \ldots, p_m\}$ is the set of places;

$T = \{t_1, t_2, \ldots, t_n\}$ is the set of transitions;

$F = (P \times T) \cup (T \times P)$ is the input function and the output function set, which is regarded as the flow relationship.

A triplet $N = (P, T, F)$ is formed to be a net structure under the practical constraints. The necessary and sufficient conditions for the triple network to form a net are as follows:

① $P \cap T = \phi$ defines that the places and the transitions are two different kinds of elements;
② $P \cup T \neq \phi$ indicates that there is at least one element in the net;
③ $F = (P \times T) \cup (T \times P)$ establishes the unilateral links from the places to the transitions and vice versa, and it is not possible to provide the direct link between similar elements.

The state of a process is modeled using tokens in places, and the transitions of states are modeled by using transitions. A token represents the state of a thing (person, goods, machine), information, condition or object; a place represents a location, a channel or a geographic location; a transition represents an event, transformation or transmission.

A process consists of a current state, reachable state, unreachable state and termination status.

The Petri net is an effective tool to investigate discrete manufacturing systems. It has the dual functions of graphical representation and mathematical description in dealing with dynamic discrete events and complex systems. It is especially suitable for modeling the concurrent behavior, and it is widely used in modeling AMHSs, modeling wafer processing systems and evaluating scheduling performances.

3.6.2 Petri Net Modeling Process for AMHSs

An agent can make decisions and control based on common goals and information. In terms of this advantage of an agent, the scheduling performance evaluation of the wafer processing system is modeled using an agent-oriented colored timed Petri net (AOCTPN) developed by Zhang Jie and Zhai Wenbin [11–13]. The modeling process is presented as follows.

In order to reduce the complexity and increase the reusability of re-entrant system model, the agents of re-entrant system with similar structure and function are aggregated to form the layered structure, which includes basic agent layer, unit layer and system layer (as shown in Figure 3.18).

1. Basic agent layer modeling
   As the basic element of a re-entrant system model, the agents in the basic agent layer include physical agents and the logical ones. Corresponding to the manufacturing resources of a re-entrant system, the physical agents include machine

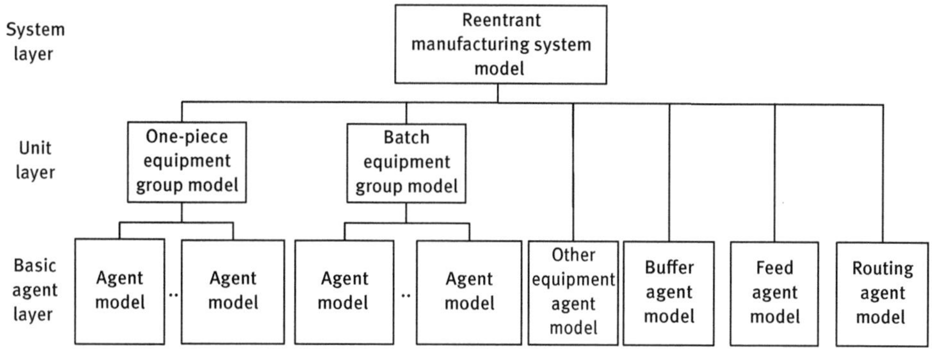

Figure 3.18: Hierarchical structure of logistics network model.

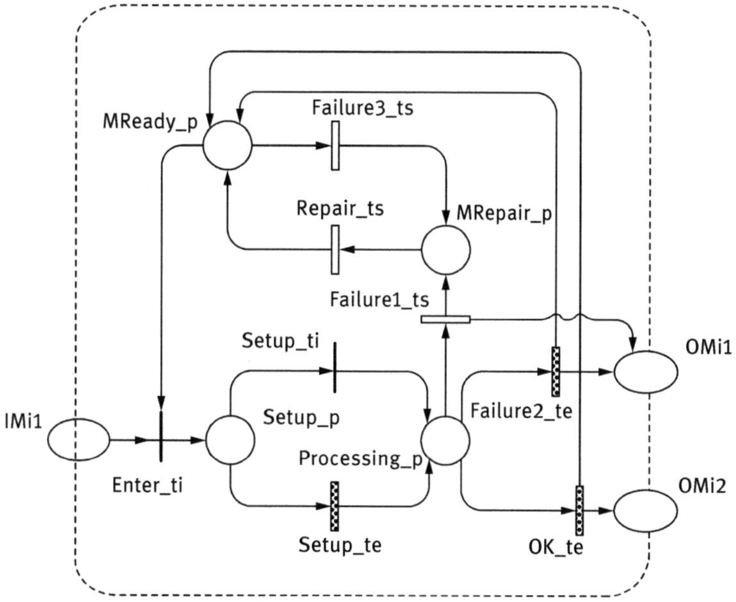

Figure 3.19: AOCTPN based machine agent.

agents, buffer agents and so on. The logical agents with dispatch or control functions include lot release agents, dispatching agents and so on.

The machine agent is the core in the machine layer, whose model characterizes the interior behavior of the agent, such as machine setup, operation or rework, failure and maintenance, and so on. According to the part types, the machines are classified into single-lot processing machines and batch processing machines. The reasoning process of the single-lot-machine-agent model is shown in Figure 3.19. The token in the input message place *IMi*1 represents the

Table 3.8: Color data type.

Data interpretation	Elements	Type name
LI	Lot NO	Natural numbers
P	Product ID	$\{p_1, \ldots, p_j, \ldots, p_m\}$
S	Lot stage	$\{S_1, NS_1, \ldots, S_j, NS_j, \ldots, S_n, NS_n\}$
PT	Product processing time	$\{PT_1, NPT_1, \ldots, PT_j, NPT_j, \ldots, PT_n, NPT_n\}$
ST	Key machine setup time	$\{ST_1, \ldots, ST_j, \ldots, ST_n\}$
KMG	Key machine group NO	Natural numbers
m	Key machine NO	Natural numbers
E	Key machine state	$\{e_{ok}, e_r, e_f\}$
L	Result of lot processing	$\{l_{ok}, lr\}$

Table 3.9: Color sets of the single lot processing machine agent.

Name	Function interpretation	Color set
IM_{i1}	Input message of lot waiting for processing	$LI \times \{P_i\} \times \{S_j\} \times KM$
MReady_p	Machine being idle	$KM \times \{e_{ok}\} \times \{P_i\} \times \{S_j\}$
Enter_ti	Transfer to machine	$LI \times \{P_i\} \times \{S_j\} \times KM \times \{e_{ok}\}$
Setup_p	Check machine setup	$LI \times \{P_i\} \times \{S_j\} \times KM \times \{e_{ok}\}$
Setup_t_i	Process without setup	$\{LP_k\} \times \{LS_m\} \times KM \times \{e_{ok}\}$
Setup_t_e	Machine setup	$\{P_i\} \times \{S_j\} \times KM \times \{e_{ok}\} \times \{ST_j\}$
Processing_p	Start lot processing	$LI \times \{l_{ok}\} \times \{P_i\} \times \{S_j\} \times KM \times \{e_{ok}\} \cup$ $LI \times \{lr\} \times \{P_i\} \times \{S_j\} \times KM \times \{e_{ok}\} \cup$ $LI \times \{lr\} \times \{Pi\} \times \{S_j\} \times KM \times \{e_f\}$
Failure1_t_s	Machine failure and lot rework	$LI \times \{P_i\} \times \{S_j\} \times KM \times \{e_f\}$
Failure2_t_e	Machine idle and lot rework	$LI \times \{P_i\} \times \{S_j\} \times KM \times \{e_{ok}\} \times \{PT_j\}$
Failure3_t_s	Machine failure	$KM \times \{e_f\} \times \{P_i\} \times \{S_j\}$
OK_t_e	Lot processing	$LI \times \{P_i\} \times \{S_{j+1}\} \times KM \times \{e_{ok}\} \times \{PT_j\}$
Mrepair_p	Start machine repair	$KM \times \{e_r\} \times \{P_i\} \times \{S_j\}$
Repair_t_s	Machine repair	$KM \times \{e_r\} \times \{P_i\} \times \{S_j\}$
OM_{i1}	Output message of rework	$LI \times \{P_i\} \times \{S_j\}$
OM_{i2}	Output message of next process stage	$LI \times \{P_i\} \times \{S_{j+1}\}$

lot-processing request messages sent by the dispatch agent. The data type of different colors is shown in Table 3.8. The color sets of the single-lot processing machine agent are shown in Table 3.9. The token in the place *MReady_p* corresponds to machine idleness. When the machine is idle and receives the lot-process-request message, the immediate transition *Enter_ti* is fired, and the lot token enters the place *Setup_p*. The color of the immediate transition *Enter_ti* is

$\{LPk\} \times \{LSm\} \times KM$, which records the last process step of the lot. The deterministically timed transition *Setup_te* represents the setup time of the machine. If the color $\{Pi\} \times \{Sj\} \times KM$ of the token in the place *Setup_p* is the same as the color of the transition *Setup_ti*, then the transition *Setup_ti* is fired and the lot token directly enters place *Processing_p* for processing. Otherwise, the transition *Setup_te* is fired and the machine starts to set up. The place *Processing_p* may fire one of three kinds of the stochastic transitions, *Failure1_ts, Failure2_te* or *OK_ts*, which respectively correspond to machine failure, lot fault and finish. According to the rework rate of a lot, whether the lot reworks or not is decided. If the lot does not rework, then the immediate transition *OK_te* is fired. After a period of process time *PTj*, the machine becomes idle and the machine token enters *MReady_p*, and the lot token in place *Processing_p* enters the output message place *OMi*2. If the lot needs rework, then machine failure is decided by the mean time to failure rate (MTTF). In the case of the machine failure, the transition *Failure1_ts* is fired and the machine starts repair. The machine token enters place *Mrepair_p*. At the same time, the lot token enters the output message place *OMi*1 to send a rework request.

If the machine is okay, the transition *Failure2_te* is fired and the machine is idle. The machine token enters place *MReady_p*. Similarly, the lot token enters output message place *OMi*1 to send rework request. Since the process time is shorter than the failure cycle of the machine, it supposes that the machine never fails during the processing period. As the machine is idle, the failure of the machine is decided by MTTF. If the machine fails, the transition *Failure3_ts* is fired. The time for machine repair is given by the mean time to repair (MTTR) rate.

The case of the batch processing machine agent is complicated, and its figure is the same as Figure 3.19. The color set of *IMi*1, *Enter_ti, Setup_p, Setup_ti, Setup_te, Failure1_ts, Failure2_te, OK_ts, OMi*1 and *OMi*2 depends on the batch size of the machine. For example, if the batch size is two, the color set of *Enter_ti* is $LI1 \times \{Pi\} \times \{Sj\} \times LI2 \times \{Pm\} \times \{Sn\} \times KM$.

The other agent models based on AOCTPN are shown in Figure 3.20. In the actual production, it is assumed that the capacity of the non-bottleneck equipment in the buffer zone is infinite. In the non-critical equipment agent model, the machining process is simplified to a fixed time delay, and the time is related to the total processing time of non-critical processes. The agent model invokes the interactive protocol network model through message bits *OMi*2 and *IMi*2. Through the multi-agent negotiation, the scheduling process of the processing tasks is completed.

2. Unit layer modeling

The unit layer agent belongs to the composite agent and consists of several processing machine agents and machine group agents with the same machining

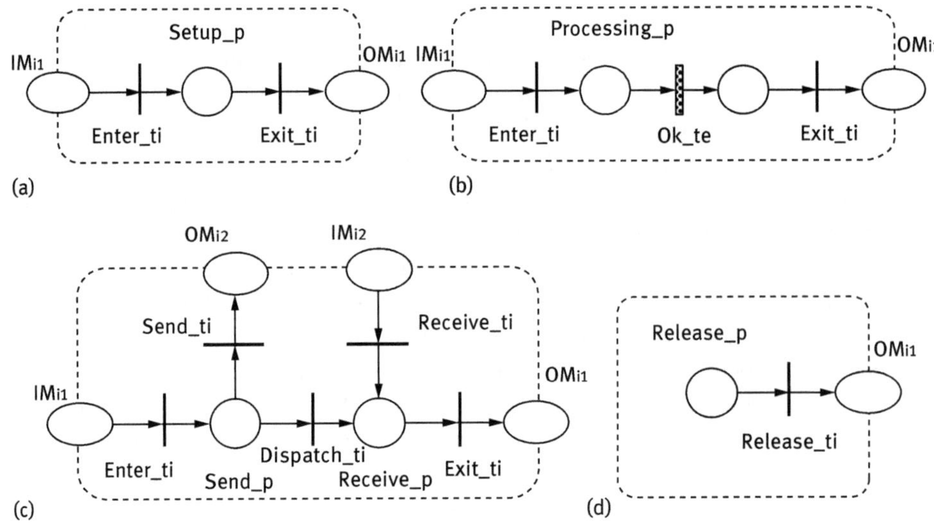

Figure 3.20: Other agent models based on AOCTPN: (a) the buffer agent model; (b) the non-key equip agent model; (c) the dispatch agent model; and (d) the lot release agent model.

function. Each basic agent is connected to form a complete AOCTPN model according to specific control logic; that is, the communication transition determines the cooperative and interactive relationship between the manufacturing resources of the unit layer.

In the specific unit layer model, the triggering mode and the triggering order of the communication transition determine the control logic of the unit layer. The triggering processes of some traffic changes can be determined by the characteristics of the Petri net (Figure 3.21(a)). However, the triggering processes of other communication changes should introduce the corresponding multi-agent interaction protocol in the model, which is regarded as the conflict when the characteristics of Petri nets cannot be determined and multi-agent cooperative decision is needed. Its corresponding multi-agent cooperative strategy is regarded as the multi-agent interaction protocol of conflict discrimination. When a collision is detected, the device agent issues a conflict resolution request to the machine group agent. The machine group agent calls the multi-agent interaction protocol to complete the conflict resolution. The conflict in the model can be classified into three classes according to the specific situation:

1. There is a conflict between the communication changes; that is, two or more communication changes have the same input bit. Therefore, it is necessary to determine which communication transition is triggered by the input bit. This is consistent with the conflict in the general Petri net. This kind of conflict usually occurs when a part in a unit chooses a processing machine, and the conflict can be solved by a multi-agent interaction protocol (Figure 3.21(b)).

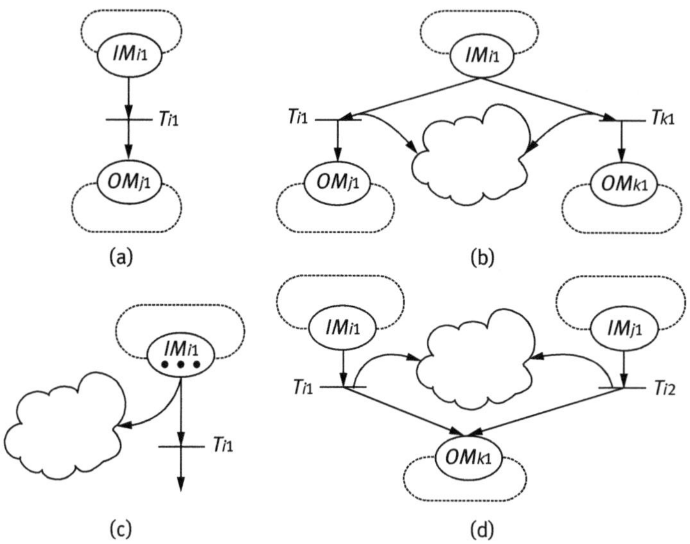

Figure 3.21: Conflicts in the AOCTPN model.

2. In collisions between tokens in bits that trigger communication transitions, where the number of tokens in the input bits of the communication transition is greater than the number of tokens required to trigger the transition, it is necessary to determine which token to trigger for the corresponding transition. This conflict usually occurs when multiple parts in the buffer are waiting to be processed, and the conflict can be resolved according to the multi-agent interaction protocol (Figure 3.21(c)).
3. Conflicts between bits trigger communication transitions, in which the input bit *IMk*1 can be triggered by a different output bit *IMi*1 or *IMj*1 (Figure 3.21(d)). The solution at this time is to determine which communication transition triggers first. This kind of conflict often occurs when parts which are processed by different machines compete with the processing machine immediately after the process, and the multi-agent interaction protocol can be used to solve the conflict.

The reasoning process of the key machine group agent is shown in Figure 3.22. The token in the input message place *IMj*1 corresponds to the lot-processing request messages sent by the buffer agent. By firing the immediate transition *ID_ti*, the lot token enters the dispatch agent place *Dispatch_ap*. Guided by the coordination mechanism, the dispatch agent optimally finds the lot token that should enter the machine agent place *KM_ap_i* and the communication transition *DK_ti_i* that should be fired. If a lot needs rework, the communication transition *KD_ti_i* is

Table 3.10: Color sets of the single lot processing machine group.

Name	Function interpretation	Color set
IM_{j1}	Input message of lot waiting for processing	$LI \times \{P_i\} \times \{S_j\}$
ID_ti	Request of dispatch	$LI \times \{P_i\} \times \{S_j\}$
Dispatch_ap	Select processing lot and machine	$LI \times \{P_i\} \times \{S_j\} \times KM$
DK_ti_i	Request of processing	$LI \times \{P_i\} \times \{S_j\} \times KM$
KM_ap_i	Lot processing	$LI \times \{P_i\} \times \{S_j\} \times KM \times \{PT_j\}$
KM_ti_i	Request of output	$LI \times \{P_i\} \times \{S_{j+1}\}$
KD_ti_i	Message of lot rework	$LI \times \{P_i\} \times \{S_j\}$
OM_{j1}	Output message of next process stage	$LI \times \{P_i\} \times \{S_{j+1}\}$

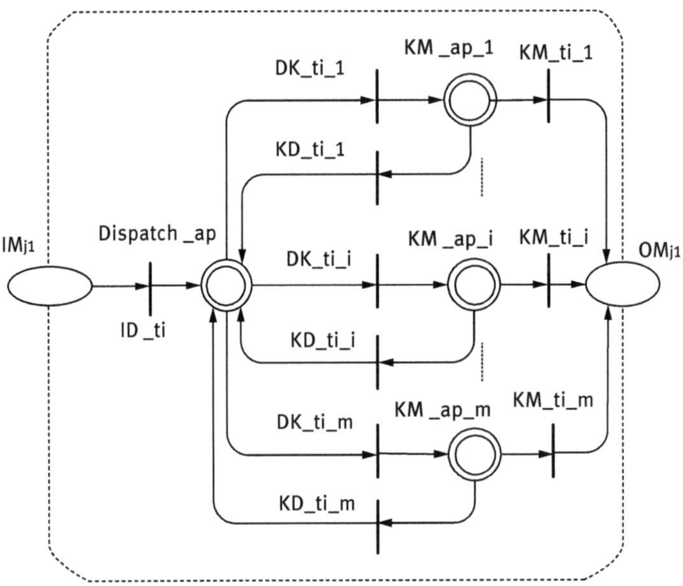

Figure 3.22: AOCTPN based machine group.

fired. Otherwise, the communication transition *KM_ti_i* is fired and the lot token enters the output message place *OMj*1. The color sets of the single-lot processing machine group agent are shown in Table 3.10. In the case of the batch-processing machine agent, the color sets of its place and transitions depend on the batch size.

3. System layer modeling
During the lot processing, the main waiting time happens in the bottleneck machine group. Hence, the bottleneck machine group is called the key

machine group. The system model focuses on the key machine group while the non-key machine group is treated as infinite possessing ability and is simplified into deterministic time delay. Since the buffer and material handling system capacity is bigger than the work-in-process (WIP) requirement, its constraints are neglected.

The re-entrant system model in the system layer is shown in Figure 3.23. The release agent place *Release_ap* sends a message to the agent place of the material handling system. After lots transfer, the lot token enters the buffer agent place *Buffer_ap_i*. When the agent place of the key machine group *KMG_ap_i* receives the processing request from the buffer agent place *Buffer_ap_i*, the lot token enters *KMG_ap_i*. After the key processing step *KMG_ap_i* is completed and the non-key processing step *NKMG_ap_i* is completed, the lot token enters check-out agent place *Check-out_ap*. If the lot has finished all steps, it enters exit agent place *End_ap*. Otherwise, the lot token reenters the *Routing_ap* for the next process step. The color sets of the system layer are shown in Table 3.11.

The AOCTPN method can describe autonomous coordination scheduling and control logic in material handling systems, but it cannot describe information and knowledge required for scheduling decision and running process. Therefore, it is difficult to effectively integrate the material handling scheduling process with the performance evaluation model.

Table 3.11: Color sets of the system layer.

Name	Function interpretation	Color set
Release_ap	Lot release	$LI \times \{P_i\} \times \{S_j\}$
RR_ti	Request of lot transfer	$LI \times \{P_i\} \times \{S_j\}$
Routing_ap	lot transfer	$LI \times \{P_i\} \times \{S_j\} \times KMG$
RB_ti_i	Request of entering buffer	$LI \times \{P_i\} \times \{S_j\} \times KMG$
Buffer_ap_i	Enter buffer	$LI \times \{P_i\} \times \{S_j\} \times KMG$
BK_ti_i	Request of lot processing by key machine group agent	$LI \times \{P_i\} \times \{S_j\} \times KMG$
KMG_ap_i	Key machine group processing	$LI \times \{P_i\} \times \{S_j\} \times KMG \times KM \times \{PTj\}$
KN_ti_i	Request of lot processing by non-key machine group agent	$LI \times \{P_i\} \times \{S_{j+1}\} \times \{NS_j\}$
NKMG_ap_i	Non-key machine group processing	$LI \times \{P_i\} \times \{S_{j+1}\} \times \{NS_j\} \times \{NPTj\}$
NC_ti_i	Request of checking exit	$LI \times \{P_i\} \times \{S_{j+1}\} \times \{NS_{j+1}\}$
Check-out_ap	Checking whether lot exit	$LI \times \{P_i\} \times \{S_{j+1}\} \times \{NS_{j+1}\}$
CR_ti	Send message of processing next stage	$LI \times \{P_i\} \times \{S_{j+1}\} \times \{NS_{j+1}\}$
CE_ti	Send message of lot exit	$LI \times \{P_i\} \times \{S_{end}\} \times \{NS_{end}\}$
End_ap	Lot exit	$LI \times \{P_i\} \times \{S_{end}\} \times \{NS_{end}\}$

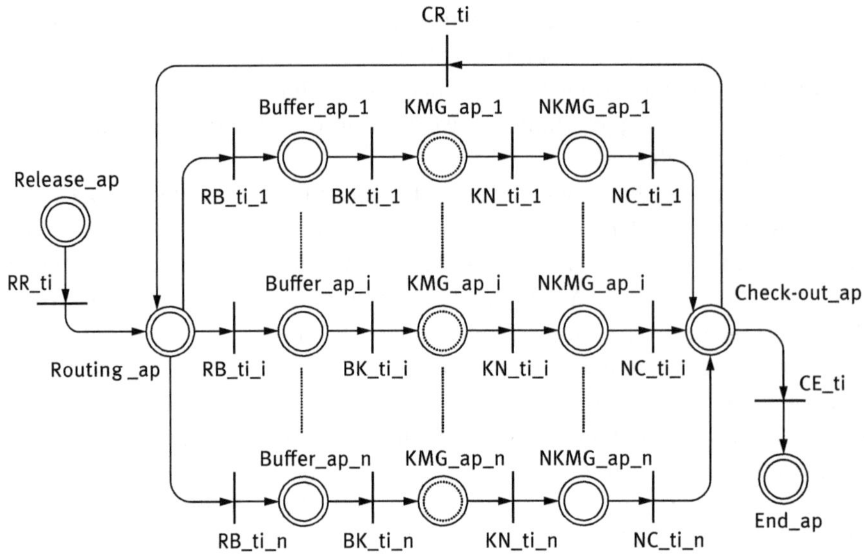

Figure 3.23: AOCTPN based system layer model.

3.7 Conclusion

This chapter presents several modeling methods of AMHSs regarding two aspects: the basic theory and its applications in scheduling problems.

The modeling methods such as network flow model, queuing theory model, mathematical programming model and Markov model are often developed on the basis of the steady state. In the modeling process, the stochastic characteristics of the material handling process are simplified, these conditions may not be satisfied in practice, and the accuracy of the analysis results cannot be guaranteed. It is difficult to obtain the solutions for large-scale problems. Both the simulation model and the Petri net model are effective tools to study the discrete manufacturing system, which can describe and analyze the operation process of an AMHS. In particular, Petri nets have the dual functions of graphical representation and mathematical description in dealing with the characteristics of dynamic discrete events and complex systems, especially for modeling concurrent behavior, and are particularly useful for modeling concurrency behavior as a powerful tool to evaluate the scheduling performances.

Different modeling methods can solve actual problems with actual demands. For example, the queuing theory and the Markov model can be used to achieve better results at lower cost to analyze the statistical properties of an AMHS. The simulation model or the Petri net model can be used to clearly describe the operations of an AMHS.

References

[1] Jiang, Q.Y., Xie, J.X., Ye J. Mathematical Model. Beijing: Higher Education Press, 2003.
[2] Chen, B.L. Theory and Algorithms of Optimization. Beijing: Tsinghua University Press, 2005.
[3] Averill, M. Law. Simulation Modeling and Analysis (Fourth Edition). McGraw Hill Higher Education, New York, 2007.
[4] Fishman, G.S. Statistical analysis for queuing simulations. Management Science, 1999, 20(3).
[5] Curry, G.L., Peters, B.A., Lee, M. Queuing network model for a class of material handling systems. International Journal of Production Research, 2003, 41: 3901–3920.
[6] Roeder, T., Govind, N., Schruben, L. A queuing network approximation of semiconductor automated material handling systems: how much information do we really need?. Proceedings of the 2004 Winter Simulation Conference. 2004, pp. 1956–1961.
[7] Nazzal, D., Mcginnis, L.F. Queuing models of vehicle-based automated material handling systems in semiconductor Fabs. Proceedings of the 2005 Winter Simulation Conference. 2005, pp. 2464–2471.
[8] Johnson, M.E., Brandeau, M.L. An analytic model for design and analysis of single vehicle asynchronous material handling systems. Transportation Science, 1994, 28: 337–353.
[9] Kobza, J.E., Yu-Cheng, S., Reasor, R.J. A stochastic model of empty-vehicle travel time and load request service time in light-traffic material handling systems. IIE Transactions, 1998, 30: 133–142.
[10] Peterson, J.L. Petri Net Theory and the Modeling of Systems. Prentice-Hall, New Jersey, USA, 1981.
[11] Zhang, J., Zhai, W., Yan, J. Multi-agent-based modeling for re-entrant manufacturing system. International Journal of Production Research, 2007, 45(13): 3017–3036.
[12] Zhai, W.B., Chu, X.N., Zhang, J., Ma, D.Z. Research on AOCTPN based modeling technology of semiconductor fabrication line. Computer Integrated Manufacturing Systems, 2005, 11(3): 326–329.
[13] Zhai, W.B., Zhang, J., Yan, J.Q., Ma, D.Z. Agent -oriented colored petri-net based interactive protocol modeling technologies of semiconductor fabrication line. Journal of Shanghai Jiaotong University, 2005, 39(7): 1150–1154.
[14] BS Koo, HS Jang. Evaluation on the impact of Lowest Bid Contracts on Site Operations in times of Severe Economic Downturn. Korean Journal of Construction Engineering and Management, 2009, 10 (6): 146–153.
[15] Kimura, Approximating the Mean Waiting Time in the GI/G/s Queue. Journal of the Operational Research Society, 1991, 42(11): 959–970.

4 Analysis of automated material handling systems in SWFSs

There have been many reports of investigation on the performance of analytical modeling approach for automated material handling systems. The most common and practical approach is the simulation model. Although the use of the simulation tool is advantageous in that all details can be taken into account, it is time-consuming and error prone to generate statistically reliable results. Especially when the dynamic system changes, e.g., the adjustment of the layout, the increase of machine or vehicles and the setup of some operational rules, the reconstruction of the simulation model is tedious. To avoid the shortcomings of simulation approaches, various mathematical models which are more efficient have been proposed to evaluate the performance of the automated material handling system (AMHS), such as queuing model, queuing network model and Markov chain model (MCM). Compared with the simulation tools, these analytical models may need to be improved in terms of the effectiveness (i.e., the accuracy); however, their runtime efficiency is much higher. This is critical in some environment with high demand for real-time decision; for example, the model is used to test various schedules. The mathematical models could evaluate the performance of schedules in several minutes while the simulation model will run a couple of hours to evaluate, which normally is not acceptable.

This chapter proposes a modified Markov chain model (MMCM) for more general AMHSs, in which the distinctive AMHS's features including shortcut rail, vehicle blockage and multi-vehicle transportation are taken into account. On the basis of the proposed model, the performance of the AMHS is efficiently analyzed and evaluated.

4.1 Running Process Description

In semiconductor manufacturing, multiple layers of miniature circuitry are built upon a wafer. For each layer, a sequence of similar processes must be undertaken repeatedly, often on the same pieces of equipment. Between two processing steps, an AMHS is used to transport wafer lots. The layout of an AMHS is similar to that of a flow shop with functional areas, and can be separated into Interbay and Intrabay AMHS (as shown in Figure 2.2). Interbay and Intrabay AMHS are connected by stockers [1]. In a typical AMHS, the distance between adjacent stockers is 20 m. The distance between the pick-up port and the drop-off port of each stocker is 5 m. And the total Interbay rail loop is 320 m long. Since the configuration and constitution of Interbay AMHS and Intrabay AMHS are similar, the Interbay AMHS is investigated in this chapter.

The Interbay AMHS typically consists of transportation rails, stockers, automated vehicles, shortcut and turntable. The transportation rails are usually monorail

https://doi.org/10.1515/9783110487473-004

systems, where automated vehicles transport wafer lots along the rails between different stockers. Each stocker has an input/output port and provides temporary storage for work in process of wafer lots. Each stocker is connected with an Intrabay AMHS. Automated vehicles are used to carry and transport wafer lots. Shortcuts are special transportation rails that are used to connect two transportation rails with opposite directions. Shortcut can shorten the vehicle's transportation distance between stockers significantly. Turntables are the intersection of shortcut and transportation rails.

Due to the space restriction in the Interbay AMHS, vehicle's movement is along a unidirectional closed loop and vehicles cannot pass each other, even when a vehicle stops to drop-off or pick-up a wafer lot at a stocker. Let $L(m)$ refer to the Interbay AMHS with m vehicles. Denote S as the set of stockers in the Interbay system, $S = \{s_i,\ i = 1, \ldots, n\}$. s_i is the ith stocker in the Interbay system. Each stocker has a pick-up station and a drop-off station where wafer lots are dropped off and picked up by vehicles, respectively. Let s_i^p and s_i^d denote the pick-up station and the drop-off station. Then, an Interbay AMHS with n stockers consists of $2n$ stations. As the pick-up and drop-off stations are represented as nodes, and transportation rails and shortcuts are represented as directed arcs, the network representation of the Interbay AMHS can be illustrated as shown in Figure 4.1.

The continuous material handling process is defined as the vehicle's discrete travel process at the pick-up/drop-off station to describe material handling logic of the Interbay AMHS and analyze its performance indices effectively. In the Interbay AMHS, the vehicles constantly move on the unidirectional loop and transport wafer lots based on first come first service (FCFS) rule. Without loss of generality, given that an empty vehicle approaches stocker s_i, it passes through the drop-off station s_h^d and travels to the pick-up station s_h^p. If there is a wafer lot waiting at s_h^p, then the vehicle stops and picks it up. After a constant loading time t^l, the vehicle moves to the destination stocker according to the transportation path $ph = (s_h^p, s_{h+1}^d, s_{h+1}^p, \ldots, s_n^d, \ldots)$ (given the destination position of the current lot is stocker s_n, $s_n \in S$). The vehicle

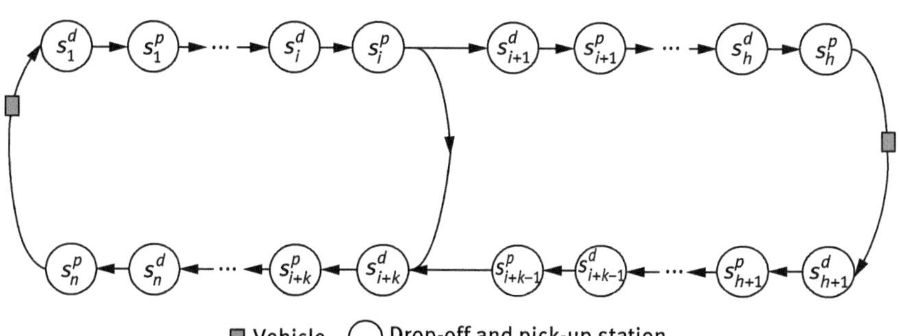

Figure 4.1: Network representation of the Interbay AMHS.

does not stop at stations s_{h+1}^d, s_{h+1}^p, ..., and s_{n-1}^p unless it is blocked by other vehicles at downstream stations. After vehicle arrives at stocker s_n^d and drops off its load, it travels to s_n^p and inspects if there are wafer lots waiting at s_n^p, and so forth until it encounters a waiting wafer lot.

4.2 Methods for Analyzing Running Characteristics

On the basis of the AMHS model, the internal changeover process and operation mechanism of an AMHS are analyzed. The static and dynamic factors that affect the performance of an AMHS are determined, and the changing rules and characteristics of an AMHS are revealed, which provide a foundation to optimize the AMHS.

As mentioned in Chapter 3, modeling is the process to abstract the system. With different mathematical or computing tools, we can establish different system models from different angles. These different models have different characteristics and application fields, while the corresponding analytical methods depend on the established system models. For example, analysis methods based on queuing network model, Markov model and perturbation analysis mainly explain the running process of an AMHS in the statistical level. Analytical methods based on formal languages and automata, Markov chains and Petri nets focus on analyzing a system's internal behavior in the logical level. Analytical methods based on a finite recursion process and a minimax algebra model study system's trajectory and characteristics in the time level. Since several common modeling methods in a wafer manufacturing system have been introduced in the previous chapter, this section mainly presents the advantages and disadvantages of the analytical methods.

The analytical method based on the network flow model generally establishes the running analysis model of an AMHS by simplifying the AMHS into a deterministic directed graph model, and regarding the material handling time as the deterministic time. Based on the model, we can study the performance of an Interbay system and the related influence factors including vehicle idle time, vehicle quantity and vehicle speed. However, this method relies on assumptions such as steady state, independence and so on. In the process of modeling, random characteristics of the material handling process are simplified, which cannot be satisfied in practice, and thus cannot guarantee the results of the analysis accuracy. Therefore, the network flow model is suitable for the statistical analysis of a stationary system, and it is difficult to describe the stochastic characteristics of the running process of an AMHS.

The analytical methods based on the queuing network model generally models the AMHS as a queuing network, and depicts different running states with Markov model. Then the model can be used to analyze the impacts of handling shortcuts location on quantity of WIP, the relationship between the speed and load of vehicle and system throughput, the impacts of vehicle number on system throughput and vehicle average transportation time, the impacts of vehicle speed and turntable system rotation speed on system's performance, the impacts of vehicle number on material waiting time, the impacts of vehicle blockage on regional control AMHS performance and so on. The queuing network model can effectively describe the random behavior of an AMHS. However, it is difficult to model the characteristics of multiple vehicles blockage and empty vehicles. Therefore, this method only applies to analyze small-scale AMHSs, while for large-scale and complex AMHSs it is easy to cause a massive explosion of model states.

The analytical method based on the mathematical programming model analyzes its running characteristics and optimizes the system's operation parameters. The method is mainly used to analyze the effects of handling shortcuts and track positions on vehicle handling time, the effects of vehicle blocking time, vehicle speed and vehicle number on vehicle empty time and vehicle load rate, and the effects of vehicle number on AMHS's throughput and production cost. Nevertheless, it is difficult to apply the method in a stochastic and large-scale environment.

The analytical method based on the Markov model considers the Markov properties of state change sequences in vehicle handling, loading and unloading wafers. The method simplifies the AMHS scheduling problem into the mathematical model by defining vehicle-related parameters, storage-related parameters and states-related parameters, and establishes the running characteristic analysis model of the AMHS. It can be used to analyze the impacts of a single vehicle's speed and load on AMHS performance, the vehicle's performance under different running speeds and loading/unloading times, and the impacts of wafer's waiting time and vehicle number on AMHS performance. However, the Markov model can easily cause the state space explosion of the model in a large-scale environment, and it is very difficult to solve the model.

To summarize, due to the complexity of the AMHS, it is difficult to apply the network flow model, the queuing theory and the mathematical programming model to the system for operation analysis. This chapter presents an improved Markov chain model for performance analysis of the AMHS. The method considering characteristics of shortcut track's layout can accurately express operation characteristics of an AMHS so as to analyze the performance of the complex AMHS [2].

4.3 Modified Markov Chain Model

So far, many approaches for AMHS performance analyzing have been proposed, such as queuing theory model, queuing network model, Markov chain model (MCM) and so on. Due to the complexity issue, few literature can be found upon modeling for the AMHS with shortcuts and blocking. In this section, an MMCM for more general

AMHSs is proposed, in which the distinctive AMHS's features including shortcut rail, vehicle blockage and multi-vehicle transportation are taken into account. On the basis of the proposed model, the performance of the AMHS is effectively analyzed and evaluated.

4.3.1 Notation and System Assumption

1. Parameters related to stockers
   S – Set of stockers in the system;
   n – Quantity of stockers in the system;
   s_i – The ith stocker in the system, $s_i \in S$, $i \in \{1, 2, ..., |S|\}$;
   s_i^p – Pick-up station of s_i;
   s_i^d – Drop-off station of s_i;
   U_i – Set of pick-up stations in the upstream of s_i;
   D_i – Set of drop-off stations in the downstream of s_i;
   p_{ij} – The probability that a moving request is picked up from s_i^p and destined to s_j^d;
   λ_i – Mean arrival rate of moving requests picked up from s_i^p;
   Λ_i – Mean arrival rate of moving requests dropped off to s_i^d.
2. Parameters related to vehicles
   m – Quantity of vehicles in the system;
   r_i – The probability that loaded vehicle drops off its load at s_i^d;
   q_i – The probability that an empty vehicle arriving at s_i^p finds a waiting moving request;
   α_i^d – Rate of loaded vehicles' arrival at s_i^d;
   α_i^p – Rate of loaded vehicles' arrival at s_i^p;
   ε_i^d – Rate of empty vehicles' arrival at s_i^d;
   ε_i^p – Rate of empty vehicles' arrival at s_i^p;
   θ_i – Arrival rate of empty and loaded vehicles to stocker s_i;
   p_i^d – The probability that the vehicle at s_{i-1}^p is blocked by another vehicle at s_i^d;
   p_i^p – The probability that the vehicle at s_i^d is blocked by another vehicle at s_i^p;
   β_i – The probability that vehicle at s_i^d is empty when it is blocked by another vehicle at s_i^p;
   γ_i – The probability that vehicle at s_i^p is empty when it is blocked by another vehicle at s_{i+1}^d;
   $t_{i,i+1}$ – The vehicle's travel time from the pick-up station s_i^p to the drop-off station s_{i+1}^d;
   t_i – The vehicle's travel time from the drop-off station s_i^d to the pick-up station s_i^p, $t_i < t_{i,i+1}$;
   t^l – The vehicle's picking-up and dropping-off time;
   t_b – The mean time that vehicle is blocked by the vehicle in the downstream station;

$\bar{\tau}_w$ – The probability that a loaded vehicle in the pick-up station s_w^p gets through the shortcut connected with stocker s_i and enters the drop-off station s_{w+k}^d, $\bar{\tau}_w = 1 - \tau_w$;

τ_w –The probability that a loaded vehicle in the pick-up station s_w^p enters the drop-off station s_{w+1}^d.

3. Parameters related to the MMCM states

   R – State transition probability matrix of the MMCM;

   W – State set of the MMCM;

   v – State visited ratio vector of the MMCM, $\mathbf{v} = \{v_w\},\ \ w \in W$;

   v_w – Visited ratio to state w of the MMCM;

   C – Mean cycle time between two visits to some reference state, e.g., $(1, p, e)$;

   T_w – The time from the instant that the vehicle enters state w until the instant that it enters the next state.

4. System assumptions

   (1) Moving requests per period in each stocker is constant.

   (2) Moving request arrived according to the Poisson distribution process.

   (3) Each vehicle transports wafer lots based on the FCFS rule.

   (4) The loading time, unloading time and running speed of vehicles are deterministic values, and acceleration and deceleration of vehicles are ignored.

   (5) The stokers have enough storage capacities.

   (6) Vehicles travel independently in the AMHS, i.e., the correlation among the vehicles is not considered.

   (7) The probability that a vehicle is blocked by the downstream stocker is proportional to the number of vehicles in the AMHS.

   (8) Since each wafer lot dropped to a stocker is also picked up from the same stocker, the mean arrival rate of unloading requests and loading requests in each stocker is equal.

4.3.2 MMCM's State Definition

According to assumption 6, the process of m vehicles transporting wafer lots is equivalent to m independent processes of single-vehicle transporting wafer lots. According to assumption 2, the time of each single-vehicle picking-up wafer lot is stochastic. That is, the state of each single-vehicle transporting wafer lot also changes randomly. Therefore, the process of each single-vehicle transporting wafer lot is analyzed as a Markov chain process, which is modeled by using the MMCM. As only those pick-up and drop-off stations are considered, a state of the MMCM for the Interbay AMHS is defined by the three-tuple structure:

$$MMCM = \{S, Kp, Ks\} = \{(s_i, i = 1, \ldots, n), (p, d), (e, f, s, b)\} \tag{4.1}$$

where Kp is the pick-up and drop-off station set where vehicles locate. Ks is the state set of vehicles at pick-up and drop-off stations. p, d indicate vehicles' location at pick-up station and drop-off station, respectively. e, f, s, b mean that the state of a vehicle at pick-up and drop-off stations is empty running, loaded running, receiving picking-up/dropping-off service and blocked, respectively. S is denoted as the set of stockers, and s_i is the ith stocker in the Interbay system.

According to the definition of the state, the whole set of states for the vehicle at the drop-off station includes (i, d, e), (i, d, f), (i, d, s) and (i, d, b), which stand for empty running, loaded running, dropping-off service and blocked, respectively. The set of states for the vehicles at the pick-up stations is the same. Therefore, for the whole Interbay system with n stockers, the set of system states for the MMCM can be denoted as

$$W = \{(1, d, e), (1, d, f), (1, d, s), (1, d, b), ..., (n, p, e), (n, p, f), (n, p, s), (n, p, b)\} \quad (4.2)$$

The number of system states is $2 \times 4 \times n = 8n$.

4.3.3 Modeling Process of the MMCM

The modeling process of the AMHS consists of the following steps. First, the state transition probability of the MMCM for a single-vehicle based material handling process is analyzed and the transition matrix R is derived. Second, the MMCM's steady state is analyzed and the visit ratio model of each state in the MMCM is inferred. Third, vehicle blocking probabilities in the AMHS are analyzed and the system blocking probability model is built. Fourth, the stability condition of the AMHS is analyzed and the system stability condition model is constructed. Fifth, the transition time of each state in the MMCM is analyzed and the mean cycle length C between two states is obtained. Finally, on the basis of the visit ratio model of each state in the MMCM, system blocking probability model, system stability condition model and the mean cycle length C, the stability state equations of MMCM are built and the visit ratio of each state in the MMCM can be solved. The performance analytical indexes can be expressed by the visit ratio of each state in the MMCM.

4.3.3.1 Transition Probabilities

The transition probability refers to a change in probability between states in the MMCM. Consider an Interbay system with n stockers; the state transition diagram is partially described in Figure 4.2. The detailed process of state changing and state transition analyzing is illustrated as follows. Denote that there is an empty vehicle that approaches the drop-off station s_i^d, i.e., the state of a model is (i, d, e). The transitions from this state depend on whether the vehicle is blocked by the vehicle at the pick-up station s_i^p. If the vehicle is blocked by the vehicle at the pick-up station

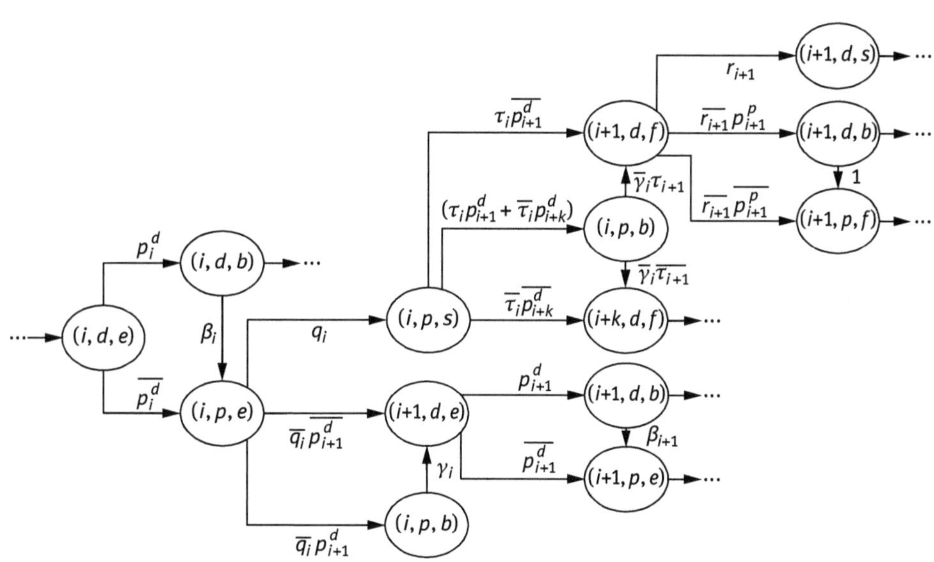

Figure 4.2: The state transition diagram of the MMCM for Interbay system.

s_i^p, then the system enters the state (i, d, b), which happens with probability p_i^p. Otherwise, if there is no vehicle at the pick-up station s_i^p, then the system enters the state (i, p, e), which happens with probability $1 - p_i^p$. When the vehicle enters the pick-up station s_i^p, the system state changes from (i, p, e) to (i, p, s), (i, p, b) or $(i+1, d, e)$ according to corresponding probability. (1) If the vehicle finds a wafer lot waiting at the pick-up station s_i^p, then the vehicle starts picking up wafer lot and the system enters the state (i, p, s), which happens with probability q_i. (2) If the vehicle does not find a load waiting at s_i^p and the vehicle is blocked by a vehicle at the drop-off station s_{i+1}^d, then system the enters the state (i, p, b) with probability $\overline{q_i}\,\overline{p_{i+1}^d}$. (3) If the vehicle does not find a load waiting at s_i^p and there are no vehicles at the drop-off station s_{i+1}^d, then the system enters the state $(i+1, d, e)$, which happens with probability $\overline{q_i} p_{i+1}^d$. The transition probability of other states of the system can be analyzed in the same way.

Based on these probabilities, the transition matrix **R**, which specifies the movement of the vehicle between the states of a system, can be identified, as described in Figure 4.3. For the Interbay system with n stockers, the dimension of the transition matrix **R** is $8n \times 8n$. In the transition matrix **R**, $\overline{\tau}_i$ presents the probability of a loaded vehicle in the pick-up station s_i^p which passes through the shortcut connected with stocker s_i and enters the drop-off station s_{i+k}^d. Denote that the Interbay system has l shortcuts, whose starting point corresponds to the stocker set $\{s_{h_1}, \ldots, s_{h_l}\}$ and terminal point corresponds to the stocker set $\{s_{h_1+k_1}, \ldots, s_{h_s+k_l}\}$. Then if $i = h_1, \ldots, h_l$, $0 < \overline{\tau_i} < 1$. Otherwise, if $i \neq h_1, \ldots, h_l$, $\overline{\tau_i} = 0$.

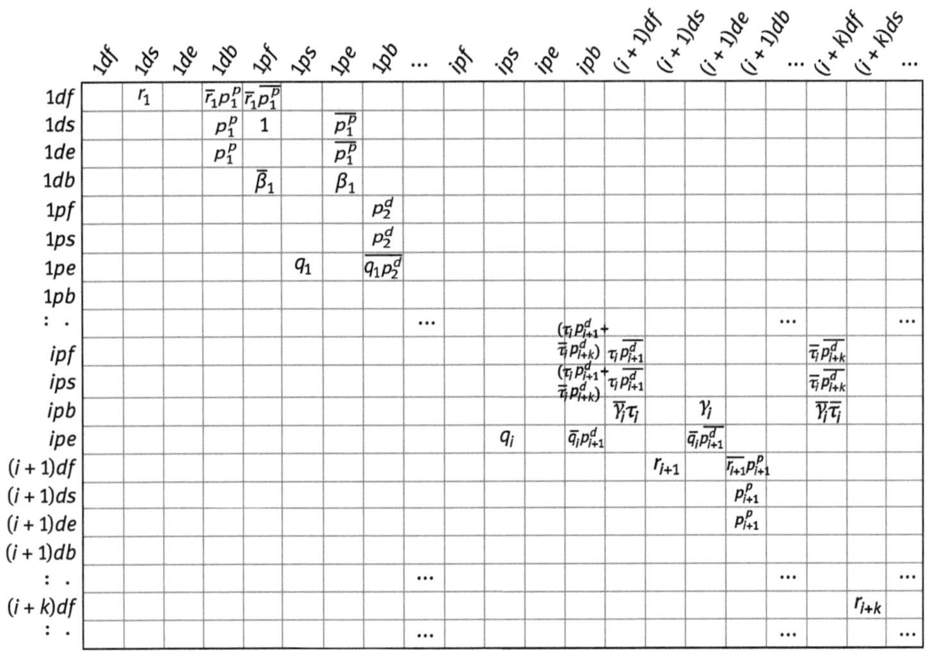

Figure 4.3: The state transition matrix **R** of the MMCM for Interbay system.

4.3.3.2 Steady-State Analysis

Since the vehicle moves continuously on the closed-loop Interbay system, the state number of the MMCM is finite, e.g., the state number of the Interbay system with n stockers is $8n$. Denote $\mathbf{v} = \{v_w\},\ \ w \in W$, where v_w means the steady-state probability of the visited ratio to state w and W is the state set of the MMCM. Without loss of generality, set $v_{(1,p,e)} = 1$. Let **R** be the state transition probability matrix of the MMCM. For a finite state, positive recurrent Markov chain, the steady-state probabilities can be uniquely obtained by solving the square system of equations.

$$\mathbf{R} \cdot \mathbf{v} = \mathbf{v} \tag{4.3}$$

$$v_{(1,p,e)} = 1 \tag{4.4}$$

Let R_{ij} $(i, j = 1,\ldots,8n)$ be the element of the transition matrix **R**, which denotes the transition probability from state i to state j in the transition matrix. Then R_{ij} can be represented as follows:

$$R_{ij} = \left(\left(q_i^{y_1} \cdot \bar{q}_i^{(1-y_1)}\right) \cdot \left(r_i^{y_2} \cdot \bar{r}_i^{(1-y_2)}\right) \cdot \left[\left(p_i^d\right)^{y_3} \cdot \overline{p_i^d}^{(1-y_3)}\right] \cdot \left[\left(p_i^p\right)^{y_4} \cdot \overline{p_i^p}^{(1-y_4)}\right] \cdot \left(\tau_i^{y_5} \cdot \bar{\tau}_i^{(1-y_5)}\right)\right)^{y_6} \tag{4.5}$$

where y_1, y_2, y_3, y_4, y_5 and y_6 are the binary variables, $y_1, y_2, y_3, y_4, y_5, y_6 \in \{0,1\}$. The R_{ij} is related with vectors $\mathbf{q} = \{q_i\}(i = 1, ..., n)$, $\mathbf{r} = \{r_i\}(i = 1, ..., n)$, $\mathbf{p^d} = \{p_i^d\}(i=1, ..., n)$, $\boldsymbol{\tau} = \{\tau_i\}(i=h_1, ..., h_l)$, $\boldsymbol{\beta} = \{\beta_i\}(i=1, ..., n)$, $\boldsymbol{\gamma} = \{\gamma_i\}(i=1, ..., n)$ and $\mathbf{p^p} = \{p_i^p\}(i=1, ..., n)$. According to the definition of probabilities β_i and γ_i, the expressions of β_i and γ_i can be described as follows:

$$\beta_i = \varepsilon_i^p / (\varepsilon_i^p + \alpha_i^p) \tag{4.6}$$

$$\gamma_i = \varepsilon_i^d / (\varepsilon_i^d + \alpha_i^d) \tag{4.7}$$

From eqs (4.1) and (4.2), the visit ratio to each state can be inferred as follows:

$$V_{(i,d,e)} = \overline{q_{i-1} p_i^p} V_{(i-1,p,e)} + \gamma_{i-1} V_{(i-1,p,b)} \tag{4.8}$$

$$V_{(i,d,f)} = \begin{cases} \tau_{i-1}(\overline{p_i^d} V_{(i-1,p,f)} + \gamma_{i-1} V_{(i-1,p,b)} + \overline{p_i^d} V_{(i-1,p,s)}) & \text{if } (i-1) = h_1, ..., h_l \\ \overline{p_i^d} V_{(i-1,p,f)} + \overline{\gamma}_{i-1} V_{(i-1,p,b)} + \overline{p_i^d} V_{(i-1,p,s)} & \text{if } (i-1) \neq h_1, ..., h_l, h_{1+k}, ..., h_{l+k} \\ \overline{p_i^d}(V_{(i-1,p,f)} + V_{(i-1,p,s)}) + \overline{\gamma}_{i-1} V_{(i-1,p,b)} + \overline{\tau_{i+k}}(\overline{p_i^d}(V_{(i+k,p,f)} + V_{(i+k,p,s)}) & \\ + \overline{\gamma}_{i+k} V_{(i+k,p,b)}) & \text{else} \end{cases} \tag{4.9}$$

$$V_{(i,d,b)} = p_i^p (V_{(i,d,s)} + V_{(i,d,e)}) + \overline{r_i} p_i^p V_{(i,d,f)} \tag{4.10}$$

$$V_{(i,d,s)} = r_i V_{(i,d,f)} \tag{4.11}$$

$$V_{(i,p,e)} = p_i^p V_{(i,d,e)} + \overline{p_i^p} V_{(i,d,s)} + \beta_i V_{(i,d,b)} \tag{4.12}$$

$$V_{(i,p,f)} = \overline{r_i p_i^p} V_{(i,d,f)} + \bar{\beta}_i V_{(i,d,k)} \tag{4.13}$$

$$V_{(i,p,b)} = \begin{cases} (\tau_i p_{i+1}^d + \overline{\tau_i} p_{i+k}^d)(V_{(i,p,f)} + V_{(i,p,s)}) + \overline{q_i p_{i+1}{}^d} V_{(i,p,e)} & \text{if } i = h_1, ...h_1 \\ p_{i+1}^d (V_{(i,p,f)} + V_{(i,p,s)}) + \overline{q_i p_{i+1}^d} V_{(i,p,e)} & \text{if } i = h_1, ...h_1 \end{cases} \tag{4.14}$$

$$V_{(i,p,s)} = q_i V_{(i,p,e)} \tag{4.15}$$

In eq. (4.15), some of these probability variables are unknown: specifically, the picking-up/dropping-off probability variables $\mathbf{q} = \{q_i\}$ and $\mathbf{r} = \{r_i\}, i = 1, ..., n$, the blocking probabilities variables $\mathbf{p^d} = \{p_i^d\}$ and $\mathbf{p^p} = \{p_i^p\}, i = 1, ..., n$, and the shortcut-selecting probabilities variables $\tau = \{\tau_i\}$, $i = h_1, ..., h_l$. The probability variables $\mathbf{r}, \mathbf{q}, \boldsymbol{\tau}$ can be obtained by system's stability condition analyzing, and the probability variables $\mathbf{p^d}$, $\mathbf{p^p}$ can be obtained based on system blocking probability analysis.

4.3.3.3 System Blocking Probabilities

System blocking probability refers to the probability that the vehicle in a pick-up/drop-off station is blocked by the vehicle of a downstream station. In the Interbay system, the distance between the drop-off station and the pick-up station of a stocker is less than the distance between stockers. Hence, a vehicle at the drop-off station s_i^d cannot be blocked by a downstream vehicle at the pick-up station s_i^p unless this downstream vehicle is traveling toward the pick-up station (i.e., the system state is in (i, p, e) or (i, p, f)). However, a vehicle at the pick-up station s_i^p will be blocked when there is a vehicle at the pick-up station s_{i+1}^d (i.e., the system state is in $(i+1, d, +)$). As the blocking probabilities increase linearly with vehicle quantity according to system assumption 7 in Section 4.3.1, the blocking probability variables p_i^d and p_i^p can be estimated as follows:

$$p_i^d = (m-1)\frac{V_{(i,d)}}{\sum_{j=1}^{|R|}\left(V_{(i,d)} + V_{(i,p)}\right)}, \quad \forall s_i^d, i = 1, \ldots, n \tag{4.16}$$

$$p_i^p = (m-1)\frac{V_{(i,p,s)} + V_{(i,p,b)} + V_{(i,p,k)}}{\sum_{j=1}^{|R|}\left(V_{(i,d)} + V_{(i,p)}\right)}, \quad \forall s_i^p, i = 1, \ldots, n \tag{4.17}$$

As the Interbay system runs stably, the following stability conditions should be satisfied. (1) For each stocker s_i, $i = 1, \ldots, n$, the mean arrival rate that moving requests dropped off to station s_i^d should be equal to the drop-off ratio of loaded vehicles at the current station. (2) For each stocker s_i, $i = 1, \ldots, n$, the mean arrival rate that moving requests picked up from station s_i^p should be equal to the pick-up ratio of empty vehicles at the current station. (3) According to from-to-table (it is a conveying relation table where the transition ratios between stockers in the Interbay system are described) of the Interbay system, vehicles in the Interbay system should pass through each shortcut by certain probabilities. Based on system stability conditions, the probability variables $\mathbf{r}, \mathbf{q}$ and $\boldsymbol{\tau}$ can be obtained.

① The probability variables $\mathbf{r}$

According to system stability conditions, for each stocker s_i, $i = 1, \ldots, n$, the mean arrival rate of moving requests dropped off at station s_i^d, Λ_i, which is equal to the drop-off ratio of loaded vehicles at the current station, which is the product

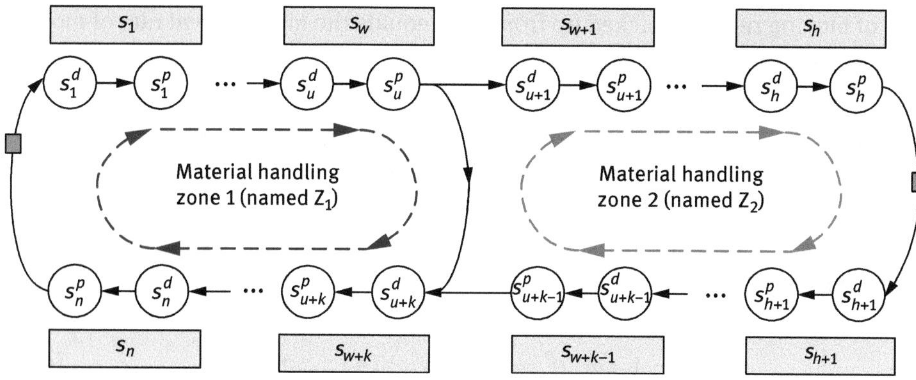

Figure 4.4: The Interbay material handling system with single-shortcut.

of the rate of loaded vehicles' arrival at s_i^d, a_i^d and the probability that loaded vehicle drops off its load at s_i^d, r_i. Then the vehicle's picking up probability r_i can be described as

$$\Lambda_i = r_i \cdot a_i^d \Rightarrow r_i = \Lambda_i / a_i^d \tag{4.18}$$

a_i^d can be obtained based on material handling flow analyzing of the Interbay system. For example, there is an Interbay system with single shortcut (Figure 4.4.), which can be separated as two material handling zones based on shortcut, labeled Z_1 (i.e., the stocker set is $\{s_1, .., s_u, s_{u+k}, ..., s_n\}$) and Z_2 (i.e., the stocker set is $\{s_{u+1}, .., s_{u+k-1}\}$). Assume that a loaded vehicle runs in Z_1 and arrives at station s_u^p; then the vehicle will decide whether it should enter in Z_2 according to the destination stocker of the wafer lot on the vehicle. If the destination stocker belongs to Z_2, then the loaded vehicle selects to enter in Z_2. Otherwise, if the destination stocker belongs to Z_1, the loaded vehicle runs in the current zone continuously. The equation of the parameter a_i^d can be described as

$$a_i^d = \begin{cases} \sum\limits_{j=i+1}^{n} \sum\limits_{r=i}^{j} \lambda_j p_{jr} + \sum\limits_{j=1}^{i-1} \sum\limits_{\substack{r=1 \\ r \neq [j, i-1]}}^{n} \lambda_j p_{jr}, \forall s_i \in \{s_1, .., s_u, s_{u+k}, ..., s_n\} \\ \sum\limits_{\substack{j=i, \\ j \neq [u+1, u+k-1]}}^{n} \sum\limits_{r=i}^{u+k-1} \lambda_j p_{jr} + \sum\limits_{j=u+1}^{i-1} \sum\limits_{\substack{r=1, \\ r \neq [j, i-1]}}^{n} \lambda_j p_{jr} + \sum\limits_{j=i+1}^{u+k-1} \sum\limits_{r=i}^{j} \lambda_j p_{jr}, \forall s_i \in \{s_{u+1}, .., s_{u+k-1}\} \end{cases} \tag{4.19}$$

The expression of a_i^d in the Interbay system with multi-shortcut can be inferred by the same way. Since the SWFS has the characteristic of multi re-entrant, the wafer lot dropped off at station s_i^d will be picked up at station s_i^p later. So, the mean arrival rate of moving requests picked up from s_i^p; λ_i equals the mean arrival rate of moving requests dropped off at s_i^d, Λ_i, i.e., $\Lambda_i = \lambda_i$.

② The probability variables **q**

$\mathbf{q} = \{q_i\}, i = 1, ..., n$, denotes the probability that a loaded vehicle picks up a waiting moving request at s_i^d. According to system stability conditions, the mean arrival rate of moving requests picked up from the station s_i^p; λ_i equals the pick-up ratio of an empty vehicle at the current station, which is the product of the rate of empty vehicles arriving at s_i^p, ε_i^p and q_i. Then q_i can be described as follows:

$$\lambda_i = q_i \cdot \varepsilon_i^p \Rightarrow q_i = \frac{\lambda_i}{\varepsilon_i^p}, \quad \forall i = 1, ..., n \tag{4.20}$$

In the mean cycle time C, the visit ratio that any vehicle enters the system state (i, p, e) is $v_{(i,p,e)}$, which can be described as follows: $v_{(i,p,e)} = \frac{\varepsilon_i^p C}{m}$, where m is the vehicle number in the Interbay system. Then, q_i can be inferred as

$$q_i = \frac{\lambda_i C}{v_{(i,p,e)} m} \tag{4.21}$$

The value of the mean cycle time C will be solved in Section 4.3.3.4.

③ The probability variables τ

$\tau = \{\tau_i\}, i = h_1, ..., h_l$, denotes the probability that a loaded vehicle at the pick-up station $s_i^p (i = h_1, ..., h_l)$ enters the downstream drop-off station s_{i+1}^d. For an Interbay system with a single shortcut (Figure 4.4.), assume that there is a moving request at the pick-up station s_u^p in Z_1. If an empty vehicle arrives at s_u^p and picks up the load, then the loaded vehicle will choose the running path according to the destination stocker of wafer lots on the vehicle. If the destination stocker belongs to Z_1, then the loaded vehicle will pass through shortcut and run along paths $\{s_u \to s_{u+k} \to \cdots \to s_n \to \cdots\}$. Otherwise, if the destination stocker belongs to Z_2, the vehicle will enter Z_2 and run along paths $s_u \to s_{u+1} \to \cdots \to s_h \to \cdots\}$. The equation of $\overline{\tau_u}$ can be described as follows:

$$\overline{\tau_u} = \frac{\sum_{i=1}^{u} \left(\sum_{j=u+k}^{n} p_{ij} + \sum_{j=1}^{i-1} p_{ij} \right) + \sum_{i=u+k+1}^{n} \sum_{j=u+k}^{i} p_{ij}}{\sum_{i=}^{u} 1 \left(\sum_{j=u+1}^{n} p_{ij} + \sum_{j=1}^{i-1} p_{ij} \right) + \sum_{i=u+2}^{n} \sum_{j=u+1}^{i} p_{ij}} \tag{4.22}$$

The expression of $\overline{\tau_u}$ in an Interbay system with multiple shortcuts can be inferred by the same way.

4.3.3.4 The Mean Cycle Time *C*

C is the mean cycle time between two successive visits to some reference state, e.g., $(1, p, e)$. Denote T_w as the time from the instant the vehicle enters state w until the instant it enters the next state. According to the definition of the mean cycle time, the expression of C can be described as

$$
\begin{aligned}
C &= \sum_{w=1}^{|W|} v_w E(T_w) \\
&= \sum_{i=1}^{n} \big(v_{(i,d,e)}E(T_{(i,d,e)}) + v_{(i,d,f)}E(T_{(i,d,f)}) + v_{(i,d,b)}E(T_{(i,d,b)}) + v_{(i,d,k)}E(T_{(i,d,k)}) \\
&\quad + v_{(i,d,s)}E(T_{(i,d,s)}) + v_{(i,p,e)}E(T_{(i,p,e)}) + v_{(i,p,f)}E(T_{(i,p,f)}) + v_{(i,p,b)}E(T_{(i,p,b)}) \\
&\quad + v_{(i,p,k)}E(T_{(i,p,k)}) + v_{(i,p,s)}E(T_{(i,p,s)})\big)
\end{aligned}
\tag{4.23}
$$

where $E(T_w)$ is the expected mean state time, which can be determined based on the state transition probabilities. The expected mean state time of each state is described as follows. (1) Consider the state $(i, d, s)/(i, p, s)$, which means vehicle dropping off/picking up wafer lot at the station s_i^d/s_i^p; the time that the vehicle spends in these states are the vehicle's picking-up and dropping-off times t^l. (2) For the state $(i, d, f)/(i, d, e)$, which means a loaded/empty vehicle arriving at some drop-off station s_i^d, the time that the vehicle spends in these states is the traveling time of a loaded/empty vehicle from the pick-up station s_{i-1}^p to the drop-off station s_i^d, $t_{i-1,i}$. (3) The time that the vehicle spends in the state $(i, p, f)/(i, p, e)$ is the traveling time of a loaded/empty vehicle from the drop-off station s_i^d to the pick-up station s_i^p, t_i. (4) The state (i, d, b) means a vehicle blocked at the drop-off station s_i^d, which happens when there is another vehicle picking up wafer lots at the station s_i^p. The time that a vehicle spends in this state is related to its previous states. (5) Consider the state (i, p, b), which means a vehicle is blocked at the pick-up station s_i^p, and the time that the vehicle spends in this state is influenced by the states in the downstream station s_{i+1}^d. The expressions of the expected transition time for each state of a vehicle at s_i^d and s_i^p are derived as follows:

$$E(T_{(i,d,f)}) = t_{i-1,i} \tag{4.24}$$

$$E(T_{(i,d,e)}) = t_{i-1,i} \tag{4.25}$$

$$E(T_{(i,d,b)}) = \frac{\beta_i \times t^l \times v_{(i,d,e)}}{2 \times (v_{(i,d,e)} + v_{(i,d,s)})} + \frac{\beta_i \times t^l}{2} \tag{4.26}$$

$$E(T_{(i,d,s)}) = t^l \tag{4.27}$$

$$E(T_{(i,p,f)}) = t_i \tag{4.28}$$

$$E(T_{(i,p,e)}) = t_i \tag{4.29}$$

$$E(T_{i,p,b}) = \\ \overline{\gamma_i}\frac{\frac{(v_{(i+1,d,f)}+v_{(i+1,d,e)})(t_{i,t+1}-t_i)+v_{(i+1,d,s)}((t_{i,t+1}-t_i)+t^l)}{(v_{(i+1,d,f)}+v_{(i+1,d,e)}+v_{(i+1,d,s)})}v_{(i,p,f)}+(t_{i,t+1}-t_i)v_{(i,p,s)}}{(v_{(i,p,f)}+v_{(i,p,s)})\times 2} \\ +\gamma_i\frac{(v_{(i+1,d,f)}+v_{(i+1,d,e)})(t_{i,t+1}-t_i)+v_{(i+1,d,s)}((t_{i,t+1}-t_i)+t^l)}{(v_{(i+1,d,f)}+v_{(i+1,d,e)}+v_{(i+1,d,s)})\times 2} \tag{4.30}$$

$$E(T_{(i,p,s)}) = t^l \tag{4.31}$$

According to analytical results from Sections–4.3.3.4, the stability state equations of the MMCM are obtained and described as follows:

$$\mathbf{R}\cdot\mathbf{v} = \mathbf{v} \tag{4.32}$$

$$v_{(s_1,p,e)} = 1 \tag{4.33}$$

$$p_i^d = \frac{(m-1)v_{(s_i,d,*)}}{\sum_{j=1}^{|W|}(v_{(s_j,d,*)}+v_{(s_j,p,*)})}, \quad \forall s_i^d, i = 1, \ldots, n \tag{4.34}$$

$$p_i^p = \frac{(m-1)v_{(s_i,p,s)}+v_{(s_i,p,b)}}{\sum_{j=1}^{|W|}(v_{(s_j,d,*)}+v_{(s_j,p,*)})}, \quad \forall s_i^p, i = 1, \ldots, n \tag{4.35}$$

$$q_i = \frac{\lambda_i Ct}{v_{(s_i,p,e)}m} \tag{4.36}$$

$$Ct = \sum_{w=1}^{|W|} v_w E(T_w) \tag{4.37}$$

The quantity of stability state equations described above is $|W|+3n+1$. The number of variables in these equations is also $|W|+3n+1$, which includes $|W|$ steady-state visit ratio variables $v_w(w \in W)$, n picking-up probability variables $q_i(i=1,\ldots,n)$, n vehicle-blocking probability variables $p_i^p(i=1,\ldots,n)$ and $p_i^d(i=1,\ldots,n)$, and a mean cycle time variable Ct.

Proposition 1. For the system with non-linear equations given in above equations, if the Interbay AMHS can handle all moving requests within the planning horizon, i.e., the Interbay AMHS is stable, then there exist unique values for the variables $q_i(i=1,\ldots,n)$, $p_i^p(i=1,\ldots,n)$, $p_i^d(i=1,\ldots,n)$ and Ct, and these values provide the sole solution to the steady-state visit ratio vector $\mathbf{v}$ and the state transition probability matrix $\mathbf{R}$ in the MMCM.

4.3.4 Model Validation

1. Experimental data

This section presents a numerical experiment study to demonstrate the effectiveness of the proposed MMCM analytical approach. The system data used for numerical experiments in this study were shared by a semiconductor manufacturer in Shanghai, China. To protect the confidentiality of the manufacturer's information, some system data presented in this chapter have been modified and are for demonstrative purpose only. There are 14 stockers in the Interbay material handling system, and each stocker connects with an Intrabay system (Figure 4.5). The distance between adjacent stockers is 20 m, the distance between the pick-up port and the drop-off port of a stocker is 5 m and the Interbay rail loop is 320 m long with two shortcuts. Each automated vehicle can load only one wafer lot at a time, and the vehicle's picking-up and dropping-off time at a stocker is 10 s. Two types of wafer lot (jobs A and B) are being handled in the system and the release ratio of jobs A and B is 1:1. The system's impact factors considered in the

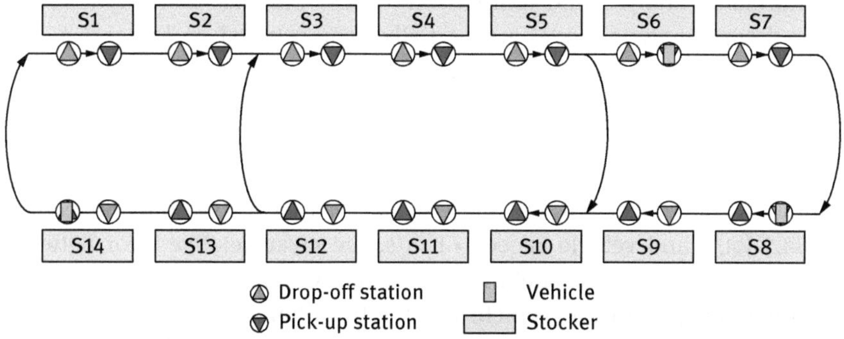

Figure 4.5: The layout of the Interbay material handling system.

Table 4.1: Investigated factors.

Factor	Levels
Vehicle quantity	3–15 vehicles
System's release ratio	9,000 wafer/month, 13,500 wafer/month, 15,000 wafer/month
Vehicle's speed	1 m/s, 1.5 m/s

experimental study include vehicle quantity, system's release ratio and vehicle's speed, as listed in Table 4.1. Totally, 72 scenarios are designed for different combinations of the selected factors.

With the actual data of the Interbay material handling system, the basic data and key system parameters of the MMCM in each scenario are identified, including from-to-table of the Interbay system, mean arrival rate of moving requests picked up from s_i^p, λ_i, mean arrival rate of moving requests dropped off to s_i^d, Λ_i, rate of loaded vehicles' arrivals at s_i^d, α_i^d, the dropping-off probability r_i and the shortcut-selecting probability τ_i. For instance, in the scenario where system's release ratio = 13,500 wafer/month and vehicle's speed = 1 m/s, the from-to-table and the key system parameters of the MMCM are calculated and shown in Tables 4.2 and 4.3, respectively.

2. Comparison and discussion

Based on the basic data and the values of key system parameters, the MMCM analytic model is comprehensively compared with discrete event simulation model in each scenario. A total of three performance indexes, system's throughput capability, vehicle's mean utilization ratio and mean arrival time interval of empty vehicle, are compared. The eM-Plant simulation software is adopted to construct Interbay's simulation model, in which moving requests arrive at the stocker according to Poisson process. The simulation model in each scenario is executed with three replications of 60 days each.

The performance analysis and comparison results of the MMCM analytic model and the simulation analytic model are partially demonstrated in Tables 4.4–4.6. Table 4.4 shows that, in those scenarios that the arrive ratio of moving requests = 56 lot/h and vehicle speeds = 1 m/s, the mean relative error between the MEMCM and the simulation model in terms of empty vehicle's mean arrival time interval, throughput capacity and vehicle's mean utilization ratio is 3.75%, 2.88% and 1.86%, respectively. Table 4.5 illustrates that, in those scenarios that arrive ratio of moving requests = 84.5 lot/h and vehicle speeds = 1 m/s, the mean relative error between the MMCM and the simulation model in terms of empty vehicle's mean arrival time interval, throughput capacity and vehicle's mean utilization ratio is 3.6%, 2.87% and 10.0%, respectively. Table 4.6 shows that, in the scenario that the number of vehicles = 7, vehicle speeds = 1 m/s and the arrive ratio of moving requests = 56 lot/h, and the

Table 4.2: The from-to-table of the Interbay system (system's release ratio = 13,500 wafer/month, vehicle speed = 1 m/s).

Stocker ID	S1	S2	S3	S4	S5	S6	S7	S8	S9	S10	S11	S12	S13	S14
S1	0	1.5	0.5	1	0	1	0	0.5	1	0	0.5	0	0.5	0.5
S2	0	0	0.5	0.5	0.5	1	0	0.5	0.5	0.5	2	0	1	0.5
S3	0.5	0	0	0	0	1	1	1	0.25	1	1	0.5	0.5	1
S4	0.5	0.5	0	0	0	0	0.5	0	1	0	0	0.5	1	1
S5	1	0	0	0	0	0	0	0	0	0	0	1	1	0.5
S6	0.5	0	0.75	0	0.5	0	0.5	0	0	0.5	1	1	0	0
S7	1	0	0.5	0	0	0.5	0	0	1	0	0	0.5	0.5	0.5
S8	0	1	0	0.5	0	0.25	0	0	0	0.75	0.5	0.5	0.5	0.5
S9	2	0	0	0	1	0	0.5	0	0	0	1	1	1	1
S10	0.5	0.5	0	1	0.5	0.5	0.5	0	0	0	1.25	1	0	1
S11	0	1.5	1.5	1	0	0.5	1	0	0.75	1	0	0	0	0
S12	0	1	0.5	0.5	0.5	0	0.5	1.5	1	0.5	0	0	0	0
S13	1	0	1.5	0	0	0	0	0	1	2.5	0	0	0	0
S14	0	1.5	2	0.5	0.5	0	0	1	1	0	0	0	0	0

Table 4.3: The key system parameter values of the MMCM (system's release ratio = 13,500 wafer/month, vehicle speed = 1 m/s).

Variable	S1	S2	S3	S4	S5	S6	S7	S8	S9	S10	S11	S12	S13	S14
λ_i	7	7.5	7.75	5	3.5	4.75	4.5	4.5	7.5	6.75	7.25	6	6	6.5
Λ_i	7	7.5	7.75	5	3.5	4.75	4.5	4.5	7.5	6.75	7.25	6	6	6.5
a_i^d	23	21.5	40.75	40.75	40.75	19.75	19.75	19.75	19.75	40.75	40.75	40.75	23	23
r_i	0.304	0.349	0.190	0.123	0.086	0.241	0.228	0.228	0.380	0.166	0.178	0.147	0.261	0.283
$\overline{\tau_w}$	/	/	/	/	0.515	/	/	/	/	/	/	0.460	/	/

Table 4.4: Performances comparison of the MMCM and the simulation-based model (arrive ratio of move requests = 56 lot/h, vehicle speeds = 1 m/s).

Vehicle number	Empty vehicle's mean arrival time interval (s)			Throughput capacity (lot)			Vehicle's mean utilization ratio (%)		
	MEMCM	Simulation model	Relative error (%)	MEMCM	Simulation model	Relative error (%)	MEMCM	Simulation model	Relative error (%)
3	200.7	212.3	5.8	68,344	70,458	3.00	60.4	62.2	2.9
4	119.5	122.2	2.3	91,151	93,953	2.98	45.7	46.3	1.3
5	85.6	86.0	0.5	113,963	117,437	2.96	36.8	37.1	0.8
6	67.0	66.5	0.7	136,782	140,920	2.94	31.0	31.1	0.3
7	55.2	54.2	1.8	159,610	164,401	2.91	26.8	26.7	0.4
8	47.2	45.9	2.8	182,450	187,907	2.90	23.6	23.5	0.4
9	41.3	39.8	3.6	205,301	211,382	2.88	21.2	20.8	1.9
10	36.8	35.2	4.3	228,167	234,846	2.84	19.2	18.9	1.6
11	33.2	31.5	5.1	251,048	258,343	2.82	17.6	17.3	1.7
12	30.4	28.6	5.9	273,947	281,873	2.81	16.3	15.9	2.5
13	28.0	26.2	6.4	296,863	305,318	2.77	15.2	14.8	2.7
14	26.1	24.2	7.3	319,800	328,835	2.75	14.2	13.8	2.9
Mean value	/	/	3.75	/	/	2.88	/	/	1.86

Table 4.5: Performances comparison of the MMCM and the simulation-based model (arrive ratio of move requests = 84.5 lot/h, vehicle speeds = 1 m/s).

Vehicle number	Empty vehicle's mean arrival time interval (s)			Throughput capacity (lot)			Vehicle's mean utilization ratio(%)		
	MEMCM	Simulation model	Relative error (%)	MEMCM	Simulation model	Relative error (%)	MEMCM	Simulation model	Relative error (%)
4	178.8	203.6	12.2	91,245	94,013	2.94	68.5	63.5	7.9
5	112.1	120.0	6.6	114,092	117,509	2.91	55.3	50.9	8.6
6	82.1	85.4	3.9	136,948	141,010	2.88	46.4	42.6	8.9
7	65.1	66.4	2.0	159,818	164,507	2.85	40.1	36.7	9.3
8	54.2	54.5	0.6	182,704	188,015	2.82	35.4	32.2	9.9
9	46.5	46.3	0.4	205,607	211,514	2.79	31.7	28.8	10.1
10	40.9	40.3	1.5	228,531	235,015	2.76	28.8	26.1	10.3
11	36.5	35.7	2.2	251,477	258,524	2.73	26.4	23.8	10.9
12	33.1	32.1	3.1	274,447	282,001	2.68	24.4	22.0	10.9
13	30.3	29.3	3.4	297,444	305,526	2.65	22.7	20.4	11.3
14	28.0	27.0	3.7	320,470	329,035	2.60	21.3	19.2	10.9
15	26.1	25.1	4.0	343,525	352,530	2.55	20.1	18.1	11.0
Mean value	/	/	3.6	/	/	2.87	/	/	10.0

Table 4.6: Empty vehicle's mean arrival time interval comparison between the MMCM and the simulation-based model.

Stocker ID	Vehicles = 7, vehicle speeds = 1 m/s, arrive ratio of move requests = 56 lot/h			Vehicles = 7,vehicle speeds = 1 m/s, arrive ratio of move requests = 84.5 lot/h		
	MEMCM (s)	Simulation model (s)	Relative error (%)	MEMCM (s)	Simulation model (s)	Relative error (%)
S1	55.5	53.7	3.24	65.5	65.2	0.46
S2	55.2	53.4	3.26	64.9	64.6	0.46
S3	54.8	53.5	2.37	64.1	64.6	0.78
S4	56.2	54.7	2.67	67.2	67.7	0.74
S5	57.0	55.6	2.46	68.9	69.6	1.02
S6	56.0	54.9	1.96	66.6	68.0	2.10
S7	55.9	55.2	1.25	66.4	68.6	3.31
S8	55.7	55.1	1.08	66.0	68.3	3.48
S9	53.8	53.4	0.74	62.1	64.6	4.03
S10	54.2	53.9	0.55	63.0	65.8	4.44
S11	54.0	53.4	1.11	62.5	64.6	3.36
S12	54.7	54.4	0.55	64.0	66.7	4.22
S13	55.2	54.3	1.63	65.1	66.4	2.00
S14	55.4	54.0	2.53	65.5	65.8	0.46

scenario that the number of vehicles = 7, vehicle speeds = 1 m/s and the arrive ratio of moving requests = 84.5 lot/h, the mean relative error of empty vehicle's mean arrival time interval between the MMCM and the simulation model is less than 4.44%. The results demonstrate that the MMCM analytical approach is feasible.

To further evaluate the effectiveness of the MMCM analytical approach, the frequency of relative error percentages of the MMCM and the simulation model in all scenarios are analyzed. Figure 4.6. illustrates that 96% of all relative error values belong to [−8%, 10%]. Meanwhile, runtime comparison between the MMCM and the eM-Plant model-based analyzing methods is given in Table 4.7 and the results show that the computation efficiency of the MMCM is improved about 600 times than that of the eM-Plant model. It demonstrated that the proposed MMCM analytical approach performed reasonably well with acceptable error percentages and was an effective performance analyzing approach for the AMHS with shortcut and blocking.

4.4 Analysis of AMHS Based on the MMCM

The stability state equations of the MMCM are solved quickly using Gauss iterative method [3, 4]. Based on the MMCM and its state visit ratio, the performance of the Interbay AMHS is analyzed and evaluated.

Table 4.7: Runtime comparison between the MMCM and eM-Plant model based analyzing methods.

Analyzing methods	Runtime (min)
eM-Plant model	≈60
MMCM	≈0.1

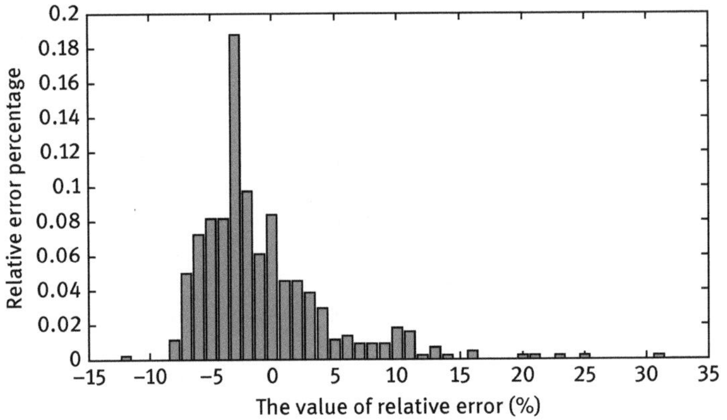

Figure 4.6: The frequency of relative error percentages of the MEMCM and the simulation model.

4.4.1 The Overall Utilization Ratio and the Mean Utilization Ratio of a Vehicle

The overall utilization ratio ρ refers to the ratio of times for a vehicle to stay in the handling wafer state, the loading state and the blocking state.

$$\rho_u = \frac{\sum\limits_{w \in X \cup F \cup K} v_w T_w}{\sum\limits_{w=1}^{|W|} v_w T_w} = \frac{\sum\limits_{w \in X \cup F \cup K} v_w T_w}{C}$$

The mean utilization ratio ρ_u means the ratio of times that a vehicle stays in the handling wafer state and the loading state.

$$\rho_u = \frac{\sum\limits_{r \in X \cup F} v_w T_w}{\sum\limits_{w=1}^{|W|} v_w T_w} = \frac{\sum\limits_{r \in X \cup F} v_w T_w}{C} \tag{4.38}$$

where X is the set of states when the vehicle is picking up or dropping off. F is the set of states when the vehicle is loaded. K is the set of states when the vehicle is blocked.

4.4.2 The Mean Arrival Time Interval of an Empty Vehicle

The mean arrival time interval of an empty vehicle T_e refers to the mean cycle time for an empty vehicle to successively visit the pick-up station s_i^p of the stocker s_i. The smaller the value of T_e, the shorter the mean waiting time of the wafer lot at a stocker. T_e can be calculated as follows:

$$T_e = \frac{1}{n}\sum_{i=1}^{n}\frac{1}{\varepsilon_i^p} = \frac{1}{n}\sum_{i=1}^{n}\frac{T_{(i,p,e)}}{v_{(i,p,e)}} \tag{4.39}$$

4.4.3 Expected Throughput Capability and Real Throughput Capability of an AMHS

The expected throughput capability of an AMHS refers to the expected maximum number of wafer lots that can be handled in a given period. Denote *Mov* as the expected throughput capability of an AMHS; the expression of *Mov* is defined as

$$Mov = m\cdot\frac{\sum_{i=1}^{n}(v_{(i,p,e)}+v_{(i,p,f)})}{n\cdot C}$$

where $v_{(i,p,e)}+v_{(i,p,f)}$ is the probability vehicle arriving at states (i,p,f) and (i,p,e).

The real throughput capability of an AMHS refers to the number of wafer lots that the AMHS can handle in a given period. Denote *TH* as the real throughput capability of an AMHS; the expression of *TH* is defined as

$$Mov = m\cdot\frac{\sum_{i=1}^{n}(v_{(i,p,e)}+v_{(i,p,f)})}{n\cdot C} \tag{4.40}$$

where n is the stocker quantity in an Interbay system and m is the number of vehicles in an Interbay system.

4.4.4 The Waiting Time of a Wafer

The waiting time of a wafer R_i refers to the total time that a wafer stays in the stocker before it is loaded to the vehicle. Assuming that moving requests arrive at the stocker according to Poisson process, R_i is the expected waiting time of wafer x at the loading point s_i^p and there are L lots at the loading point s_i^p before wafer x arrives. Then the expected waiting time of wafer x is defined as

$$R_i = \sum_{L=0}^{\infty} R_i(L)P_i(L) \tag{4.41}$$

where $P_i(L)$ is the probability that there are L lots to be transported at loading point $s_i^p, L \in Z^+$; $R_i(L)$ is the expected waiting time of wafer x at the loading point s_i^p when there are L lots at the loading point s_i^p. $R_i(L)$ consists of two parts: the expected waiting time of the first lot at the loading point s_i^p and the total waiting time of the remaining lots.

Then $R_i(L)$ can be given by

$$R_i(L) = R_i(0) + L \cdot T_e^i \tag{4.42}$$

where T_e^i is the average time of the empty vehicle to the stocker. According to Little's law:

$$WIP_{s_i^d} = \lambda_i \cdot R_i = \sum_{L=0}^{\infty} L \cdot P_i(L) \tag{4.43}$$

where λ_i is the arrival rate of moving requests at the loading point and $WIP_{s_i^d}$ is the expected WIP at the loading point s_i^p. Then it can be further defined as

$$\begin{aligned} R_i = \sum_{L=0}^{\infty}(R_i(0) + L \cdot T_e^i) \cdot P_i(L) = \sum_{L=0}^{\infty} R_i(0) \cdot P_i(L) + T_e^i \sum_{L=0}^{\infty} L \cdot P_i(L) = R_i(0) \\ + T_e^i \sum_{L=0}^{\infty} L \cdot P_i(L) = R_i(0) + T_e^i \cdot \lambda_i \cdot R_i, \quad \Rightarrow \quad R_i = R_i(0)/(1 - T_e^i \cdot \lambda_i) \end{aligned} \tag{4.44}$$

According to the system's hypothesis, the arrival time of an empty vehicle to the loading point s_i^p is random, and the average time of the empty vehicle to the stocker is T_e^i. The computation formula of remaining waiting time, and the expected waiting time of the first lot can be denoted by

$$E(R_i(0)) = \frac{E(T_e^i) + (C_{Ti}^2 + 1)}{2} = \frac{T_e^i}{2} = \frac{1}{2\varepsilon_i^p} \tag{4.45}$$

where C_{Ti}^2 is the variance of the cycle time vehicle arriving at the loading point s_i^p. Then the expected waiting time of wafer x can be defined as

$$R_i = \frac{T_e^i}{2(1 - T_e^i \cdot \lambda_i)} \tag{4.46}$$

4.4.5 Vehicle Blockage-Related Indicators

Vehicle blockage-related indicators include vehicle blockage rate β, vehicle efficient utilization rate ρ_R, system's material handling capability reduction rate ρ_{mov} and empty vehicle arrival time interval elongation rate ρ_T. Each performance indicator is presented as follows.

1. Vehicle blockage rate β.

Vehicle blockage rate refers to the ratio of the vehicle blocking time to the overall handling time in the material handling process. The vehicle blockage rate β based on the MMCM can be calculated by

$$\beta = \frac{\sum_{r \in B \cup K} v_r T_r}{\sum_{r=1}^{|R|} v_r T_r} = \frac{\sum_{r \in B \cup K} v_r T_r}{C} \tag{4.47}$$

where K is the set of states in which a vehicle is fully loaded and B is the set of states in which vehicle e is unloaded and blocked.

2. Vehicle efficient utilization rate ρ_R.

Vehicle efficient utilization rate mainly describes the influence of the vehicle blockage on the vehicle utilization rate, and ρ_R is the ratio of the effective utilization to the overall vehicle utilization, which can be given by

$$\rho_R = \frac{\rho_u}{\rho} \times 100\% \tag{4.48}$$

3. System's material handling capability reduction rate ρ_{mov}.

System's material handling capability reduction rate mainly describes the influence of the vehicle blockage on the whole handling capacity of an AMHS, and can be defined by

$$\rho_{mov} = \frac{TH/\rho_u - Mov}{TH/\rho_u} \times 100\% \tag{4.49}$$

4. Empty vehicle arrival time interval elongation rate ρ_T.

Empty vehicle arrival time interval elongation rate mainly describes the influence of the vehicle blockage on the average arrival time interval of an empty vehicle, and can be calculated by

$$\rho_T = \left(\frac{n \times \beta \times C}{m \times \sum_{i=1}^{n} v(i, p, e)} \right) / T_e \times 100\% \tag{4.50}$$

4.4.6 Case Study

In this section, practical data collected from a 300 mm wafer fabrication factory in Shanghai are analyzed. The AMHS in the factory consists of one Interbay subsystems and 22 Intrabay subsystems. They are connected by stockers, with a total of 496 devices distributed in 22 Intrabay subsystems. The analysis process is presented as follows:

1. Input system handling track (including shortcut track) layout information, stocker location and loading/unloading port position information, as well as vehicle information.
2. Set parameters of the modified Markov model. This includes set wafer arrival rate, processing route of different wafers and their corresponding processing areas in the modified Markov model.
3. Construct modified Markov model. According to the input parameters of the modified Markov model, establish wafer arrival rate model of each stocker, system conservation and stability condition model, state transition probability model and vehicle blockage probability model, and obtain the steady-state probability of the modified Markov model.
4. Analyze the running characteristics of an AMHS based on the modified Markov model including the impacts of vehicle number, vehicle speed and vehicle blockage on the system's performance. The results are shown in Figures 4.7–4.9.

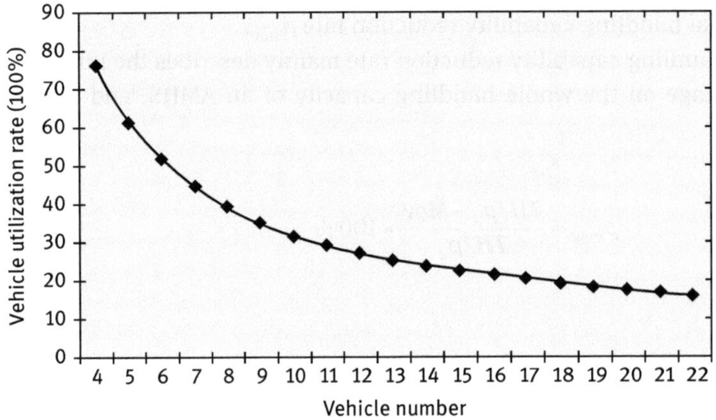

Figure 4.7: Impact of the vehicle number on the vehicle utilization rate.

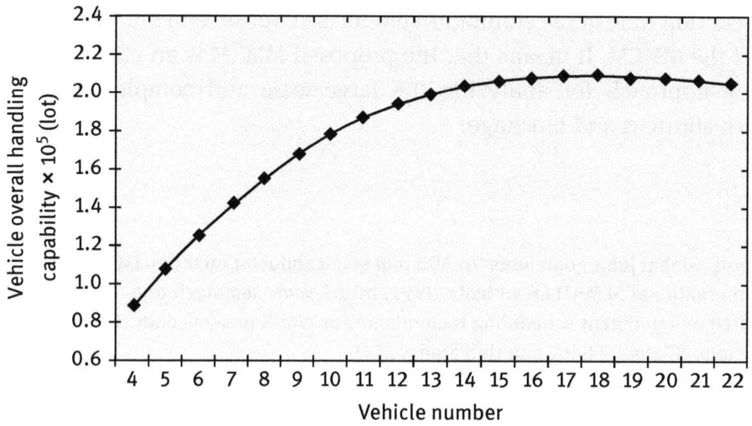

Figure 4.8: Impact of the vehicle number on the vehicle overall handling capability.

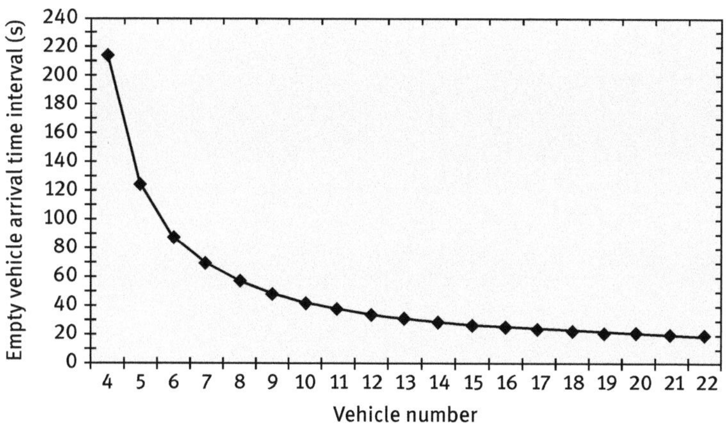

Figure 4.9: Impact of the vehicle number on the empty vehicle arrival time interval.

4.5 Conclusion

AMHS is indispensable and critical in 300 mm SWFSs. In order to analyze and evaluate the performances of AMHS with shortcut and blockage effectively and efficiently in system's design stage, an MMCM has been proposed. With real production data, the proposed MMCM has been compared with the simulation model. The results have demonstrated that the proposed MMCM and the simulation approach have small relative errors in terms of system's throughput capability, vehicle's mean utilization ratio and empty-vehicle's mean arrival time interval. The 96% of all relative error percentages range from −8% to 10%. In the aspect of computational efficiency, the execution time of the simulation model is about several hundred times more than that of the MMCM. It means that the proposed MMCM is an effective and efficient modeling approach for analyzing the large-scale and complex AMHS's performances with shortcut and blockage.

References

[1] Sematech Report. Global joint guidelines for 300 mm semiconductor factories I300I/J300. Available at: International SEMATECH website. 1997, http://www.sematech.org.
[2] Wu, L.H.Research on intelligent scheduling technologies of AMHS in semiconductor wafer fabrication system, Shanghai Jiao Tong University, 2011.
[3] Sampoorna, M., Bueno, J.T. Gauss-Seidel and successive overrelaxation methods for radiative transfer with partial frequency redistribution. Astrophysical Journal, 2010, 712(2): 1331–1344.
[4] Chatterjee, C., Roychowdhury, V.P., Chong, E.K.P. A nonlinear Gauss-Seidel algorithm for noncoplanar and coplanar camera calibration with convergence analysis. Computer Vision and Image Understanding, 1997, 67(1): 58–80.

5 Scheduling methods of automated material handling systems in SWFSs

In order to respond quickly to changes in market demands, improve on-time delivery rate and shorten the production cycle time, it is necessary to schedule wafer fabrication systems and AMHS. The overall optimization objectives are achieved by reducing transport time, waiting time and WIP quantity. Scheduling in AMHSs refers to meeting the material handling constraints and arranging the sequence and time of wafer carriers handled by vehicles in order to optimize the material handling performance. The scheduling problem in AMHSs is a typical NP-hard problem due to its large-scale [1], stochastic, real-time and multi- objective features. It is ineffective to solve such a complex problem by using traditional mathematical programming methods. In addition, the dynamic scheduling problem of complex production processes in the large-scale uncertain environment cannot be solved by using the traditional scheduling theory. Therefore, the scheduling problem in AMHSs is intensively studied by domestic and foreign scholars, using the common methods including heuristic rules, operation research methods and artificial intelligence algorithms.

5.1 Heuristic Rules-Based Scheduling Methods

The heuristic approach refers to a method of finding an optimal schedule based on empirical rules. It solves problems by using past experiences and proven methods rather than using determinate steps to find the answer. The heuristic method reduces the number of attempts in the limited search space and solves the problem quickly [2]. Many major discoveries of scientists are obtained by using simple heuristic rules. In the information processing theory of the cognitive psychology, heuristic is an important way to solve the problem of human thinking. In artificial intelligence, heuristic methods are often used to design computer programs in order to simulate human thinking activities. It has been proved that this is an effective way to solve practical problems.

5.1.1 Heuristic Rules

The method based on the single heuristic rule is to use a kind of heuristic rule as the optimization strategy. It is usually applied to optimize the single performance index of the AMHS, such as the shortest wafer processing cycle, the shortest wafer waiting time, the minimum number of wafer products and the highest machine utilization

https://doi.org/10.1515/9783110487473-005

Table 5.1: Dispatching rules.

	RULE	DATA	ENVIRONMENT
1	SIRO	_	_
2	ERD	r_j	$1 \mid r_j \mid \mathrm{Var}(\sum(C_j - r_j)/n)$
3	EDD	d_j	$1 \parallel L_{\max}$
4	MS	d_j	$1 \parallel L_{\max}$
5	SPT	p_j	$P_m \parallel \sum C_j$; $F_m \mid p_{ij} = p_j \mid \sum C_j$
6	WSPT	w_j, p_j	$P_m \parallel \sum \omega_j C_j$
7	LPT	p_j	$P_m \parallel C_{\max}$
8	SPT-LPT	p_j	$F_m \mid \text{block}, p_{ij} = p_j \mid C_{\max}$
9	CP	p_j, prec	$p_m \mid \text{prec} \mid C_{\max}$
10	LNS	p_j, prec	$p_m \mid \text{prec} \mid C_{\max}$
11	SST	$s_{j,k}$	$1 \mid s_{jk} \mid C_{\max}$
12	LFJ	M_j	$p_m \mid M_j \mid C_{\max}$
13	LAPT	p_{ij}	$O2 \parallel C_{\max}$
14	SQ	_	$P_m \parallel \sum C_j$
15	SQNO	_	$J_m \parallel \gamma$

rate. However, it is difficult to use the single heuristic rule-based method to optimize the multi-objective AMHS. Commonly used heuristic rules are presented in Table 5.1.

For a number of heuristic rules, each rule has a different nature; hence, there is no uniform measure. There is a contradiction between the rules; the objective value under one rule is improved while the objective value under another rule is compromised. The process that uses a variety of prioritized strategies and considers various heuristic rules to optimize is regarded as a heuristic rule composition. The method based on the compound heuristic rules that consider multiple system parameters, such as idle vehicle delivery time, wafer waiting time, the number of wafer cards waiting to be delivered, the distance between the wafer lot and the vehicle, and the current vehicle location, is able to optimize the multi-objective AMHS.

5.1.2 Heuristic Rules Applied in AMHS Scheduling

For the scheduling problem in AMHSs, the system parameters such as the idle vehicle delivery time, the wafer waiting time and the number of the wafer lots waiting for transportation are taken into consideration. According to the actual demand of the manufacturing system, the heuristic search is implemented to quickly obtain the search results in order to determine the material handling program in AMHSs [3–9].

The performance of an AMHS is evaluated in terms of wafer production efficiency and the satisfaction degree of production strategies. The production efficiency concerns movement efficiency, throughput, use of processing tools and reduction in cycle time; production strategy issues include delivery due dates and manufacturing priorities. Several common vehicle dispatch rules are presented as follows:

1. HP rule: highest priority lot first
   The wafer with the highest manufacturing priority is delivered first in order to complete all of its processing steps in higher priority.

2. ERT rule: earliest release time first
   The earliest released wafer is moved first.

3. LWT rule: longest waiting time lot first
   The wafer with the longest waiting time is moved first in order to reduce the delivery waiting time.

4. NJF rule: nearest job first
   Select a recent wafer waiting to be delivered.

5. ECD rule: earliest commit due date first
   The wafer with the earliest delivery date is delivered first.

6. MNJF rule: modified nearest job first
   It is a combination of NJF and LWT rules. The implementation is depicted in Figure 5.1.

7. FNJF rule: factorized nearest job first
   It is another combination of NJF and LWT rules.

   Based on the above description, eight scheduling rules are tailored according to the production strategy and the manufacturing efficiency, but they cannot meet the demands of both strategy and efficiency. A fuzzy logic- based multi-mission-oriented vehicle scheduling method, which aims to improve efficiency and meet the policy requirements, is shown in Figure 5.1.

   The control architecture of the fuzzy logic-based multi-mission-oriented scheduler is illustrated in Figure 5.2. The fuzzy logic- based multi-mission-oriented scheduler is used to determine an appropriate scheduling rule so that potentially risky wafer cards can meet the production strategy. The procedure of establishing the fuzzy scheduling model can be divided into defining input, constructing membership function, establishing fuzzy rule table and inverse fuzzification. The inputs to the fuzzy-based scheduler are based on the latency statistics and the degree to which the policy conforms to the priority and lead time requirements.

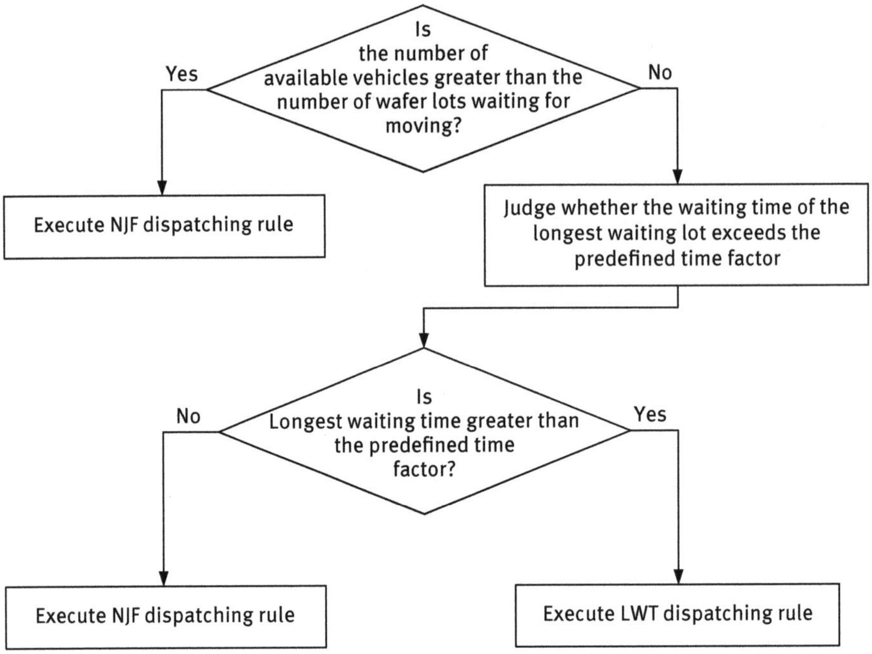

Figure 5.1: MNJF dispatching rule execution scenario.

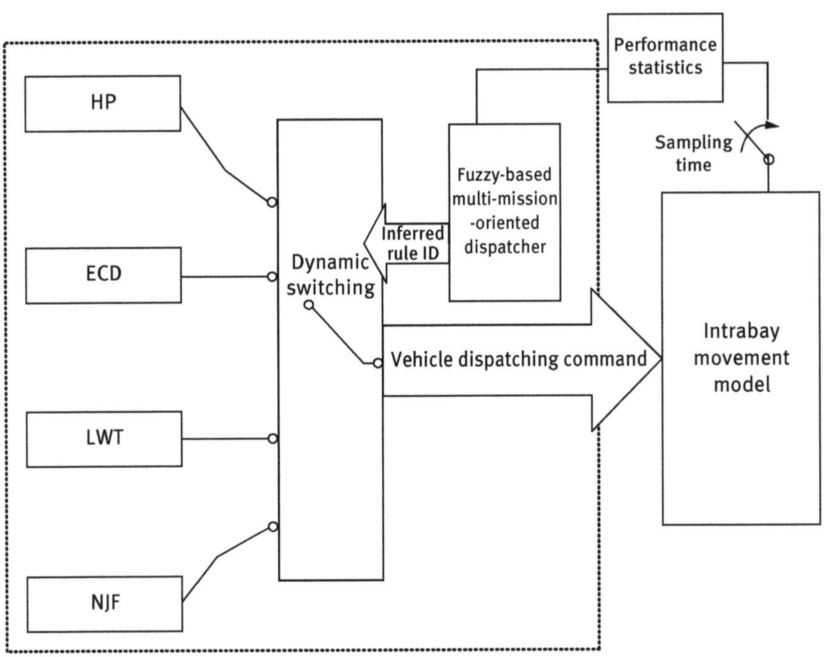

Figure 5.2: The control architecture of the fuzzy-logic-based multi-mission-oriented vehicle dispatcher.

In fact, in the actual wafer manufacturing material handling process, it is difficult to use the single heuristic rule-based method to optimize the multi-objective AMHS. The weight of each parameter in the traditional compound heuristic rule is usually set to a fixed value, and it is only applicable to the static environment. How to adjust the scheduling strategy according to the change-over of system state and interference factors is a key problem to effectively schedule the material handling system in AMHSs under the dynamic and stochastic material handling environment.

5.2 Operation Research Theory-Based Scheduling Methods

Operation research methods solve various systematic optimization problems by using statistics, mathematical models and algorithms so as to find the best or near optimal solution in complex problems. It has been widely applied in the field of scheduling in AMHSs.

5.2.1 Operation Research Theory

The general procedure of applied operational research includes the following main steps: (1) define the nature and extent of the problem; (2) establish the mathematical model of the problem; (3) define an objective function as a scale for comparing various possible actions; (4) determine the specific value of each parameter in the model and (5) solve the model to find the optimal solution to make the objective function reach the maximum value (or minimum value). In the field of wafer material handling scheduling, operation research methods include linear programming, integer programming, dynamic programming and other methods.

5.2.1.1 Linear Programming

The linear programming method is used to reasonably arrange resources under certain conditions so as to achieve the optimal economic effect.

1. Simplex method

   Applicable problem: all constraints are ≤; the right side are all nonnegative. There is no requirement for the coefficients of the objective function. For example,

$$\begin{aligned} \min z &= 3x_1 - 2x_2 \\ \text{s.t.}\ \ x_1 + 2x_2 &\le 12 \\ 2x_1 + x_2 &\le 18 \\ x_1, x_2 &\ge 0 \end{aligned}$$

The following steps are used to solve this problem:
Step 1: Standardize the linear programming problem.
Step 2: Check the initial feasible solution: if there is, go to the fourth step; otherwise, go to the third step.
Step 3: Construct the auxiliary problem, and use the two-stage method to solve the auxiliary problem. If the objective function value of the optimal solution is greater than 0, there is no feasible solution and the algorithm terminates. Otherwise, go to the fourth step.
Step 4: Write a simplex table, in which the coefficient of the base variable in the constraints is set to the unit matrix and the coefficient of the base variable in the objective function is set to 0. Go to the fourth step.
Step 5: If the number of all non-base variables is negative or 0, the optimal solution is obtained and the algorithm terminates. Otherwise, a non-base variable whose number of tests is positive and whose absolute value is the largest is selected as the base variable. Go to the sixth step.
Step 6: If the coefficients of the base variables in the constraint condition are all negative or 0, the objective function is unbounded and the algorithm terminates. Otherwise, the base variables are determined according to the minimum ratio of the right and positive coefficients. Go to the seventh step.
Step 7: Transform the row of the simplex table; the principal element becomes 1 and the other elements become 0. Go to the fifth step.

2. Dual simplex method
   Applicable problem: at least one of the constraints is "≥," the corresponding right constant is nonnegative and the coefficients of the objective function are all nonnegative.
   The following steps are used to solve this problem:
   Step 1: In the first step, the initial basis B of the original problem (L) is determined, and the initial simplex table is established for all test numbers, namely, the dual feasible solution.
   Step 2: In the second step, the values of the base variables are checked; if ≥ 0, the optimal solution has been obtained, then terminate the search process; otherwise, L line in the simplex table corresponding to the base variable is determined as the spin-out variable.
   Step 3: If ≥ 0, then there is no feasible solution for the original problem, so terminate the search process; otherwise, the corresponding base variable is determined as the spin-in variable.
   Step 4: In the fourth step rotation transform for the new simplex table is established; go to step 2. It can be shown that the iterative solution is always the dual feasible solution.

5.2.1.2 Integer Planning

When the variables in the plan (some or all) are limited to be integer, it is called an integer plan. The main methods include the following:

1. Cutting plane method

   In this method, it is possible to cut the original problem by adding new constraints in order to shrink the feasible region. The optimal solution of the original problem is gradually exposed and tends to the position of the feasible pole. It is possible to use the simplex method to find the optimal solution.

   The following steps are used in this method:

   Step 1: Solve the relaxation problem using the simplex method to obtain the optimal simplex table.

   Step 2: Seek a cutting plane equation, add it to the optimal simplex form and continue to solve the problem using the dual simplex method.

   Step 3: If integer optimal solutions are not obtained, then continue to cut the plane equation and go to the second step.

2. Hungarian method

   In the real life, there are assignment problems of various natures. Assignment problems are also important issues for integer planning. For example, there are *n* tasks that need to be assigned to *n* individuals (or departments) to complete; there are *n* contracts need to select *n* bidders to contract; there are *n* classes need to be arranged in the classroom and so on. The basic requirements for these problems are to optimize the overall effectiveness of the assignment scheme when the specific assignment requirements are satisfied.

   The following steps are used in this method:

   Step 1: Transform the efficiency matrix so that the coefficient matrix of the assignment problem is transformed and zero elements appear in each row and column. This is done by subtracting the minimum nonzero elements of the row from the rows of the efficiency matrix, and subtracting the smallest nonzero element of the columns from the resulting coefficient matrix. The new resulting matrix will have zero elements in every row and column.

   Step 2: Use the circle 0 method to find the independent zero elements in the matrix C1. After the first step of transformation, the coefficient matrix of each row and each column has a separate zero element. If *n* is small, it can be directly observed to find *n* independent zero elements; while *n* is large, it must follow certain steps to find zero elements. (1) From a row (or column) with only one zero element, start by adding 0 to the element. This means that for the person represented by this line, there is only one task that can be assigned, and then the other elements of the column (row) are marked as ф, which means that the task represented by this column has been assigned without having to consider other people. (2) Add a circle to the zero element with only one zero element column (row). (3) Repeat steps (1) and (2) until each column has no unmarked zero elements or at least two unmarked zero elements.

Step 3: Test assignment. If case (1) occurs, assignment can be made: let the decision variable in the position of circle 0 be 1 and let the other decision variables be 0, and get an optimal assignment scheme and stop the calculation. In this example, after C2 is obtained, case (1) occurs. Let $x14 = x22 = x31 = x43 = 1$ and the remaining $xij = 0$ is the best assignment scheme. If case (2) occurs, then for each row, each column has two unmarked zero elements (choose one, and add the tag); that is, circle the zero element. And then mark "×" to other not marked zero elements of the same row, the same column. Case (1) or (3) may occur. If case (3) appears, go to the next step.

Step 4: Make at least a straight line covering all zero elements.

5.2.1.3 Dynamic Programming

Dynamic programming is a branch of algorithm design and operational research, which is an effective way to solve multi-stage optimization problems. In the middle of the twentieth century, the American mathematician Chad Berman divided the decision-making problem into multiple stages and solved the problem of each stage. He proposed the "optimization principle" to establish a large branch of operational research. Dynamic programming is the global solution for each set, and a scheme is chosen such that the objective function is minimized to find the optimal solution of the problem. Since the emergence of the dynamic programming theory, the decision-making process is not a linear decision-making process, it fully considers different circumstances of the decision-making process and it provides an effective way for the solution. In other words, it cannot be replaced by using an existing algorithm. According to the actual background of the given problem, a new plural is selected by using continuous decision- making process until the optimal solution is found; hence, it is called "dynamic."

Dynamic programming is established on the basis of the global optimization, to make local interest decision for each stage in order to ensure that the final global optimal is obtained. A large, complex and multidimensional variable problem is decomposed into a number of simple, small- scale variables so as to reduce the computational complexity.

One of the most common shortest paths in dynamic programming is shown in Figure 5.3.

In the initial stage, the optimal principle is introduced from the simple logic, the basic equation is derived on the premise of the optimal strategy and then the optimal strategy is solved by this equation. In its application process, it is found that the optimal principle is not universal for any decision-making process, and it is not unconditionally equivalent to the basic equation, and there is no definite implication

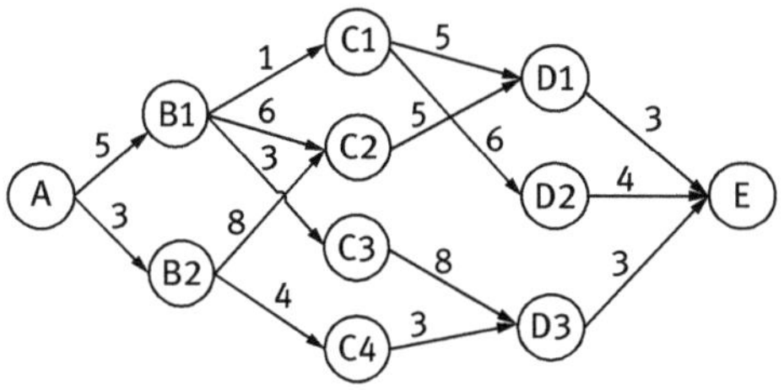

Figure 5.3: Shortest path.

relationship between them. The basic equation plays a more essential role in the dynamic programming.

For the initial state $x_1 \in X_1$, the necessary and sufficient condition of the optimal strategy $p^*_{1,n} = u^*_1, \ldots, u^*_n$ is for any k, $\mathbf{1} \le k \le n$, there is

$$V_{1,n}(x_1, p^*_{1,n}) = \varphi\left(opt_{p_{1,k-1} \in p_{1,k-1}(x_1)}[V_{1,k-1}(x_1, p_{1,k-1})], opt_{p_{k,n} \in p_{k,n}(x_k)}[V_{k,n}(x_m, p_{k,n})]\right) \tag{5.1}$$

If $p^*_{1,n} \in p_{1,n}(x_1)$ is the best solution, for any k $(1<k<n)$, its sub-strategy $p^*_{k,n}$ is also optimal for x_1 and the kth to nth sub-procedures starting from x^*_k determined by $p^*_{1,k-1}$. The abovementioned inference is called the optimization principle, which gives the necessary condition for the optimal strategy and is generally described as follows: regardless of the past state and decision, the remaining decisions must constitute the optimal strategy for the current state of the previous decision.

According to the inference of the basic theorem, the basic equation of dynamic programming is presented as follows:

$$\begin{cases} f_k(x_k) = opt\{\phi(v_k(x_k, u_k), f_{k+1}(x_{k+1}))\}, x_{k+1} = T_k(x_k, u_k), k = 1, 2, \ldots, n \\ f_{n+1}(x_{n+1}) = \delta(x_{n+1}) \end{cases} \tag{5.2}$$

where $f_{n+1}(x_{n+1}) = \delta(x_{n+1})$ is the terminal condition of the decision process and δ is a known function. It is called a fixed terminal when x_{n+1} only takes a fixed state, and it is called a free terminal when x_{n+1} can be changed in the terminal set X_{n+1}. The final optimal objective function satisfies (5.3):

$$opt\{V_{1,n}\} = opt_{x_1 \in X_1}\{f(x_1)\} \tag{5.3}$$

In the recursive formula (5.2), you can directly use the formula to obtain the optimal recursive formula. However, in practical applications, the recursive formula is changed to a recurrence formula, which will be more efficient. The numerical solution is quite complicated when the dynamic programming method is used, and the memory required by the computer is quite large. The above difficulties are faced by solving dynamic planning problems; they are the limitations in the development of dynamic planning methods.

5.2.2 Operation Research Theory Applied in Scheduling in AMHSs

Here a case of the hybrid CP (constrained programming)/MIP (hybrid integer programming)-based approach is used to introduce how to use operation research methods to optimize tasks assignment and conflict-free vehicle routing.

Considering numbers of AGVs and a series of transport requests, it is necessary to determine (1) the number of AGVs required to complete a given task; (2) the allocation of handling requests to the vehicle; (3) the location of each AGV in the flexible manufacturing system and (4) a scheduling scheme for tasks picked up or delivered by the cart.

The hybrid CP/MIP method can effectively utilize the powerful scheduling capability of CP, which can deal with nonlinear constraints. On the other hand, MIP is used to optimize the vehicle routing. The hybrid algorithm flow is shown in Figure 5.4. In this decomposition method, the CP method is based on the shortest path to solve the distribution of the handling requests and the scheduling of AGVs. The MIP method finds the collision-free path in the main problem solutions

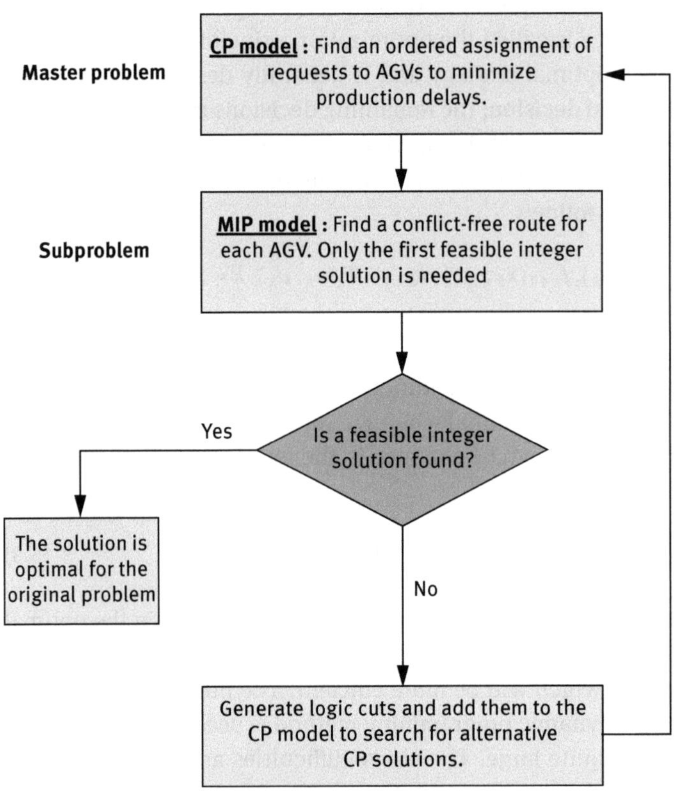

Figure 5.4: The hybrid CP/MIP method.

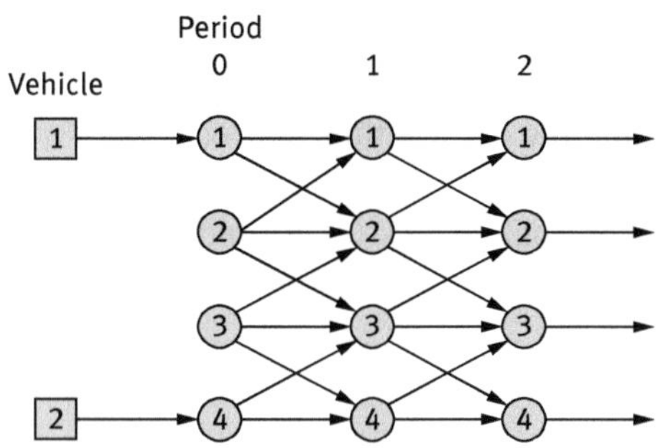

Figure 5.5: Time- space network.

obtained by the CP method. If there is no solution (that is, there is no collision-free path), the logic reduction is triggered and returned to the main problem.

The CP model determines which material handling request will be processed by which vehicle at a time when the total delivery delay is minimized. The total delay is determined by calculating the difference between the actual earliest delivery time and the planned earliest delivery time. The distance matrix is determined by computing the shortest path between the nodes. Therefore, the delay calculation (regardless of possible collisions) provides a lower bound for the actual delay. Shipping requirements consist of pickup and delivery tasks. Thus, the starting position of the window and the ending position of the window can be considered as nodes of the current and next virtual tasks of each AGV. The set is defined as the set of all tasks, including virtual initiation.

As shown in Figure 5.5, each node defines the end of a time unit and a path. A node of a given time interval is connected to an adjacent node of the next section by an arc. An arc indicates that the AGV moves between adjacent endpoints of two paths. The nodes of each time interval are connected by arcs to the same associated node in the next time interval (e.g., a horizontal arc), which means waiting for a unit time at this node. There is no arc connection for the same time interval. Each AGV has a start node at time zero. This problem resembles a network flow model for multiple goods, where each item represents an AGV. The solutions of the main problem restrict a particular node that each flow passes through at a particular time.

1. Index and parameters
   - N the set of nodes
   - V the set of AGVs
   - A^+ the set of all arcs
   - A the set of arcs
   - T^+ the set of periods but the first period
   - T the set of periods
   - $A^O[.]$ the opposite arc of each arc
   - $A^w[.]$ the waiting arc associated with each node

$A^t[.]$	the set of arcs entering in node i
$A^f[.]$	the set of arcs from node i
W	record the nodes of each task and the earliest time matrix
n^v	the starting node of AGV v
R	the set of requests: each request is a pair of pick-up (r^p) and delivery (r^d)
$R^+[.,t]$	for every pair (node i, period t), the service node for loading and handling task r is i
n_i	the node for task i
$t_r \in T$	the start time of request r

2. Variables

$X^t_{k,a}$ Boolean variable = 1 if AGV$_k$ goes through the directional arc a at period t; otherwise, $X^t_{k,a}$ is equal to 0.

The constant objective function is presented as follows.

Min 1

s.t.

$$\sum_{a \in A^f[n^k]} X^0_{k,a} = 1, \forall k \in V \tag{5.4}$$

$$\sum_{a \in A^+} X^t_{k,a} = 1, \forall t \in T, k \in V \tag{5.5}$$

$$\sum_{a \in A^f[i]} X^t_{k,a} + \sum_{a \in A^f[i]} X^t_{k,a} = 0, \forall i \in N, k \in V, t \in [1...(M-1)] \tag{5.6}$$

$$\sum_{k \in V} X^t_{k,a} + \sum_{k \in V} X^t_{k,A^0[a]} \leq 1, \forall t \in T, a \in A \tag{5.7}$$

$$X^{T_{r^p}}_{V_r, A^w[n_r,p]} = 1, \forall r \in R \tag{5.8}$$

$$X^{T_{r^d}}_{V_r, A^w[n_r,d]} = 1, \forall r \in R \tag{5.9}$$

$$\sum_{k \in V, a \in A^t[i]} X^t_{k,a} \leq 1 + \left(\sum_{r \in R^+[t-1,i]} 1 \right) \times \left(\sum_{r \in R^+[t,i]} 1 \right), \forall i \cap N, t \in T^+ \tag{5.10}$$

Constraints (5.4) make sure that every AGV starts at the initial position in each time interval. Constraints (5.5) ensure that each AGV is in a unique position in each time interval. The constraint (5.6) is a flow conservation constraint. Constraints (5.7) are anti-collision constraints. Constraints (5.8) indicate that each load operation must be completed by the correct vehicle at the correct time. Constraints (5.9) indicate that each handling operation must be completed by the correct vehicle at the correct time. Constraints (5.10) are the capacity constraints of a node, that is, only one AGV can be on the same node at any time, except that one is leaving and the other is entering the same node.

Since the scheduling model has many different solutions for the same production delay, the purpose of the logic reduction is to prohibit not only solutions that are not currently feasible (i.e., conflicts), but also many solutions with other undesirable characteristics. Logical reductions are triggered when a feasible path is not found. Logical reductions are allocated on the basis of the starting time of the delivery process and the AGV request. When two AGV cars conflict, at least one of them will not have enough time to complete its task, and thus a logical reduction approach is adopted.

In the scheduling process in AMHSs, material handling scheduling methods based on operation research methods are capable of obtaining the global optimal solution or suboptimal solutions while considering factors such as vehicles' temporary blockage. However, due to large calculation number and long computing time of the mathematical programming model, it is usually suitable for a small-scale scheduling problem. And in the large-scale, complex and stochastic wafer manufacturing environments, it is often difficult to meet the requirements of a real-time scheduling process.

5.3 Artificial Intelligence-Based Scheduling Methods

The use of the artificial intelligence theory and technology to solve the production scheduling problem is regarded as intelligent scheduling. These methods mainly include the expert system based on symbol intelligence, neural network based on computational intelligence, fuzzy rule and genetic algorithm (GA) and distributed production scheduling based on multi-agent cooperative solution technology. Intelligent scheduling has become a hotspot in the field of material handling scheduling in AMHSs in the past decade.

5.3.1 Artificial Intelligence Algorithms

5.3.1.1 Genetic Algorithm

The operation of the GA begins with the generation of a population that represents the potential solution of the problem. A population consists of a number of individuals, each of which is actually a chromosome with a characteristic entity. Chromosomes are the main carriers of genetic material, and their internal manifestations are a combination of genes that determine the external manifestations of individual traits, such as the characteristics of black hair, which are determined by a combination of genes that control this feature in chromosomes. Therefore, at the outset, it is necessary to implement mapping from phenotype to genotype, that is, coding. After the initial population is generated, it is evolved according to the survival of the fittest principle, and better and better approximate optimal solutions are obtained. The evolution process is a specific genetic operation. First, the individuals are selected according to the fitness of the individuals in the population, and the genetic operators such as crossover and mutation are used to generate the new population representing the new solution set. This evolutionary process will make the fitness of the individuals in the population become more and more close to the approximate optimal solution of the problem.

GAs have the following characteristics: (1) self-organization, self-adaptation and self-learning. When the coding scheme and the fitness function are determined, the algorithm will use the information obtained in the evolutionary process to organize the search. This self-organizing, self-adapting feature enables it to automatically discover the characteristics and laws of the environment according to environmental changes without describing all the features of the problem in advance. (2) The parallelism: the GA searches the population in parallel, rather than on a single point. Its parallelism is manifested in two aspects. One is the internal parallelism, which can make the evolution calculation of the independent population by using hundreds or even thousands of computers. When all the machines are finished, the communication is compared and the best individual is selected. The second is the inclusion of parallelism. By using the population search, multiple regions are searched within the solution space, and the information is exchanged. (3) The GA does not need other auxiliary knowledge; it only needs the objective function and the fitness function that influence the search. (4) GAs emphasize probability conversion rules rather than deterministic transformation rules. (5) The GA can be more directly applied as it is simple and easy to understand.

Generally, the GA is designed according to the following steps:

Step 1: Determine the coding scheme for the problem. Because GA usually does not directly affect the solution space of the problem, some coding representations of the solution are adopted to evolve. A reasonable coding mechanism has a great impact on the quality and efficiency of the algorithm.

Step 2: Determine the fitness function. Because genetic operations are operated on the basis of the fitness value, a reasonable fitness value of the individual can reflect the good/bad degree of the individual, and adapt to the evolution process of the algorithm.

Step 3: Select algorithm parameters. Usually they include the population number, crossover probability, mutation probability, iteration and so on.

Step 4: Design genetic operators. They include initialization, selection, crossover, mutation and replacement operations.

Step 5: Determine the termination condition of the algorithm. Termination criteria should be developed according to the nature of the problem; it should balance the quality and the efficiency.

The key parameters and operations of the GA are presented as follows:

1. Coding

   Coding uses a code to represent the solution of the problem so that the state space of the problem corresponds to the code space of the GA. This largely depends on the nature of the problem and will affect the design of genetic operations. Because the optimization process of a GA is not directly applied to the problem parameters, it is carried out in the code space corresponding to a certain coding mechanism. Hence, the choice of coding is an important factor affecting the performance and efficiency of the algorithm.

   For the function optimization, the coding technology mainly includes binary coding, decimal coding, real coding and so on.

2. Fitness function

   The fitness function is used to evaluate the individual and is the basis of the optimization process.

   For a simple minimization problem, the objective function can usually be transformed directly to a fitness function. For complex optimization problems, it is often necessary to construct an appropriate evaluation function to improve the optimization efficiency of a GA.

3. Algorithm parameters

   The standard GA has four key parameters, namely, population size (M), crossover probability, mutation probability and generation gap (G).

   (1) Population size (M)

   The population size affects the final optimization performance and the algorithm efficiency. In the optimization process, the population size is allowed to change, and a larger population can be decomposed into several populations to evolve.

   (2) Crossover probability

   When the crossover probability is too large, the strings are updated quickly, and the individuals with higher fitness values are destroyed soon.

   (3) Mutation probability

   The mutation probability is used to increase the diversity of the population. Usually a lower mutation probability is sufficient to prevent any gene in the entire population from remaining unchanged. Moreover, the mutation probability cannot be too high; otherwise, the genetic search process will behave like a random search process.

   (4) Generation gap (G)

   The generation gap is used to control the replaced proportion of population in each generation.

4. Genetic operations
   The survival of the fittest is the basic idea of designing GA, which should be reflected in the genetic operations, such as selection, crossover, mutation and population substitution.
   (1) Population initialization
       In view of the probabilistic convergence properties of a GA, the initial population is generated randomly. However, considering the search efficiency and quality, it is necessary to distribute the initial population in the solution space as far as possible. On the other hand, some simple methods or rules can be used to quickly generate some solutions as initial individuals.
   (2) Selection operation
       The selection operation is used to avoid the loss of effective genes to make the high-performance individuals survive with greater probability so as to improve the global convergence and computational efficiency. The most common methods are proportional selection, ranking-based selection and tournament selection.
   (3) Crossover operation
       Crossover operations are used to combine new individuals, perform efficient searches in the solution space and reduce the probability of failure for the active mode. Binary coding GA typically uses single-point crossover and multi-point crossover.
   (4) Mutation operation
       When the fitness of the offspring produced by the crossover operation does not evolve and does not reach the optimum, it means the premature convergence of the algorithm. This phenomenon is rooted in the defect of effective gene; to a certain extent, the mutation operation is adopted to overcome this situation in order to increase the diversity of the population.
   (5) Replacement strategy
       The population replacement strategy is usually replaced by whole or partial. The whole replacement will completely cover the original population, while the partial replacement will replace part of old individuals with some new individuals.

5.3.1.2 Agent-Based Distributed Algorithm

Since the 1990s, with the development of computer networks and computer communications technology, agent technology has not only become a hotspot for distributed artificial intelligence research, but also become a hot field of information technology. It is noted that research achievements in various fields of artificial intelligence should be integrated into an agent of intelligent behavior. The nature of artificial intelligence is social intelligence. "Collaboration," "competition" and "negotiation" are the main manifestations of human intelligent behavior. The prerequisite for the application of the agent is to achieve its intelligent behavior by building an agent, guiding its interaction with the surrounding environments and communicating with other agents, which play an important role in forming new computing and problem-solving norms and establishing an agent technology-based distributed collaborative model with certain autonomy.

In terms of the theoretical basis, the structure of an agent can be classified into three major classes: (1) Thinking agent (cognitive or deliberative agent): In this type, the information processing unit consists of a variety of behavioral knowledge, domain knowledge and decision rules. The information processing unit is able to implement complex logical reasoning in order to make decisions. (2) Reactive agent: In this type, the information processing unit does not include any domain knowledge and behavior knowledge of the environment. Complex reasoning mechanisms are not adopted by the information processing unit, and decision-making is performed by using the predefined rules inside the information processing unit. (3) Hybrid agent: It is a new structure that is formed by the combination of the above two structures.

The "perception-action" mappings are preset in the internal part of a reactive agent; when certain conditions are met by the environmental information, an agent directly calls the preset perception-action mappings to generate its corresponding action output. The basic structure of a reactive agent is shown in Figure 5.6. In the figure, information perception and actions are combined by the perception-action relationship base, which interacts with the environment through sensors and actuators.

In contrast to a thinking agent, a reactive agent responds quickly; the main reason for this is that it does not contain logical reasoning modules and does not

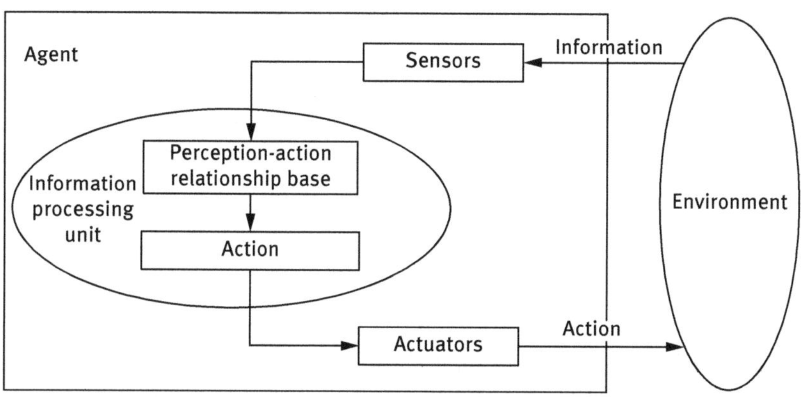

Figure 5.6: The basic structure of a reactive agent.

adopt complex reasoning systems. Experimental results indicate that the processing speed of a reactive agent is faster than that of a thinking agent when processing a limited number of tasks in the real world. However, there are some problems for a reactive agent to handle tasks requiring knowledge of the environment; for this knowledge must be obtained from memory or by inference, not by perception of sensors. In addition, a reactive agent has a relatively poor reasoning ability, it is not able to learn, and each behavior of a reactive agent is required to be encoded separately, which also leads to poor system scalability.

From the intelligence point of view, an agent is an expert system; agency means that an agent can be used to represent the role of a man while mobility means that an agent can move to run on a different machine on the Internet. An agent should have the following properties:

1. Autonomy: An agent can control its behavior and internal state by itself, and it cannot be controlled by others. This is used to differentiate an agent from other concepts such as process and object.
2. Reactive: An agent can feel the environment and respond appropriately to the environment-related events by its behavior.
3. Sociality: An agent is in a social environment constituted by multiple agents. These agents exchange information with each other in some interactive ways. These agents collaborate with each other to solve different problems and help other agents complete related activities. Agents exchange information with each other by a communication language.
4. Initiative: The reaction of an agent to the environment is a goal-directed initiative behavior. In some cases, the behavior of the agent is triggered by its own requirement. The reactive behavior is a kind of positive behavior or an active communication with the environment.
5. Adaptability: An agent can respond to environmental changes, adopt a goal-oriented action at the appropriate time and learn from their own experience, the environment and the interaction process with other agents.
6. Interoperability: An agent can work with other agents to complete complex tasks, which is a social behavior.
7. Learning ability: An agent can learn from results of the surrounding environment and cooperated experiences so as to improve its own capability.
8. Evolutionary: An agent can improve itself through learning, reproducing and following Darwin's natural selection rule "survival of the fittest."
9. Honesty: An agent does not intend to deceive users.
10. Intellectuality: The action taken by an agent and its consequences will not harm its own interest and other agents' interests.
11. Persistent: An agent is ongoing, not temporary; its status should be consistent, which is not in contradiction with property (8).
12. Mobility: An agent should have the ability to move independently in the network, and its status remains unchanged.

13. Reasoning: An agent can reason and forecast in a rational manner according to the accumulated past knowledge, the states of the current environment and other agents.

5.3.1.3 Ant Colony Algorithm

The ant colony algorithm is inspired by the foraging behavior of real ants in nature and the artificial ant is defined as the abstraction of real ants. Owing to simulating the foraging behavior of ants in nature, the ant colony algorithm is an application of mechanism. Therefore, it is necessary to abstract real ants and it is neither possible nor necessary to fully reproduce real ants. Its purpose is to more effectively portray the mechanism of the real ant colony that could be used by algorithms and to abandon the factors unrelated to the algorithm model. According to the principle of bionics, the prototype of ant colony algorithm consists of the following three major components:

1. When a path is explored, the amount of pheromone is deposited on the path.
2. When an ant is at a starting node, the next node is selected according to its respective selection probability.
3. Set the Tabu list, ants are not allowed to select passed nodes or nodes unsatisfied with executable conditions.

Suppose that $b_i(t)$ denotes the quantity of ants at node i at the moment t, and $\tau_i(t)$ denotes the amount of pheromone on the path (i, j). n represents the scale of problem, and m is the total number of ant colony. Then,

$$m = \sum_{i=1}^{n} b_i(t) \tag{5.11}$$

$$T = \{\tau_{ij}(t) | c_i, c_j \subset C\} \tag{5.12}$$

Here, T is a set of the amount of remaining pheromone on the path l_{ij} in set C at the moment t. At the initial moment, the amount of pheromone on all paths is equal and is expressed by the following equation:

$$\tau_{ij}(0) = const \tag{5.13}$$

In the movement process of Ant k (1, 2, . . . K), its moving direction is determined by the amount of pheromone on each path. The taboo list $\Gamma^k(1, 2, ..., K)$ is used to record the visited points. Γ^k is dynamically adjusted as the ant move on. In the search process, the state transition probability of ants is calculated in accordance with the amount of pheromone and heuristic information on each path. At the moment t, the state transition probability that ant k chooses to move from node i to node j is

$$p_{ij}^k(t) = \begin{cases} \dfrac{[\tau_{ij}(t)]^\alpha \cdot [\eta_{ij}(t)]^\beta}{\sum\limits_{s \notin \Gamma^k} [\tau_{ij}(t)]^\alpha \cdot [\eta_{ij}(t)]^\beta} & if\, j \notin \Gamma^k \\ 0 & if\, j \in \Gamma^k \end{cases} \tag{5.14}$$

where α is the information heuristic factor that indicates the relative importance of pheromone accumulated in the movement process and changes the dependence of ant on the existing pheromone. The larger the α value, the greater the impact of path information on the decision-making of ants and the more the probability selection consists of the collaboration among ant colony. β is the expectation heuristic factor that represents the path selection and the dependence on visibility (such as path distance). The larger the β value, the greater the impact of the path distance on ant selection and the more the probability selection consists of the greedy rule.

$\eta_{ij}(t)$ is the value of the heuristic function, which is calculated by using the following formula:

$$\eta_{ij}(t) = \frac{1}{d_{ij}} \tag{5.15}$$

In formula (5.15), d_{ij} denotes the distance l_{ij} between two adjacent nodes. For ant k, the smaller the d_{ij}, the larger the $\eta_{ij}(t)$. Excessive residual pheromone trail on certain paths directly dispatches the probability selection so as to cause undesirable effects. In order to avoid excessive residual pheromone, the residual pheromone is globally updated after each ant has visited all the nodes (n nodes). At the moment $t+n$, the pheromone on the path (i, j) is updated according to formula (5.16):

$$\tau_{ij}(t+n) = (1-\rho) \cdot \tau_{ij}(t) + \Delta\tau_{ij}(t) \tag{5.16}$$

$$\Delta\tau_{ij}(t) = \sum_{k=1}^{m} \Delta\tau_{ij}^k(t) \tag{5.17}$$

where ρ is the pheromone evaporating rate and $1-\rho$ is the residual pheromone factor. In order to prevent unlimited accumulation of pheromone, ρ is to be $\rho \subset [0, 1]$. $\Delta\tau_{ij}(t)$ is the pheromone increment on the path (i, j) in this cycle, and $\Delta\tau_{ij}(0) = 0$ at the initial time. $\Delta\tau_{ij}^k(t)$ is the pheromone increment of ant k on the path (i, j) in this cycle.

According to the pheromone update strategy, Dorigo proposed three different basic ant colony models, respectively, called Ant-Cycle model, Ant-Quantity model and Ant-Density model. Their difference lies in how to achieve $\Delta\tau_{ij}^k(t)$.

(a) In Ant-Cycle model:

$$\Delta\tau_{ij}^{k}(t)=\begin{cases}\dfrac{Q}{d_{ij}} & \text{if ant } k \text{ traverses } (i,j) \text{ in this cycle}\\ 0 & \text{other}\end{cases} \tag{5.18}$$

where Q is the pheromone intensity that affects the convergence speed of the ant colony algorithm L_k is the total length of the path visited by ant k in this cycle.

(b) In Ant-Quantity model:

$$\Delta\tau_{ij}^{k}(t)=\begin{cases}\dfrac{Q}{d_{ij}} & \text{if ant } k \text{ traverses } (i,j) \text{ between time } t \text{ and time } t+1\\ 0 & \text{other}\end{cases} \tag{5.19}$$

(c) In Ant-Density model:

$$\Delta\tau_{ij}^{k}(t)=\begin{cases}Q & \text{if ant } k \text{ traverses } (i,j) \text{ between time } t \text{ and time } t+1\\ 0 & \text{other}\end{cases} \tag{5.20}$$

In the Ant-Quantity and Ant-Density models, the local information is used by the pheromone, that is to say, whenever an ant completes a path, the pheromone trail on the path it has visited will be updated. In the Ant-Cycle model, the global information is used by the pheromone, that is to say, after all ants have completed all paths (a cycle), the pheromone trails on all paths they have visited will be updated. Among these three models, it is noted that the Ant-Cycle model achieves good results in application. In general, it is regarded as the basic model in the ant colony algorithm.

5.3.2 Intelligent Algorithms Applied in AMHS Scheduling

Intelligent algorithms have been extensively studied in scheduling in wafer fabrication and processing systems [10–13], and the research works on scheduling in material handling systems are relatively scarce. A multi-agent algorithm is introduced as an example to present the intelligent algorithms in scheduling in AMHSs. Figure 5.7 shows a single-ring slot Interbay handling system for a semiconductor fab.

When the AGV is in the single-loop cassette, the movement direction of the vehicle is clockwise. Stockers are used to store the processing wafer lots. Each stocker has sufficient storage space. Each stocker is associated with a manufacturing unit in which there are various production facilities. The input/output robot transfer mechanism (RTM) is connected to the hopper and the track. When a vehicle loaded with a wafer lot found many other wafer lots waiting to be delivered in the stocker, the vehicle needs to decide which wafer lot is the next. Once the wafer lot has been determined, the source stocker moves the selected wafer lot to the output RTM. When

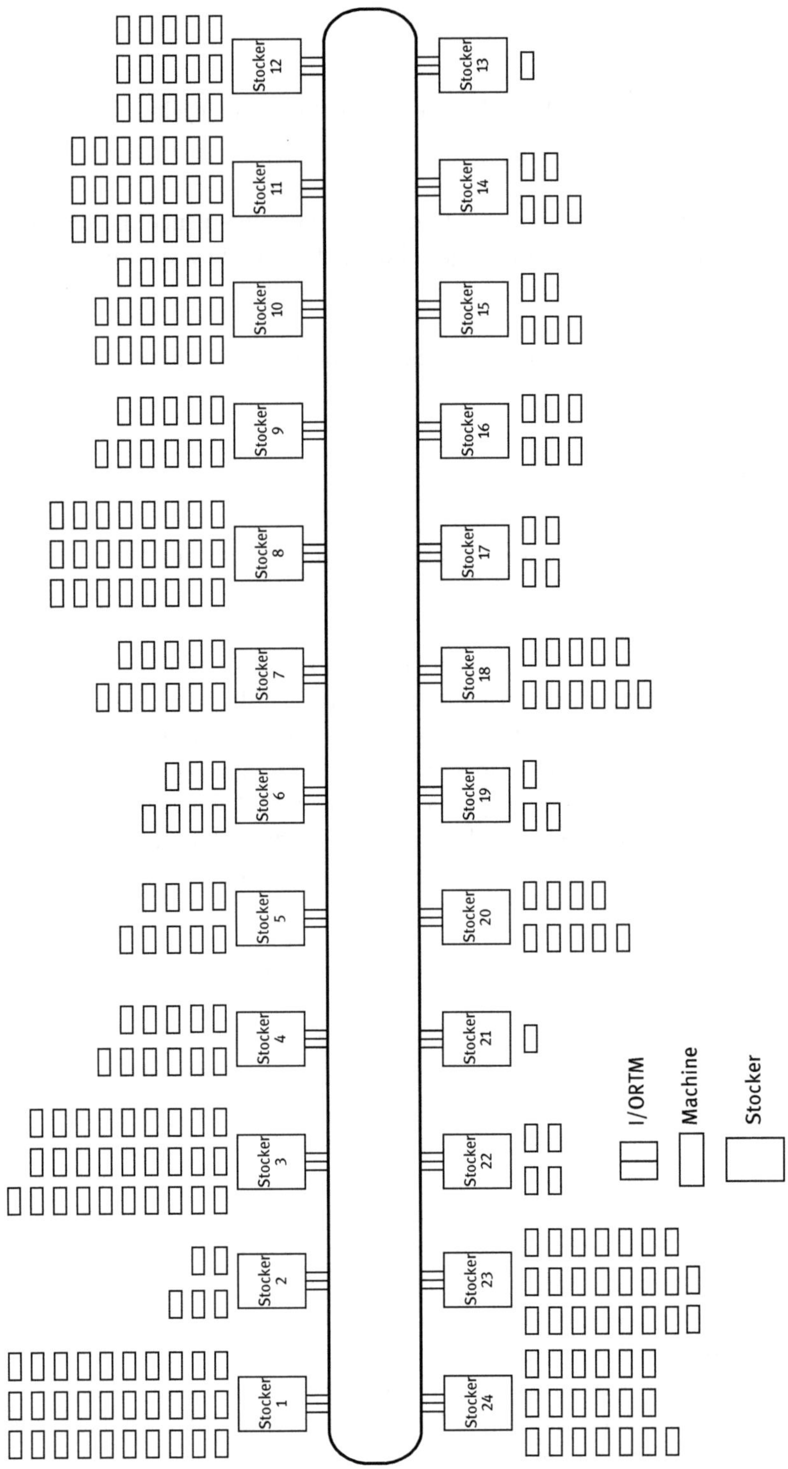

Figure 5.7: An Interbay system layout.

the vehicle arrives at the source storage depot, the output RTM places the wafer lot in the vehicle. After the vehicle is delivered to the destination stocker via the single-track track, the wafer lot is delivered from the vehicle to the storage location via the input RTM. This is the end of a basic shipping cycle.

The Interbay material handling system consists of wafer cards, machine stations, vehicles and stockers, where the wafer card is the resource requester (acting as a mobile agent) and the rest are the resource agents. Figure 5.8 lists the distributed control architectures.

Although the first structure is very scattered, rather than the rest, it is necessary to adopt more agents to develop its control system. These agents are defined by the following structure: manufacturing unit agent, vehicle agent and wafer card agent. A distributed Interbay control architecture is adopted to facilitate the control manufacturing and handling process in an Interbay (as shown in Figure 5.9). The structure includes m wafer card (CST) agents, q manufacturing unit agents and n

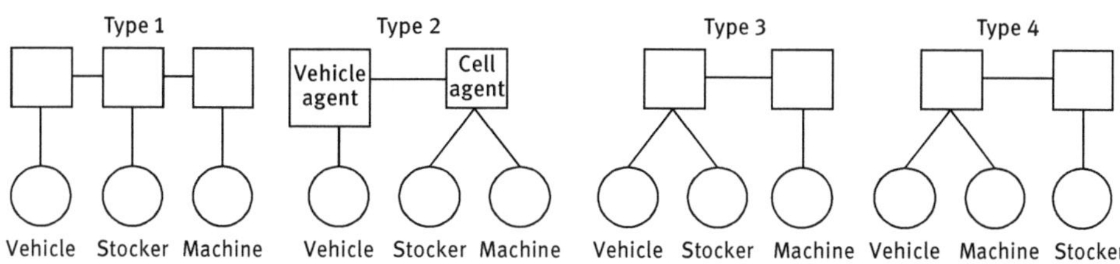

Figure 5.8: The distributed control architectures of the resource agents in an Interbay system.

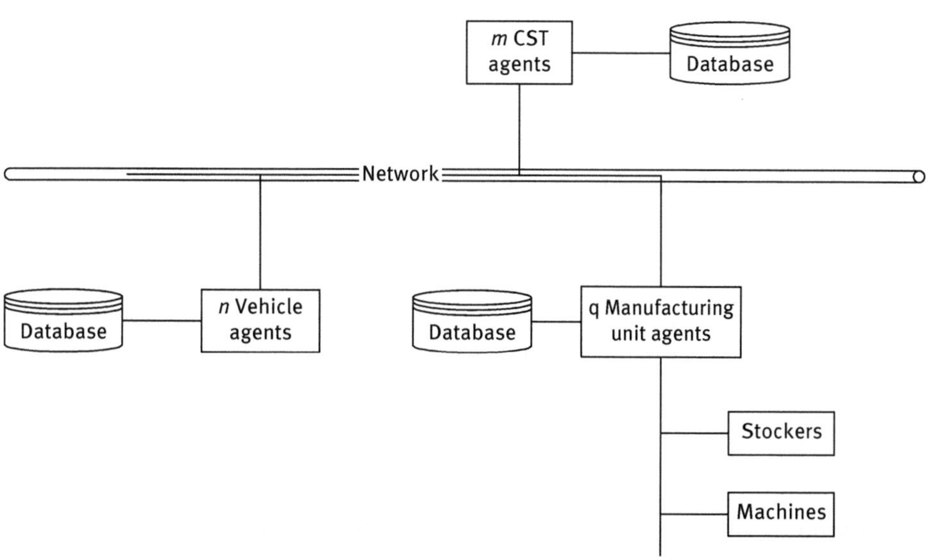

Figure 5.9: The control structure of a distributed Interbay.

vehicle agents, which connects the local vehicle database, the manufacturing unit database and the wafer card sequence database with the manufacturing execution system. Agents communicate with each other through a contract network agreement.

Some agents must act independently and must not be controlled by other agents. Other agents are only allowed to act under an agent's permission. The agent can access the information encapsulated in the other agents. The agent should share the geographic location information. This information can be obtained by request; the agent should use modular software and hardware.

In the establishment of an agent-based manufacturing system, it is necessary to consider the following functions: multi-task operating system environment; distributed database management system; distributed reasoning; and communication equipment and decision-making model.

Next, we consider the distributed Interbay control coordination mechanism driven by the vehicle. A bid strategy refers to the handling requirements or resource availability of the agent. It plays a key role in a successful agent-based system. The construction of a bidding strategy significantly affects resource allocation and system performance. Three types of bidding strategies used by a vehicle coordination mechanism based on a contract network protocol are presented as follows:

1. One-time bidding (OTB);
2. Wafer look-ahead bid (CLAB);
3. Vehicle running time bid (VTTB).

The value of each bid is calculated using the expected times of a wafer lot in the manufacturing and handling processes.

The concept of three bidding strategies can be further described as follows. A manufacturing unit uses OTB to estimate the earliest completed wafer as the object of the next operation. The bid value consists of three parts: wafer processing time, estimated waiting time and estimated handling time. The processing time is the time required for the wafer to be processed in the manufacturing unit. A manufacturing unit agent considers its workload and estimates the processing time of the wafer. The estimated waiting time is the time that a wafer is waiting in the manufacturing unit until it can begin processing. This time is calculated by allocating the total processing time of the existing wafer card in the processing waiting area of the manufacturing unit to all the machines in the manufacturing unit. In addition, a manufacturing unit agent uses the location information of the vehicle and the wafer to calculate the distance between the vehicle and the wafer in order to estimate the vehicle handling time. The expected handling time is the sum of the time taken to wait for a free vehicle to travel to the source stocker and deliver the wafer to the destination stocker. The manufacturing unit agent uses CLAB to estimate the total processing time for subsequent operations of a wafer. The wined wafers with the CLAB rule have the least overall estimated processing time, while the wined wafers with the OTB rule have only minimal processing time for a single subsequent operation. VTTB is a simplified version of the OTB, which only estimates the vehicle traveling time. VTTB rules are designed to reduce the traveling times of vehicles.

On the basis of the agent-based distributed architecture given in this case, a coordination mechanism is developed. This mechanism is triggered when the vehicle is idle. The process consists of two stages: the negotiation phase and the implementation phase. During the negotiation phase, the idle car communicates with the wafers waiting in the handling RTM in the stocker and picks up a winning candidate wafer. During the implementation phase, the winning wafer card is moved by the requested vehicle from its current location to the destination stocker. Figure 5.10 depicts the activation process between agents. The negotiation process consists of three steps: request for delivery, construction and submission of tenders, evaluation of tenders and confirmation of delivery. Specific steps are presented as follows:

1. Request for delivery
   An idle vehicle starts using the following message format to request for delivery tasks: vehicle identification, request starting time and current position of the vehicle. Once the information sent by the vehicle has been received, the wafer agent waiting at RTM sends its next processing request to the manufacturing unit using the following message format: vehicle identification, current location of the vehicle, wafer identification, current location of the wafer and the next processing operation of the wafer. If there is no waiting wafer, the corresponding

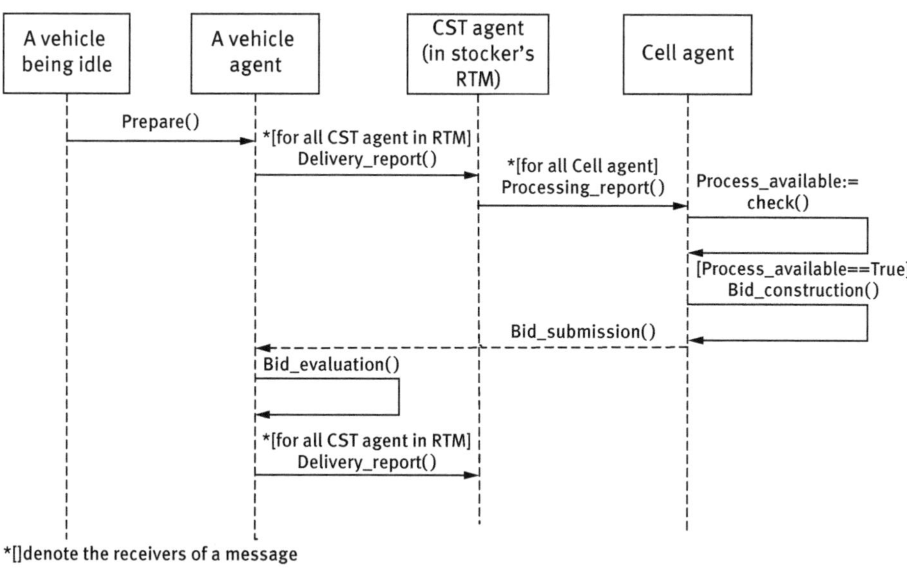

Figure 5.10: Interaction among agents.

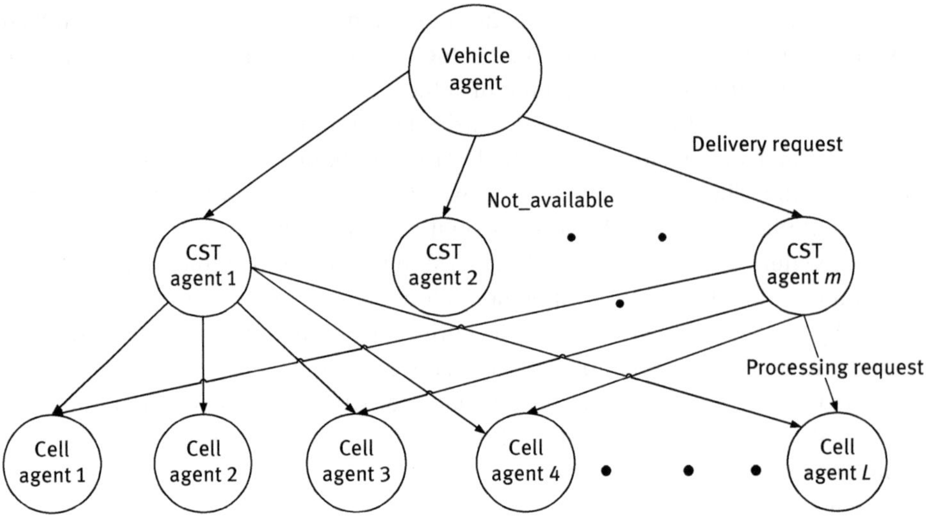

Figure 5.11: Request propagation.

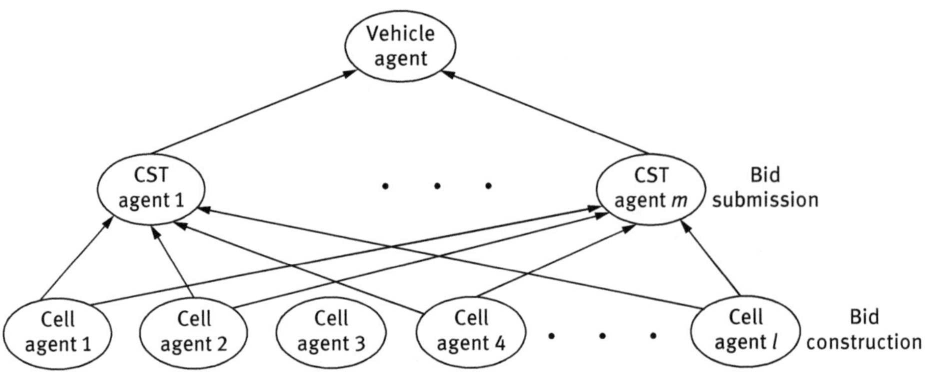

Figure 5.12: Bid construction and submission.

manufacturing unit will send back an unusable message to the activated vehicle. Figure 5.11 illustrates the propagation process.

2. Construct and submit bids.
   When a manufacturing unit agent receives a request of a wafer agent, the manufacturing unit agent checks whether the processing type is the same. If they are the same, the bid construction process starts. Figure 5.12 depicts the process of constructing and submitting a proposal.

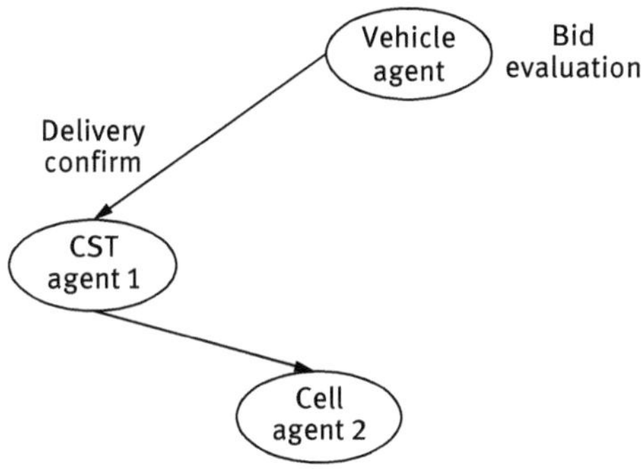

Figure 5.13: Bid evaluation and delivery confirmation.

3. Evaluation and delivery confirmation.
   Once the bid deadline is reached, the vehicle agent will evaluate all bids, and the bids with the lowest value will be picked as winners. The vehicle agent, which issues the delivery request, will then send the message to the winning wafer agent to confirm the delivery process in the following message format: vehicle identification, CST destination and delivery confirmation identification. The evaluation and delivery confirmation procedures are shown in Figure 5.13.

   In recent years, intelligent algorithms have been widely used in scheduling in AMHSs, which has become a hotspot in academia and industry. The scheduling method based on the intelligent algorithms can dynamically optimize the material handling system. However, for the complex large-scale scheduling problem, these methods are often difficult to effectively guarantee the efficiency of the scheduling method. On the other hand, most heuristic scheduling methods based on GA and swarm intelligence do not consider factors such as jamming of multi-handling vehicles and the time constraints of wafer handling on scheduling performances; hence, it is difficult to meet the multi-objective optimization requirements of scheduling in AMHSs.

5.4 Conclusion

This chapter presents three methods commonly used in scheduling in AMHSs: heuristic rules, operation research methods and artificial intelligence algorithms. These three methods have their own distinct characteristics, and their scopes of application are also different.

The heuristic rules are developed on the basis of past experiences in order to select an efficient method in a limited searching space, which greatly reduces the number of attempts and can very quickly solve the scheduling problem. This method is simple, practical and widely used in the actual production activities. However, there is no solid theoretical foundation for this method, and there is a possibility of failure. How to design efficient and reliable heuristic rules has been the focus of this study. On the contrary, operation research methods have a rigorous theoretical basis, and can find the best or near-best solution for complex problems. However, as the problem size increases, the solving efficiency declines sharply. Obviously, the operation research methods are only suitable for solving small-scale problems and doing scientific studies, and seldom applied in actual production activities. The disadvantages of heuristic rules and operation research methods do not exist in the intelligent scheduling algorithms, which can make a reasonable trade-off between solving speed and accuracy. For this reason, intelligent scheduling algorithm has become a hotspot in academic research works.

References

[1] Jiang, Z.B. Scheduling Control System Modeling and Optimization for Semiconductor Wafer Manufacturing System. Shanghai: Shanghai Jiao Tong University Press, 2010, p. 4.
[2] Chen, B.L. Theory and Algorithms of Optimization. Beijing: Tsinghua University Press, 2005.
[3] Bozer, Y.A., Yen, C.K. Intelligent dispatching rules for trip-based material handling systems. Journal of Manufacturing Systems, 1996, 15(4): 226–239.
[4] Yamashita, H. Analysis of dispatching rules of AGV systems with multiple vehicles. IIE Transactions, 2001, 33: 889–895.
[5] De-Koster, R., Le-Anh, T., Van-der-Meer, J.R. Testing and classifying vehicle dispatching rules in three real-world settings. Journal of Operations Management, 2004, 22(4): 369–386.
[6] Liao, D.Y., Wang, C.N. Differentiated preemptive dispatching for automatic materials handling services in 300 mm semiconductor foundry. International Journal of Advanced Manufacturing Technology, 2006, 29(9–10): 890–896.
[7] Bilge, B., Esenduran, G., Varol, N., et al. Multi-attribute responsive dispatching strategies for automated guided vehicles. International Journal of Production Economics, 2006, 100: 65–75.
[8] Kim, B.I., Oh, S.J., Shin, J.J., et al. Effective of vehicle reassignment in a large-scale overhead hoist transport system. International Journal of Production Research, 2007, 45(4): 789–802.
[9] Lin, J.T., Wang, F.K., Chang, Y.M. A hybrid push/pull dispatching rule for a photobay in a 300 mm wafer fab. Robotics and Computer Integrated Manufacturing, 2006, 22(1): 47–55.
[10] Wang, K.J., Lin, J.T., Weigert, G. Agent based interbay system control for a single-loop semiconductor manufacturing fab. Production Planning & Control, 2007, 18(2): 74–90.
[11] Song, Y., Zhang, L., Zhang, M. et al., ACO algorithm for machine conversion reduction in semiconductor assembly manufacturing. ISSM 2005-IEEE International Symposium on Semiconductor Manufacturing, Conference Proceedings, 2005:339–342.
[12] Liang,J., Qian, S.S., Ma, L. Two-level ant algorithm for the furnace batch scheduling in semiconductor furnace operation. Systems Engineering-Theory & Practice, 2005, 12: 96–101.
[13] Ma, H.M., Ye, C.M. Particle swarm optimization algorithm for batch scheduling in semiconductor furnace operation. Computer Integrated Manufacturing Systems, 2007, 13(6): 1121–1126.

6 Scheduling in Interbay automated material handling systems

An Interbay material handling system is responsible for wafer transportation among different processing areas and the stockers are used as a temporary storage warehouse. Whether it is the dominant segmented spine layout in the 300 mm wafer fab or the integrated spine layout in the 450 mm wafer fab, the Interbay material handling system plays a central role and its operation process directly influences the performance indicators, such as yields, overall productivity, cycle times and equipment utilization. Therefore, it is of significant importance to improve the operation efficiency of an Interbay material handling system through scheduling methods.

6.1 Interbay Automated Material Handling Scheduling Problems

As shown in Figure 6.1, an Interbay AMHS typically consists of three subsystems: transportation rails, stockers and automated vehicles. The transportation rails are usually monorail systems, where automated vehicles move wafer lots along the rails among different stockers. Each stocker system acts as an input/output port and provides temporary storage of wafer lots for corresponding Interbay systems. Automated vehicles are used to transport wafer lots among stockers.

During the production process of wafer fabrication, the Interbay AMHS is used to transport the wafer cassettes between different bays in accordance with the processing requirements. Once a cassette needs to move to the destination stocker, a vehicle is requested. The cassette faces a decision point to determine which vehicle has the highest priority and then the source stocker's crane retrieves the cassette and moves it onto the output RTM. An appropriate vehicle is then assigned based on the dispatching rule to execute this transportation task. When the allocated vehicle arrives at the source stocker's horizontal transfer, the output RTM moves the cassette onto the vehicle. After the vehicle transports the cassette via the loop to the destination stocker, the cassette is transferred from the vehicle to a storage location via the input RTM and stocker's crane. This completes a simple transportation cycle from the cassette's point of view when vehicles are available. Such dispatching of cassettes and vehicles can significantly affect the transportation performance of the Interbay material handling system and is studied in the present chapter. The scheduling process diagram of an Interbay material handling system is shown in Figure 6.2.

The purpose of scheduling for Interbay material handling is to allocate different vehicles to wafer cassettes so that the movement efficiency can be optimized while the production efficiency is still kept. The scheduling problem of Interbay material

https://doi.org/10.1515/9783110487473-006

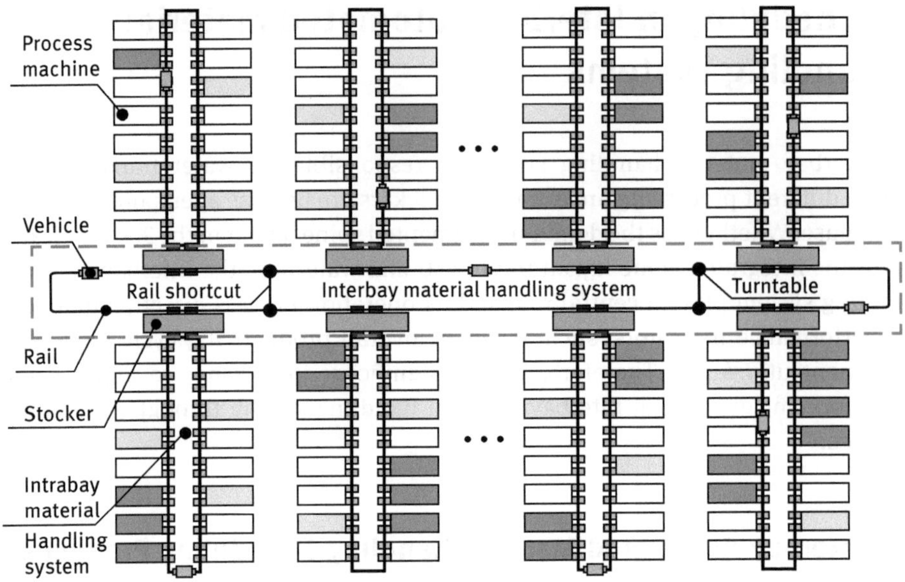

Figure 6.1: Layout of an Interbay material handling system.

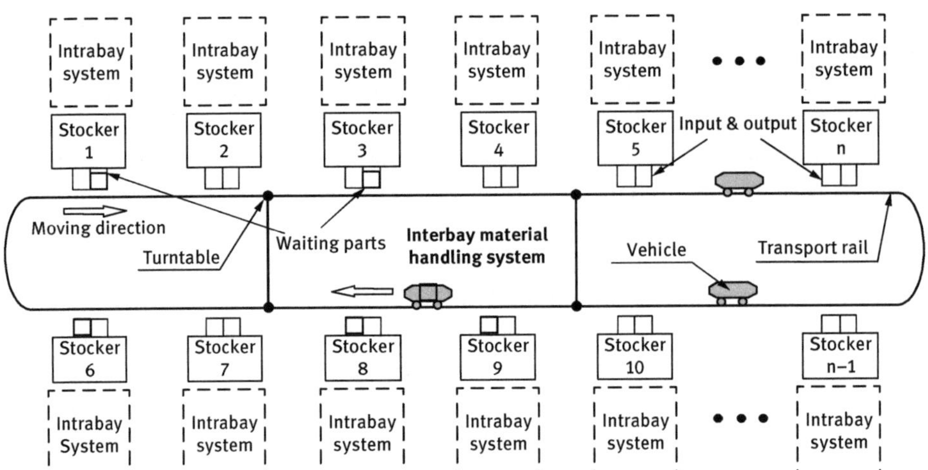

Figure 6.2: The scheduling process diagram of Interbay material handling system.

handling is actually a special dispatching problem where at any scheduling time t, there are $|R|$ waiting wafer cassettes and $|V|$ vehicles. The set of waiting cassettes R and the set of vehicles V constructed a complete bipartite graph K ($|R|$, $|V|$), and the optimal scheduling problem is then converted to an optimal dispatching problem. Edmonds and Karp classified the scheduling problems of Interbay material handling into three types: (1) the number of waiting cassettes is greater than the number of idle vehicles ($|R| > |V|$), while this case is called task-prioritized dispatching; (2) the number of waiting cassettes is smaller than the number of idle vehicles ($|R| \leq |V|$), while this case is called vehicle-prioritized dispatching; and (3) the buffer of stocker is limited in the system, the vehicles have to wait until the stocker has enough storage space and this case is called buffer-prioritized dispatching.

6.2 AMPHI-Based Interbay Automated Material Handling Scheduling Methods

Although the existing research works on Interbay scheduling problems which focused on heuristic rules have shown good efficiency and usability, it is hard to implement them to meet multi-objective optimization. Some scheduling methods based on intelligent algorithms or mathematical programming can describe the optimization target more accurately and can realize multi-objective optimization; however, it is difficult to implement them in a real-time way. In this section, an AMPHI (adaptive multi-parameter and hybrid intelligence)-based Interbay material scheduling method is proposed to implement production scheduling in a real-time way by combining a modified Hungarian algorithm (MHA) and a fuzzy-logic-based controlling method [1, 2, 3].

6.2.1 AMPHI-Based Interbay Automated Material Handling Scheduling Model

A mathematical model is established to describe the characteristics of an Interbay material handling system, which includes the performance optimization indicators related to normalized wafer lots' delivery time, wafer lots' total delivery time, transport amount, vehicle utilization, wafers' throughput, cycle time, WIP quantity and wafer's delivery due date satisfactory rate.

1. Notations

   T_u – Smallest time unit in the material handling process (vehicle speed, loading/unloading time and wafer delivery request arrival time are integer multiples of T_u);

   T – Cycle time period in the Interbay material handling system, $T \in \{T_u, 2T_u, 3T_u, \ldots\}$;

   Y – Set of tracks and shortcuts in the Interbay material handling system;

   S – Set of stockers in the Interbay material handling system;

   e_{yj}^t – $e_{yj}^t = 1$ denotes that vehicle j is located at track or shortcut y at time t, $y \in \{1, 2, \ldots, |Y|\}$;

   α_u – Weighted coefficient of system performance indexes, $\alpha_u \in (0, 1]$, $u = 1, \ldots, 8$;

   PD^s – Set of pick-up/drop-off events at stocker s, $PD^s = \{\ldots, (k, i), \ldots\}$, $s \in S$;

   $Pr(i)$ – Product type of wafer i;

$De(i)$ – Position of destination stocker or processing machine of wafer i;
$Re(i)$ – Position of stocker or processing machine of waiting wafer i;
w – Within cycle time T, the quantity of wafers waiting for transportation in the Interbay system;
n – Within cycle time T, the quantity of wafers delivered by vehicles;
m – The quantity of vehicles in the Interbay system;
$T^d_{\max}$ – Maximum time in which a vehicle can transport a wafer lot;
$T^t_{\max}$ – Maximum value of wafer lots' total transportation time;
$T^C_{\max}$ – Maximum value of wafer lots' total cycle times;
$L_{\max}$ – Maximum value of wafer lots' throughput;
t^i_w – The moment that wafer lot i sends out a transportation request;
t^i_p – The moment that wafer lot i is loaded;
t^i_d – The moment that wafer lot i is unloaded;
h – The time that the vehicle spends in loading/unloading a wafer lot;
t^i_r – The moment that wafer lot i is released to the Interbay system, t^i_r subject to certain probability distribution;
t^i_{out} – The moment that wafer lot i leaves the Interbay system after finishing all processing steps;
q – Within T, the quantity of wafer lots entering the Interbay system;
p – Within T, the quantity of wafer lots leaving the Interbay system;
l – Within T, the quantity of wafer lots leaving the Interbay system after finishing all processing steps;
F^i – Due date relaxation coefficient of wafer lot i;
T^i_p – Total processing time of wafer lot i;
T^i_D – The shortest delivery time for wafer lot i to move from the current stocker to the destination stocker;
J_w – Set of water lots at stocker w;

2. Decision variables
X^t_{ij} – A 0–1 decision variable where X^t_{ij} is equal to 1, if wafer lot i is transported by vehicle j at moment t; otherwise, X^t_{ij} is equal to 0.

3. Objective function

$$Obj = Min\left(\alpha_1\frac{\sum_{i=1}^{n}(t^i_d - t^i_p)}{n\times T^d_{\max}} + \alpha_2\frac{\sum_{i=1}^{n}(t^i_d - t^i_w)}{n\times T^t_{\max}} + \alpha_3\frac{(w-n)}{w} + \alpha_4\frac{\sum_{i=1}^{n}(t^i_d - t^i_p + 2h)}{m\times T}\right.$$
$$\left. + \alpha_5\frac{(L_{\max}-l)}{L_{\max}} + \alpha_6\frac{\sum_{v=1}^{l}(t^v_{\text{out}} - t^v_r)}{l^\star T^C_{\max}} + \alpha_7\frac{(q-p)}{q} + \frac{\alpha_8}{l}\sum_{v=1}^{l}\frac{(t^v_{\text{out}} - t^v_r)}{F^v\times T^v_p}\right) \tag{6.1}$$

4. Constraints

$$\sum_{j=1}^{m} X_{ij}^{t} = 1,\ X_{ij}^{t} \in \{0,\ 1\},\ \ \forall t \in [0, T],\ i = 1, \ldots, n \tag{6.2}$$

$$\sum_{i=1}^{n} X_{ij}^{t} = 1,\ X_{ij}^{t} \in \{0,\ 1\},\ \ \forall t \in [0, T],\ j = 1, \ldots, m \tag{6.3}$$

$$t_d^k X_{kj}^{t_d^k} + h + 1 \le t_p^i X_{ij}^{t_p^i} \tag{6.4}$$

$$t_p^i X_{ij}^{t_p^i} + T_D^i + h \le t_d^i X_{ij}^{t_d^i} \tag{6.5}$$

$$t_p^k X_{kj}^{t_p^k} + h \le t_d^i X_{ir}^{t_d^i},\ \forall\ Re(i) = De(k),\ \ (k, i) \in PD^s \tag{6.6}$$

$$t_r^i < t_w^i < t_p^i \tag{6.7}$$

$$t_p^k + h \le t_p^i,\ \forall\ i,\ k\ \in J_W : (t_r^k > t_r^i) \tag{6.8}$$

$$\sum_{j=1'}^{m} e_{yj}^{t} \le 1, \forall\, t\ \in [0, T],\ \ \forall\, y \in Y \tag{6.9}$$

where formula (6.1) is the objective function to minimize the weighted summation of normalized wafer lots' delivery time, wafer lots' total delivery time, transport amount, vehicle utilization, wafers' throughput, cycle time, WIP quantity and wafer's due date satisfactory rate; formula (6.2) ensures that each wafer lot can be transported by only one vehicle; formula (6.3) makes sure that a vehicle can transport a wafer lot only at a time; formula (6.4) ensures that the vehicle cannot start next transportation task unless it has completed current task; formula (6.5) makes sure that the loading and unloading times of the same wafer lots must satisfy certain constraints; formula (6.6) ensures that the loading and unloading times of a wafer lot at a stocker must satisfy certain constraints; formula (6.7) makes sure that the times that a wafer lot enters the Interbay material handling systems, makes transportation request and is transported by a vehicle should meet certain time constraints; formula (6.8) ensures that loading times of different wafer lots in the same stocker must meet chronological constraints and formula (6.9) ensures that each track, shortcut or turntable can only carry one vehicle at any time.

6.2.2 Architecture of AMPHI Interbay

The architecture of the proposed AMPHI-based Interbay automated material handling scheduling system is shown in Figure 6.3. First, multiple parameters including

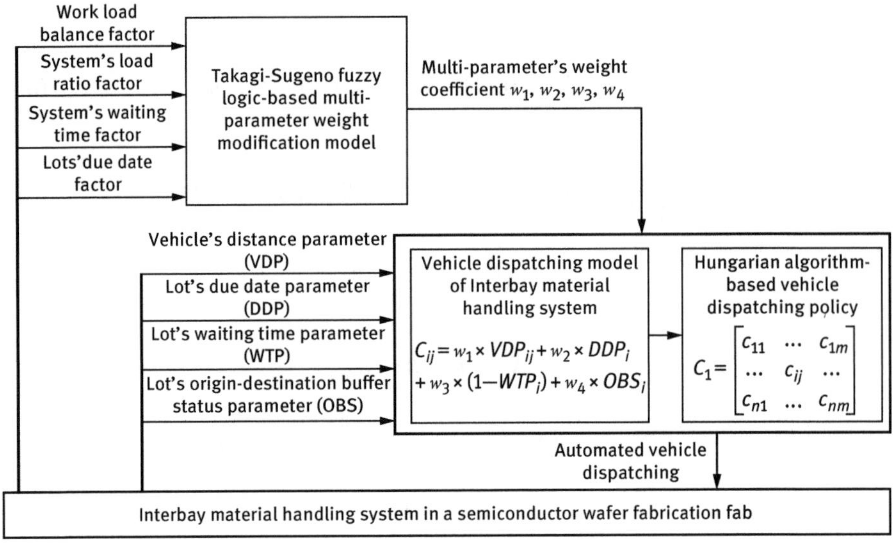

Figure 6.3: Architecture of the AMPHI-based Interbay AMHS.

Vehicle's transportation distance, lot's due date, lot's waiting time and lots' origin-destination buffer status are considered to develop a multi-parameter wafer handling cost model in order to meet the multi-objective scheduling requirements of an Interbay and balance the WIP in each processing district. Second, to reduce the temporary blockage of an Interbay material handling system, waiting wafer sets are optimally matched to empty vehicles in order to obtain a real-time material handling assignment scheme. Finally, according to the dynamic environment of an Interbay material handling system, a weight modification model is proposed based on fuzzy logic theory to adaptively adjust the weight values of these model parameters in order to optimize the multi-objective Interbay scheduling problem.

6.2.3 Vehicle Dispatching in the Interbay Material Handling System

A mathematical model is developed to solve scheduling problems in the Interbay material handling system. Suppose at time t when the cassette arrives at the Interbay material handling system or the vehicle becomes idle, there are n waiting cassettes and m available vehicles. This scheduling problem in the Interbay material handling system is formulated as follows:

$$\text{Min}\left(\sum_{i=1}^{n}\sum_{j=1}^{m} c_{ij}X_{ij}\right) = \text{Min}(\boldsymbol{C}\cdot\boldsymbol{X}) \tag{6.10}$$

$$\sum_{i=1}^{n} X_{ij} = 1, \sum_{j=1}^{m} X_{ij} = 1 \tag{6.11}$$

$$X_{ij} \in \{0,1\}, i=1, 2, \ldots, n, \; j=1, 2, \ldots, m \tag{6.12}$$

$$C = \begin{bmatrix} cc_{11} & \ldots & c_{1m} \\ \ldots & c_{ij} & \ldots \\ c_{n1} & \ldots & c_{nm} \end{bmatrix} \tag{6.13}$$

where X is the matrix of matching factor. $X_{ij} = 1$ indicates that the vehicle j is assigned to deliver the cassette i; otherwise $X_{ij} = 0$. C is the matrix of delivery cost. c_{ij} is the cost value of assigning vehicle j to cassette i.

1. Hungarian algorithm-based vehicle dispatching policy
   In the Interbay material handling system, multiple parameters including Vehicle's transportation distance lot's due date, lot's waiting time and lots' origin-destination buffer status are taken into consideration, and c_{ij} is calculated by the below-weighted multi-parameter wafer handling cost function:

$$c_{ij} = w_1 \cdot VDP_{ij} + w_2 \cdot DDP_i \; + w_3 \cdot (1 - WTP_i) + w_4 \cdot OBS_i \tag{6.14}$$

In eq. (6.14), c_{ij} is calculated as the weighted sum of parameters for vehicle j's distance to lot i (VDP_{ij}), lot i's due date (DDP_i), lot i's waiting time (WTP_i) and lot i's origin-destination buffer status (OBS_i). w_1, w_2, w_3 and w_4 are weight coefficients for parameters VDP_{ij}, DDP_i, WTP_i and OBS_i, respectively. w_1, w_2, w_3 and w_4 are adaptively adjusted by the Takagi-Sugeno fuzzy-based multi-parameter weight modification model. Multiple variables including Vehicle's transportation distance and waiting time are considered to assign vehicle, and in order to integrate these different variables, they are first normalized as ratio (parameters) by dividing itself to the maximum possible value. For example, VDP_{ij} denotes the normalized travel by dividing the distance between the current position of vehicle j to the position of job i to the maximum travel distance of vehicle. Similarly, WTP_i denotes the normalized waiting time parameter for job i. DDP_i denotes the due date parameter of candidate job i. DDP_i can be calculated by $DDP_i = RT_i/(RP_i \times Fac_{\max})$, where RT_i is the remaining time before the due date of wafer lot i, RP_i is the remaining processing time of wafer lot i and $Fac_{\max}$ is the maximum due date slack coefficient of completed wafer lots. OBS_i denotes the origin-destination buffer status parameter of job i, which is transported between two stockers. OBS_i can be calculated by $OBS_i = B^i_{\text{in}} \times (1 - B^i_{\text{out}})$, where B^i_{in} is the normalized buffer level of the input buffer in the destination stocker and B^i_{out} is the normalized buffer level of the output buffer in the source stocker.

The solution procedure using the Hungarian method is described as follows.

Step 1. According to the weighted multi-parameter wafer handling cost function (eq. (6.14)), a cost matrix C_1 is first constructed.

Step 2. For the original cost matrix C_1, if the number of available vehicles is more than that of candidate jobs, that is, $m > n$, $(m \times n)$ dummy candidate jobs are added to form an $m \times m$ cost matrix C_2. Otherwise, if the number of candidate jobs is more than that of available vehicles, that is, $m \times n$, $(n \times m)$ dummy vehicles are added to form an $n \times n$ cost matrix C_2. All dummy cost values are $c_{ij} = \max\{_{cij}\}$. Let k denote the order of cost matrix C_2, $k = \max(m, n)$.

Step 3. A reduced cost matrix C_3 is obtained by subtracting from all elements in each row of C_2, and the minimum element of that row is c_{1j} ($j = 1,\ldots,k$).

Step 4. Another reduced cost matrix C_4 is obtained by subtracting from all elements in each column of C_3, and the minimum element of that column is c_{i1} ($i = 1,\ldots,k$).

Step 5. Cross out rows or columns with element $c_{ij} = 0$ in C_4. Let k^m be the minimum number of lines needed to cross out all zero elements in C_4. If $k^m = k$, then the optimal dispatching solution is found and the procedure is stopped. Otherwise, go to Step 6.

Step 6. Let d be the minimum element in the remaining (not crossed out) elements of C_4. A new reduced cost matrix C_4 is obtained by subtracting d from all remaining elements in C_4 and adding d to elements being double crossed out (elements at the intersection of crossed-out row and column) in C_4. Go to Step 5.

2. Modified Hungarian algorithm-based vehicle dispatching policy
   In MHA-based vehicle dispatching policy, four major parameters are taken into account when calculating the cost value of assigning a vehicle to a wafer cassette: delivery time *TP_factor*, cassette waiting time *WT_factor*, cassette due date *Due_factor* and cassette processing factor *PF_factor*. Suppose the vehicle j is designated to deliver cassette i, the cost value c_{ij} is defined as follows:

$$\begin{aligned} c_{ij} = w_1 \times TP\_factor(i,j) + w_2 \times WT\_factor(i) + \\ w_3 \times Due\_factor(i) + w_4 \times PF\_factor(i,n,c) \end{aligned} \tag{6.15}$$

where *TP_factor(i, j)* is the linear normalized delivery time of assigning vehicle j to cassette i, and $TP\_factor(i,j) = dis(i,j)/(\arg\max_{i \in L, j \in K}(dis(i,j)))$.

WT_ factor(i) is calculated from linear normalization of cassette i's waiting time in the Interbay material handling system, $WT\_factor(i) = (WT^i_{\max} - WT_i)/WT_{\max}$.$WT^i_{\max}$ is the aging time for cassette i and WT_i is the time period for which cassette i has been waiting in the Interbay.

Due_factor denotes the due date parameter of candidate job i, and $Due\_factor(i) = \frac{(DUE_i - t_R^i)}{(APT_i - PT_i)} / \arg\max_{i \in L} \frac{(DUE_i - t_R^i)}{(APT_i - PT_i)}$, DUE_i is the due date of waiting cassette i.t_R^i is the remaining time to the due date of cassette i.APT_i is the sum of processing time of all operations. PT_i is the sum of processing time of all finished operations. Smaller *Due_factor* indicates that the job is more urgent.

PF_factor is designed to measure the buffer level of source and destination stockers for cassette i, and $PF\_factor(i, n, c) = (Q_{\max} + Q_c^i - Q_n^i)/2Q_{\max}$, $Q_{\max}$ is the maximum of Q_n^i and Q_c^i. Q_n^idenotes the queue length in the in-buffer of destination stocker n while Q_c^i denotes the queue length in the out-buffer of source stocker c.

In eq. (6.15), w_1, w_2, w_3 and w_4 are weight coefficients for parameters *TP_factor* (i, j), *WT_factor*(i), *Due_factor*(i) and *FF_factor* (i, n, c), respectively. w_1, w_2, w_3 and w_4 are adaptively adjusted by the Mamdani-fuzzy-logic-based multi-parameter weight adjusting method.

The Hungarian algorithm was first proposed by the Hungarian mathematician Egervary and then improved by Edmonds on the basis of the Berge Theorem and Hall's Theorem. The basic Hungarian method is a combinational optimization algorithm, which effectively optimizes solutions for assignment problems. It can be used to solve the maximal matching problem that whether there exists saturated X or Y in a bipartite graph $G = (X, Y)$. As a stable polynomial-time algorithm, it ensures to get the optimal solution, in which when $|X| = n$ and $|Y| = m$, the time complexity of the algorithm is $O(v^3)$.

The MHA is proposed to solve the maximal matching problem for a weighted bipartite graph. It is essentially a labeling method with the theoretical basis of feasible vertex labeling method. In detail, suppose in a weighted bipartite graph $G = (X, Y)$, X is the set of cassettes and Y is the set of vehicles. $\forall i \subset X$ and $\forall j \subset Y$, each vertex is given a label denoted by $l(i)$ and $l(j)$. For any edge,

$$e = (i, j), \text{ } if \text{ } l(i) - l(j) \le w(i, j) = c_{ij} \tag{6.16}$$

this labeling is called a feasible vertex labeling.

An edge can be defined as

$$e_l = \{(i, j) \in E(G) | l(i) - l(j) = w(i, j)\} \tag{6.17}$$

in which $E(G)$ denotes the set of all edges in a weighted bipartite graph $G = (X, Y)$. The sub-graph with e_l as its edges is called the equal-sub-graph G_l which is depicted in Figure 6.4. According to the Berge Theorem and Hall's Theorem, if there exists a perfect match M^* in the equal-sub-graph G_l (each vertex in G_l is related to M^*), then M^* is the minimum weight perfect match.

According to the above mentioned theoretical basis, the solution procedure using the proposed MHA is described as follows.

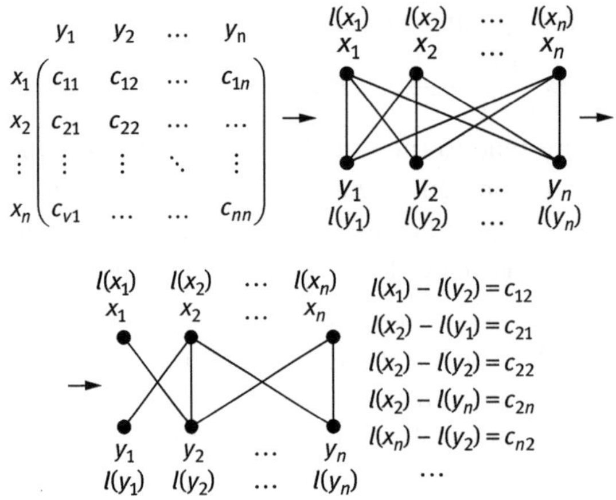

Figure 6.4: The relationship diagram of feasible vertex labeling and equal subgraph.

Step 1. The initial feasible vertex labels are given by eq. (6.18). Based on these labels, establish the equal-sub-graph G_l. Search the perfect match M^* in G_l. α indicates one of the waiting cassettes and β indicates one of the vehicles.

$$l(\alpha) = \begin{cases} \min\limits_{\forall \beta \in Y}(w(\alpha, \beta)), \alpha \in X \\ 0 \qquad\qquad\qquad\quad \alpha \in Y \end{cases} \tag{6.18}$$

Step 2. If there exists a perfect match M^* in G_l, then M^* is the minimum weight perfect match and the algorithm terminates. Otherwise, the Hungarian algorithm terminates at $S \subset X$, $T \subset Y$ and $N_{G_l}(S) = T$. S and T denote the set of matched X(cassettes) and Y(vehicles). $N_{G_l}(S)$ stands for the set of neighbor vertices for S in G_l.
Step 3. Let $Slack(l) = \min\{w(i,j) - l(i) + l(j) | i \in X, j \in Y - T\}, \forall l \in Y$, for any $j \in Y - T$, the vertex labels are modified to$l'(j) = l(j) + Slack(l)$; while the others are fixed. Such labeling is still feasible and at least there exists one edge which is included into a new equal-sub-graph G_l', which indicates that G_l' is scalable.
Step 4. Repeat Steps 2 and 3 until an equal-sub-graph G_l' with a perfect match is obtained. Denote the weight matrix of G_l' as Cr.
Step 5. If there is more than one perfect match, this means there exists more than one optimal solution. Further procedures are as follows:

(a) Let $\sigma_{ij}^2 = \dfrac{(C_{ij} - \mu)^2}{(C_{\max} - C_{\min})^2}$, $\mu = \sum\limits_{i=1}^{m}\sum\limits_{j=1}^{n} C_{ij}$, $C_{\max}$ and $C_{\min}$ denote the maximal and minimal elements in the initial cost matrix C, and σ_{ij}^2 is the normalized variance;
(b) Modify the initial cost matrix. For those elements in which Cr is valued 0, add σ_{ij}^2 to them; for those non-zero elements, reset them to $C_{\max}$. The new cost matrix C_{New} is constructed as

$$\boldsymbol{C}_{\text{New}} = \begin{pmatrix} Co_{11} + \sigma_{11}^2 & C_{\max} & \cdots & Co_{1n} + \sigma_{1n}^2 \\ C_{\max} & Co_{22} + \sigma_{22}^2 & \cdots & C_{\max} \\ \vdots & \vdots & \ddots & \vdots \\ C_{\max} & Co_{n2} + \sigma_{n2}^2 & \cdots & Co_{nn} + \sigma_{nn}^2 \end{pmatrix} \tag{6.19}$$

Go to Step 1.

When there are multiple optimal matches, the above method can be used to obtain a solution with smaller variance, and the whole process is convergent.

6.2.4 Fuzzy-Logic-Based Weight Adjustment

When assigning a vehicle to a wafer cassette, we must first determine the weight relationship between the four parameters in the wafer handling cost model, and improper parameter weighting will decrease the actual operation efficiency. If the weight coefficient w_j is set to a fixed value, the Interbay material handling task assignment method mentioned in section 6.2.3 applies only to the material handling system in the static environment. This section describes fuzzy-logic-based weight adjustment methods, which dynamically adjust arrangement of Interbay material handling tasks according to the system status and interference factors in dynamic stochastic environments.

Fuzzy logic inference is in essence a process that maps a given input space to a specific output space by using a fuzzy logic method. According to fuzzy input and fuzzy rules, fuzzy outputs are obtained by using good reasoning methods. There are three common fuzzy inference methods: pure fuzzy logic reasoning, Mamdani fuzzy inference and Takagi-Sugeno fuzzy inference. This section introduces two methods to adjust parameters of wafer handling cost model: Mamdani fuzzy inference and Takagi-Sugeno fuzzy inference.

1. Takagi-Sugeno fuzzy-based weight modification model

   In this section, a weight modification model is developed based on the Takagi-Sugeno fuzzy approach to define the weight coefficients in cost function of eq. (6.10). The main procedures for developing such fuzzy-based weight modification model include (1) defining input variables, (2) constructing membership functions and fuzzy rule set and (3) deploying Takagi-Sugeno's fuzzy inference.

   The principle of the proposed fuzzy-based model is to dynamically adjust the weights to be put on different criteria for assigning vehicle to individual jobs based on the overall system loading and job completion status. Therefore, four fuzzy input variables, namely, the Interbay system's load ratio factor, the lots' due date factor, the system's waiting time factor and the system's workload balance factor, are defined to represent the overall system dynamic status.

1) Input variables
   (1) System's load ratio factor (LRF). The LRF is used to measure the current transportation load of the Interbay material handling system. A high LRF score indicates that the Interbay system is heavily loaded. The *LRF* is calculated as

$$LRF = \sum_{s=1}^{n_1} WS_s / NV \tag{6.20}$$

   where WS_s is the number of wafer lots waiting in the output buffer of stocker *s*, *N* is the total number of stockers and *NV* is the total number of automated vehicles in the Interbay system.
   (2) Lots' due date factor (*DDF*). The *DDF* is used to measure the current wafer lots' due date satisfaction. A larger *DDF* value indicates a higher probability of satisfying the committed due date. *DDF* is calculated as

$$DDF = \sum_{i=1}^{m_1} \frac{RT_i}{RP_i \times Fac} \times \frac{1}{m_1} \tag{6.21}$$

   where RT_i is the remaining time before the due date of wafer lot *i*, RT_i is the remaining processing time of wafer lot *i*, m_1 is the total number of waiting wafer lots in the Interbay system and *Fac* is the due date slack coefficient of completed wafer lots. *Fac* can be calculated as follows: $Fac = \sum_{q=1}^{h} \frac{CT_q}{PT_q} \times \frac{1}{h}$, where PT_q is the overall processing time of wafer lot *q*, CT_q is the cycle time of wafer lot *q* and *h* is the number of wafer lots that have completed all processes in an SWFS.
   (3) System's waiting time factor (*WTF*). The *WTF* is used to measure the lateness of wafer lots in the Interbay system. A higher *WTF* value indicates that more wafer lots have a longer than expected waiting time. *WTF* is calculated by

$$WTF = Nwt / m_1 \tag{6.22}$$

   where *Nwt* is the number of wafer lots whose waiting time is greater than the expected average waiting time *AW* in the Interbay system. *AW* is calculated as $AW = \sum_{q=1}^{h} CW_q / h$, where CW_q is the mean waiting time of wafer lot *q*.
   (4) System's workload balance factor (*WBF*). The *WBF* is used to measure the starvation or blockage probability of the Interbay system. A higher *WBF* value means a higher chance that the stockers in the Interbay system are blocked or starved. *WBF* is calculated by

$$WBF = \frac{1}{n_1}\sum_{s=1}^{n_1}(B_s - \bar{B})^2 \tag{6.23}$$

where $B_s = \exp(1-(BS_s/Ns)^2)$ is the load factor of the stocker buffers, $\bar{B}$ is the mean load factor of the Interbay system, BS_s is the buffer level of stocker buffers and N_s is the buffer capacity of stock buffers.

2) Membership function and fuzzy rule table
The membership functions of the above-defined input variables are represented in the form of triangular and trapezoidal fuzzy sets in the proposed fuzzy-based weight modification model. Table 6.1 lists the membership functions of the four fuzzy input variables.
Let $\beta_{w_1}, \beta_{w_2}, \beta_{w_3}$ and β_{w_4} be the exact values of fuzzy output variables w_1, w_2, w_3 andw_4 in the fuzzy rule. By Takagi-Sugeno fuzzy inference, the weight coefficient w_r (r = 1,...,4) of eq. (6.10) can be evaluated and the fuzzy rule table is established, as shown in Table 6.2. Both Tables 6.1 and 6.2 are obtained by consulting veterans in the field of AMHSs and SWFSs, and conducting extensive simulations based on historical production data using design of experiments (DOE).

3) Takagi-Sugeno's fuzzy inference
The proposed fuzzy-based weight modification model adjusts the weight coefficients w_1, w_2, w_3 and w_4 in the cost function (eq. 6.10) by evaluating the input variables using a set of fuzzy rules. The fuzzy rule has the following generic form:

Table 6.1: The membership function of the input variables of weight modification model.

Input variables	Fuzzy sets (linguistic descriptions)	fuzzy set membership functions
LRF	Low load (LL)	$(-\infty, 0.0, 0.7, 1.0)$
	Medium load (ML)	$(0.7, 1.0, 1.3)$
	High load (HL)	$(1.0, 1.3, 5.0, +\infty)$
DDF	Urgent due date (UDD)	$(-\infty, 0.0, 0.8, 1.0)$
	Normal due date (NDD)	$(0.8, 1.0, 1.5)$
	Slack due date (SDD)	$(1.0, 1.5, 6.0, +\infty)$
WTF	Short waiting time (SWT)	$(-\infty, 0.0, 0.2, 0.35)$
	Normal waiting time (NWT)	$(0.2, 0.35, 0.5)$
	Long waiting time (LWT)	$(0.35, 0.5, 1.0, +\infty)$
WBF	Good balance (GB)	$(-\infty, 0.0, 0.2, 0.35)$
	Normal balance (NB)	$(0.2, 0.35, 0.5)$
	Bad balance (BB)	$(0.35, 0.5, 1.0, +\infty)$

Table 6.2: Fuzzy rule table of the multi-parameter weight modification model.

Fuzzy rule ID	Input linguistic variables				Crisp output variables			
	LRF	DDF	WTF	WBF	β_{w_1}	β_{w_2}	β_{w_3}	β_{w_4}
FR1	HL	UDD	/	GB	49	49	1	1
FR2	HL	UDD	/	NB	33	33	1	33
FR3	HL	UDD	/	BB	25	25	1	49
FR4	HL	/	LWT	GB	49	1	49	1
FR5	HL	/	LWT	NB	33	1	33	33
FR6	HL	/	LWT	BB	25	1	25	49
FR7	ML	UDD	/	GB	49	49	1	1
FR8	ML	UDD	/	NB	25	49	1	25
FR9	ML	UDD	/	BB	1	49	1	49
FR10	ML	/	LWT	GB	49	1	49	1
FR11	ML	/	LWT	NB	25	1	49	25
FR12	ML	/	LWT	BB	1	1	49	49
FR13	LL	/	/	GB	97	1	1	1
FR14	LL	/	/	NB	49	1	1	49
FR15	LL	/	/	BB	1	1	1	97

IF LRF_j is $\tilde{Q}_1^j$ AND DDF_j is $\tilde{Q}_2^j$ AND WTF_j is $\tilde{Q}_3^j$ AND WBF_j is $\tilde{Q}_{m_j}^j$, THEN $\beta_{W_r}^j = a_r^j (r = 1, 2, 3, 4)$

where $\tilde{Q}_i^j$ $(i = 1, \ldots, 4, j = 1, 2, \ldots, N_r)$ are fuzzy sets; LRF_j, DDF_j, WTF_j and WBF_j are input variables; j is the rule number and N_r is the total number of fuzzy rules. $\beta_{W_r}^j$ $(r = 1, \ldots, 4)$ is the output variables of the jth fuzzy rule and a_r^j $(r = 1, \ldots, 4)$ is the output coefficient of the jth fuzzy rule. By Takagi-Sugeno fuzzy inference, the weight coefficient w_r $(r = 1, \ldots, 4)$ of eq. (6.10) can be evaluated as a weighted mean of $\beta_{W_r}^j$ for all fuzzy rules.

$$w_r = \frac{\sum_{j=1}^{15} v^j \times \beta_{W_r}^j}{\sum_{j=1}^{15} v^j}, \quad r = 1, \ldots, 4 \tag{6.24}$$

where the weight v^j is the degree of relevance of the premises of the jth fuzzy rule for input variables LRF, DDF, WTF and WBF. In the Takagi-Sugeno fuzzy inference, v^j is calculated as

$$v^j = \mu_{\tilde{Q}_1^j}(LRF) \times \mu_{\tilde{Q}_2^j}(DDF) \times \mu_{\tilde{Q}_3^j}(WTF) \times \mu_{\tilde{Q}_4^j}(WBF) \tag{6.25}$$

where $\mu_{\tilde{Q}_i^j}(\cdot)$ are the membership functions of input variables LRF, DDF, WTF and WBF.

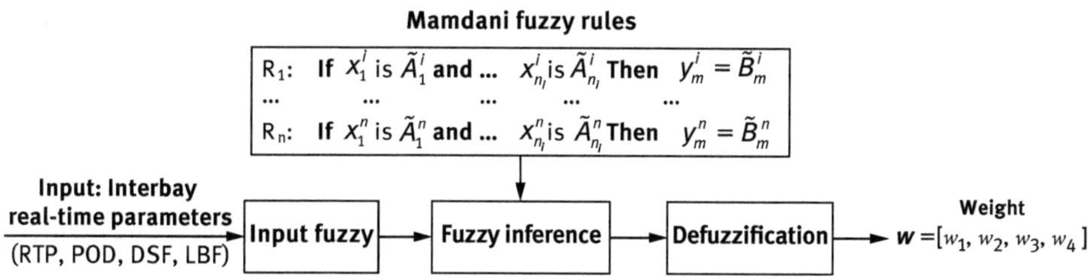

Figure 6.5: The framework of fuzzy logic-based weight-adjusting method.

2. Mamdani fuzzy-based weight modification model
   In this section, a Mamdani fuzzy-logic-based method is proposed to adjust the weight coefficients in eq. (6.15). The framework of this fuzzy-logic-based weight-adjusting method is shown in Figure 6.5. The main procedures include (1) define input variables, (2) determine the domain of discourse and membership function, (3) construct a fuzzy rule set and (4) synthesization and defuzzification.

   (1) Input and output variables
      (1) Ratio of transport and processing load (RTP)
         The *RTP* is used to measure the current transportation load of the Interbay material handling system. *RTP* is calculated as

$$RTP = \sum_{j=1}^{m} \sum_{i=1}^{n} \exp((1/TD_{ij})\, R) \tag{6.26}$$

         Among the above equations, n is the total number of waiting wafer cassettes in the current Interbay material handling system. m is the total number of available vehicles. TD_{ij} denotes the expected shortest time for which vehicle j arrives at the stocker where the cassette i locates. R is the adjustment factor which is the mean ratio of the actual delivery time for all finished job DT_i and the next operation processing time IP_i, $l'(j) = l(j) + Slack(l)$. A high *RTP* indicates that the Interbay material handling system is heavily loaded, and the mean delivery time is too long.

      (2) Due date satisfaction rate (DSF)
         The *DSF* is used to measure the current wafer lots' due date satisfaction. *DSF* is calculated as follows:

$$DSF = \sum_{i=1}^{n} \frac{RT_i}{RP_i \times Fac} \times \frac{1}{n} \tag{6.27}$$

Among the above equations, $RT_i = DUE_i - t$ is the remaining time before the due date of wafer lot i at time t; $RP_i = AP_i - PT_i$ is the sum of processing time for remaining operations. *Fac* is the adjustment factor which is the ratio of the actual processing cycle time for finished wafer lots and the total processing time for all operations. G_l'. A larger *DSF* value indicates a higher probability that the lots cannot meet the committed due date.

(3) Percent of delay (POD)
The *POD* is used to measure the lateness of wafer lots in the Interbay system. *POD* is calculated as

$$POD = \frac{1}{n}|\{i|i \in M, WT_i > AW\}| \tag{6.28}$$

Among the above equations, *AW* is the mean waiting time for all waiting wafer lots. $AW = \sum_{i=1}^{n} CW_i \times \frac{1}{n}$. CW_i is the waiting time of cassette i. A higher *POD* value indicates that more wafer lots have a longer than expected waiting time.

(4) Load balance factor (LBF)
The *LBF* is used to measure the balance of stockers in the Interbay material handling system. *LBF* is calculated as

$$LBF = \frac{1}{n-1}\sum_{i=1}^{n}(B_i - \bar{B})^2 \tag{6.29}$$

Among the above equations, B_i is the load factor of the buffer in Stocker i. $\bar{B} = \frac{1}{n}\sum_{i=1}^{n} B_i$, $B_i = 1/(1 + e^{-\lambda\left(\frac{BS_i}{Cap} - c_0\right)})$. BS_i is the number of waiting cassettes. B_i is a bell-shaped curve function, and λ and c are shape control parameters. A higher *LBF* value indicates the imbalance of current system load.

The output of the proposed fuzzy logic control (FLC) is $\boldsymbol{w}$ = [$w1$, $w2$, $w3$, $w4$]. $w1$, $w2$, $w3$ and $w4$ are weight coefficients for the proposed four input variables.

2) Domain of discourse and membership function
The domains of discourse for the input variables are determined by orthogonal design test method. In each case, simulations are carried out for different combinations of discourse domains. The best domains are obtained as U_{RTP} = [0.5, 1.5], U_{DSF} = [1.5, 2.5], U_{POD}= [0.3, 0.7] and U_{LBF} = [0.6, 1.4]. For each input variable, there are three fuzzy sets (Table 6.3).

Table 6.3: The fuzzy set of the input variables of FLC.

Input variable	Fuzzy set		
RTP	Low rate (LR)	Medium load (MR)	High load (HR)
DSF	Slack due date (LDS)	Normal due date (MDS)	Urgent due date(HDS)
POD	Short waiting time (SD)	Normal waiting time (MD)	Long waiting time (LD)
LBF	Poor balance (NLB)	Normal balance (MLB)	Good balance (GLB)

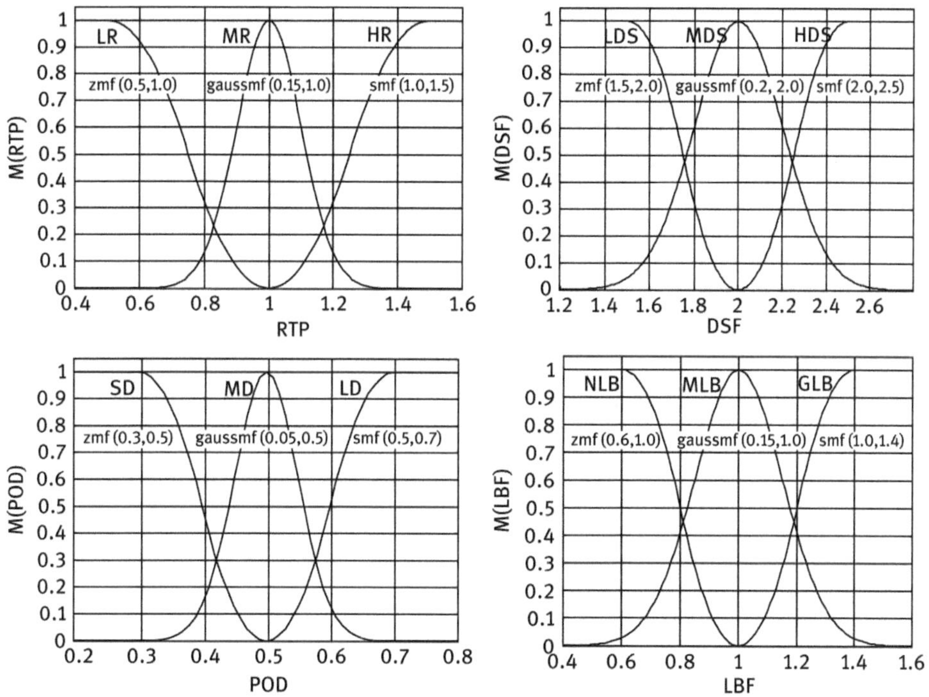

Figure 6.6: The membership–function curves of the input variables of FLC.

The membership functions of the above-defined input variables are represented in the form of Gaussian function, Z-function and S-function in the proposed fuzzy-logic-based weight adjusting method, as shown in Figure 6.6.

The domain of discourse for the output $w_i(i = 1, 2,..., 4)$ is $U = [0, 1]$; the membership function is shown is Figure 6.7.

3) Fuzzy rules

The Mamdani-based weight adjusting method modifies the weight coefficients *w*1, *w*2, *w*3 and *w*4 in the cost function eq. (6.15) by evaluating the

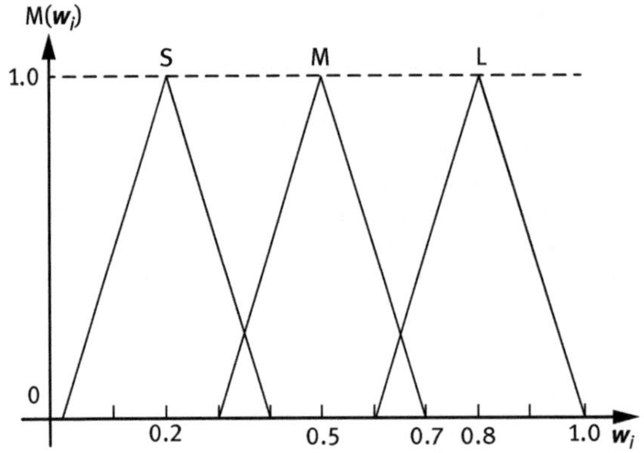

Figure 6.7: The membership–function curve of the output parameter of FLC.

Table 6.4: The fuzzy rules table of the Mamdani-based inference method.

Rule	Input linguistic variables				Output variables			
	RTP	*DSF*	*POW*	*LBF*	α_1	α_2	α_3	α_4
1	HR	LDS	/	/	L	L	S	S
2	HR	MDS	/	/	L	M	L	L
3	HR	HDS	/	/	L	S	S	S
4	HR	/	LD	/	L	S	L	S
5	HR	/	MD	/	L	S	M	S
6	HR	/	SD	/	L	S	S	S
7	HR	/	/	GLB	M	L	S	S
8	HR	/	/	NLB	M	M	S	S
9	HR	/	/	BLB	M	S	S	S
10	MR	LDS	/	/	M	S	L	S
11	MR	MDS	/	/	M	S	M	S
12	MR	HDS	/	/	M	S	S	S
13	MR	/	LD	/	M	L	L	S
14	MR	/	MD	/	M	S	M	S
15	MR	/	SD	/	M	S	S	S
16	MR	/	/	GLB	M	S	S	S
17	MR	/	/	NLB	M	S	S	M
18	MR	/	/	BLB	M	S	S	L
19	LR	LDS	/	/	S	L	S	S
20	LR	MDS	/	/	S	M	S	S
21	LR	HDS	/	/	S	S	S	S
22	LR	/	LD	/	S	S	L	S
23	LR	/	/	BLB	S	S	S	L

input variables using a set of fuzzy rules. The output of the Mamdani inference is $Z_i = [y_1^i, y_2^i, y_3^i, y_4^i]$, $i = 1, 2, \ldots, n_r$, with n_r being the total number of fuzzy rules. The fuzzy rule has the following generic form:

If $x^i{}_1$ is $\tilde{A}^i{}_1$ and...$x^i_{n_i}$ is $\tilde{A}^i{}_{n_i}$ then $y^i{}_m = \tilde{B}^i_m (m = 1, 2, 3, 4)$

$x^i_k (k = 1, 2, ..., n_i)$ is the kth input variable of rule i. $A^i_k (k = 1, 2, ..., n_i)$ is the corresponding fuzzy set for the input variable. $\widetilde{B^i_k}$ is the fuzzy set for the mth output variable y^i_m of rule i. The fuzzy implication of A→B is calculated as

$$R_c = A \rightarrow B = \int_{X \times Y} \frac{\mu_A(x) \wedge \mu_B(y)}{(x, y)} \tag{6.30}$$

There are four input variables and three fuzzy sets for each variable. Twenty-three among those 81 combinations are chosen. Table 6.4 lists the fuzzy rules adopted in the proposed weight adjusting method.

4) Synthesization and defuzzification
Mamdani inference adopts the Max method to carry out the synthesization for all output fuzzy sets in coincidence with all fuzzy rules. The final output is

$$C' = C_1' \cup C_2' \cup \ldots \cup C_{n_r}' = \cup_{i=1}^{n_r} C_i' \tag{6.31}$$

C_i' is the output fuzzy set for rule i. The centroid method is applied to realize the defuzzification. Suppose $C_m{}'$ is the final output fuzzy set of variable w_m on its discourse domain U_m; then w_m is

$$w_m = \frac{\int_{U_m} \mu_{C_m'}(x).x dx}{\int_{U_m} \mu_{C_m'}(x) dx} \doteq \frac{\sum_{k=1}^{|U_m|/\Delta x} \mu_{C_m'}(x_k).x_k}{\sum_{k=1}^{|U_m|/\Delta x} \mu_{C_m'}(x_k).x_k}, m = 1, 2, 3, 4 \tag{6.32}$$

$\mu_{C_m{}'}(x)$ is the value of the membership function for x on U_m.

6.2.5 Simulation Experiments

1. Hungarian algorithm and Takagi-Sugeno fuzzy-logic-based Interbay scheduling method
This section presents a simulation experiment to evaluate the effectiveness of the proposed AMPHI dispatching strategy and provides a comprehensive comparison to traditional material handling strategies. Discrete event simulation software eM-Plant is used to model the 300 mm SWFS and the dispatching strategies.
The system data used for simulation modeling in this study was shared by a semiconductor manufacturer in Shanghai, China. In the system, there are 216 tools belonging to 54 distinct machining tool groups. There are 22 stockers in the

Interbay material handling system, and each stocker connects with an Intrabay system. The distance between adjacent stockers is 20 m and the Interbay rail loop is 440 m long with four shortcuts and eight turntables. Each automated vehicle can load only one wafer lot at a time. The speed of the automated vehicle is 1.0 m/s and the average loading/unloading operation time is 5 s. Three types of wafer lot (jobs A, B and C) are being processed in the system. Totally, six scenarios were designed for different combinations of loading ratios (90% or 100% of the specified system capacity) and automated vehicle numbers (8, 10 or 12 vehicles). Each scenario was replicated in three simulation runs, and each run simulated a production period of 120 days with a transient period of 20 days. The lead times of job A, B and C in each scenario were set as 30, 28 and 31 days, respectively.

Simulation experiments were carried out to comprehensively compare the effectiveness of the proposed AMPHI approach and five traditional single-attribute and multi-attribute heuristic dispatching policies. These traditional dispatching approaches including Hungarian-algorithm-based overhead transporter reassignment (HABOR), cassette-look-ahead bid (CLAB), modified first come first served (M-FCFS), short transport distance (STD) and longest waiting time first (LWT) have been adopted in vehicle dispatching of the Interbay system in SWFSs and been approved to be feasible and effective methods. The HABOR dispatching approach takes the Hungarian algorithm to assign empty vehicle to wafer lots, in which wafer lot's waiting time and position information are considered. In the CLAB approach, wafer lot with minimal total estimated processing time in several succeeding operations wins the vehicle being idle. In M-FCFS, the available vehicles are dispatched to the wafer lots that have the earliest request. If there is another wafer lot (called new call) waiting to be transported at the same stocker when one wafer lot (called old call) at a certain stocker is assigned to a vehicle, then the corresponding request time associated with the new call is set equal to the time when the old call was assigned. The STD dispatching rule means that the wafer lot whose position is the closest to the designated empty vehicle is assigned to the vehicle. In the LWT-based dispatching rule, the wafer lots with the longest waiting time in stockers are assigned to empty vehicles first.

A total of eight performance indexes – wafer lot's cycle time, system throughput, wafer lot's due date satisfaction rate, system WIP, wafer lot's movement, delivery time, transportation time and vehicle's utilization rate – of the AMHS and the SWFS were compared. In the simulation study, AMPHI ranked first among the six dispatching approaches in terms of cycle time, throughput, due date satisfaction rate, WIP level, job movement and transportation time in all six scenarios. AMPHI ranked first in five out of six scenarios in terms of delivery time, and ranked first in four out of six scenarios in terms of vehicle utilization. For clarity and brevity, only the results with 100% loading ratios are listed in Tables 6.5 and 6.6. To further verify the effectiveness of AMPHI approach, the desirability function $D(X)$ is computed. Using the desirability function, the

Table 6.5: Performance comparison of different dispatching policies (part 1).

Scenario	Dispatching Approach	Throughput (lot)	Cycle time (s)	Movement (lot)	Delivery time (s)
Loading ratio = 100%, vehicles = 8	AMPHI	A (2523)	A (1794486)	A (589712)	A (521.7)
	HABOR	B (2215)	B (2206627)	A (586410)	A (521.2)
	CLAB	C (2147)	C (2527311)	D (413480)	C (623.2)
	M-FCFS	B (2215)	B (2220757)	A (586410)	D (770.6)
	STD	D (2026)	D (2631832)	B (563111)	BC (601.3)
	LWT	C (2144)	C (2511691)	C (476486)	B (579.5)
Loading ratio = 100%, vehicles = 10	AMPHI	A (2581)	A (1700622)	A (585098)	B (415.59)
	HABOR	B (2218)	C (2200471)	A (586403)	A (402)
	CLAB	C (2152)	B (2124585)	C (503901)	C (450.5)
	M-FCFS	B (2221)	C (2212919)	A (586511)	BC (426.28)
	STD	D (2016)	D (2612980)	B (560315)	D (499.47)
	LWT	B (2218)	C (2205469)	A (586468)	E (611.68)
Loading ratio = 100%, vehicles = 12	AMPHI	A (2543)	A (1733301)	A (597908)	A (328.93)
	HABOR	B (2415)	B (1831976)	A (596433)	A (321.22)
	CLAB	C (2358)	C (1989665)	B (584596)	C (476.85)
	M-FCFS	D (2220)	D (2206790)	B (586490)	B (346.98)
	STD	F (2028)	F (2574916)	C (563404)	D (492.18)
	LWT	E (2142)	E (2318331)	B (586538)	B (348.79)

Table 6.6: Performance comparison of different dispatching policies (part 2).

Scenario	Dispatching approach	Transport time (s)	Due date satisfaction (%)	WIP in Interbay (lot)	Vehicle utilization (%)
Loading ratio = 100%, vehicles = 8	AMPHI	A (191.5)	A (0.9841)	A (874)	A (0.768)
	HABOR	A (192.4)	B (0.6433)	B (1179)	A (0.7614)
	CLAB	A (196.4)	C (0.6040)	C (1248)	E (0.5342)
	M-FCFS	A (191.3)	B (0.6379)	B (1182)	D (0.5688)
	STD	B (209.7)	D (0.5237)	D (1367)	B (0.728)
	LWT	A (191.8)	B (0.6378)	C (1255)	C (0.6167)
Loading ratio = 100%, vehicles = 10	AMPHI	A (196.63)	A (0.9895)	A (812)	B (0.6207)
	HABOR	A (196.27)	C (0.6456)	B (1177)	A (0.6456)
	CLAB	B (200.87)	B (0.6780)	B (1248)	D (0.5325)
	M-FCFS	A (193.54)	C (0.6380)	B (1176)	A (0.638)
	STD	A (195.91)	D (0.5278)	C (1384)	C (0.5957)
	LWT	A (197.16)	C (0.6411)	B (1175)	B (0.6237)
Loading ratio = 100%, vehicles = 12	AMPHI	A (200.23)	A (1.0000)	A (851)	A (0.5247)
	HABOR	A (199.86)	B (0.7521)	B (986)	A (0.5236)
	CLAB	B (205.54)	C (0.7307)	C (1035)	A (0.5268)
	M-FCFS	A (196.57)	D (0.6387)	D (1179)	B (0.5183)
	STD	A (201.49)	F (0.5350)	E (1369)	B (0.51)
	LWT	A (200.79)	E (0.5901)	DE (1251)	B (0.5191)

Table 6.7: ANOVA of the AMPHI and traditional HABOR, CLAB, M-FCFS, STD, and LWT dispatching approach.

No.	Proposed approach	Traditional approaches	Sum of squares	Degrees of freedom	Mean squares	*F*-value	Significant differences level
1	AMPHI	HABOR	0.48908	35	0.09314	8.0	$p < 0.01$
2		CLAB	1.22688	35	1.0621	219.15	$p < 0.01$
3		M-FCFS	0.7646	35	0.38126	33.82	$p < 0.01$
4		STD	2.13051	35	2.06919	1147.16	$p < 0.01$
5		LWT	0.92152	35	0.706	111.37	$p < 0.01$

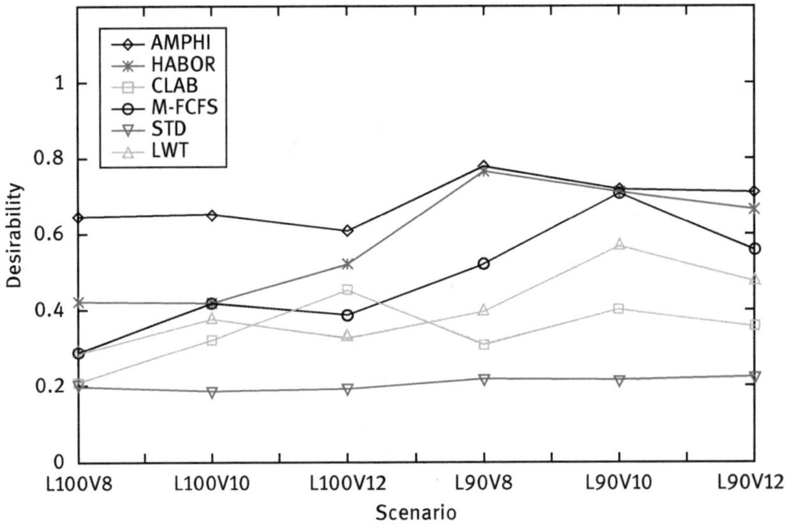

Figure 6.8: Comparison of the desirability values of the AMPHI, HABOR, CLAB, M-FCFS, STD and LWT dispatching approaches.

desirability values of the AMPHI, HABOR, CLAB, M-FCFS, STD and LWT dispatching approaches in different scenarios are calculated. Figure 6.8 illustrates that the proposed AMPHI approach clearly outperforms the other traditional dispatching approaches in terms of comprehensive performance in all six scenarios.

A statistical analysis of variance (ANOVA) was tested on the desirability values. Table 6.7 shows that AMPHI is significantly different ($p \leq 0.01$) from other HABOR, CLAB, M-FCFS, STD and LWT dispatching approaches.

The reasons for suggesting the proposed dispatching approach for complex semiconductor wafer fabrication are briefly summarized as follows. (1) The

Vehicle's transportation distance lot's due date, lot's waiting time and lot's origin-destination buffer status are thoroughly considered and used as the key parameters in the wafer handling cost function model. The weight coefficients of these parameters are adaptively adjusted according to the dynamic material handling environment. Therefore, wafer lots with worse cost value can be given priority to transport first in real time. (2) The proposed fuzzy-based Hungarian algorithm is proven as a rapid and effective approach to obtain the optimal or near-optimal vehicle dispatching solutions. The vehicle blockage phenomenon in Interbay material handling can be largely reduced. (3) The proposed fuzzy-based Hungarian algorithm can respond to events such as arrival and transport of the wafers in real time in order to dynamically respond to the dynamic environment of the Interbay material handling system.

2. Modified Hungarian algorithm and Mamdani fuzzy-logic-based Interbay scheduling method
To evaluate the MHA and Mamdani fuzzy-logic-based Interbay scheduling method, discrete event simulation models are constructed using the eM-Plant software. The layout of the 300 mm wafer fab is shown in Figure 6.9. There are 14 bays and 127 sets of equipment for 23 types of process (Table 6.8).
The layout parameters of the Interbay material handling system are summarized as follows: the Interbay rail loop is 370 m; the length of the vertical guide is 5.5 m; the vehicle speed is 1 m/s; the time on turntable is 10 s and the loading/unloading time is 5 s. Three types of wafer lot (jobs A, B and C) are being processed in the system, and the quantity is equal to each other. Figure 6.10 shows the eM-Plant simulation models.

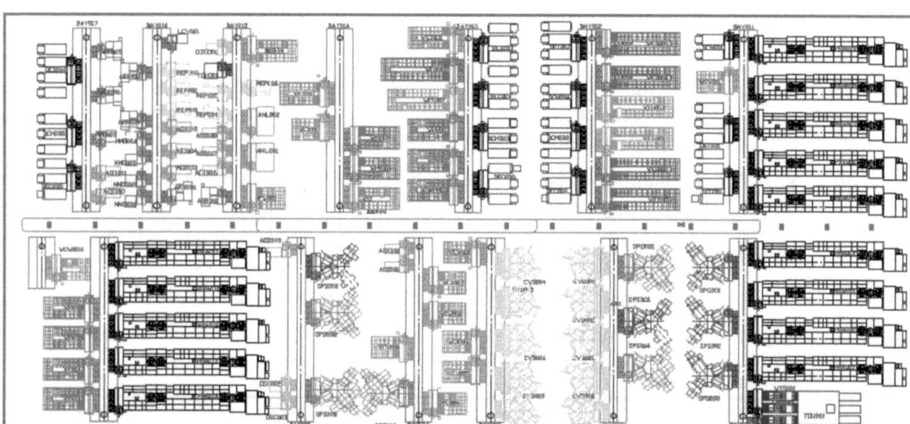

Figure 6.9: The layout of the 300 mm Wafer Fab.

Table 6.8: The fabrication data and key equipment of bays.

Stocker	Bay	Process	Number of critical equipment	Processing time	Capacity (lot/ cassette)
Stoker-1	Bay-1	Cleaning WIS	3	20 min	2
		Photo-etching	4	73 min	3
		Sputtering 1	3	35 min	2
Stoker-2	Bay-2	Short circuit test	3	44 min	1
		Sputtering 2	1	28 min	2
		Sputtering 3	1	26 min	2
		Chemical vapor deposition	5	43 min	2
Stoker-3	Bay-3	Cleaning WCW	3	19 min	2
		Chemical vapor deposition	5	41 min	3
Stoker-4	Bay-4	Auto-appearance	2	48 min	1
		Cleaning WCW	4	21 min	3
Stoker-5	Bay-5	Auto-appearance	1	48 min	1
		Sputtering 2	1	28 min	2
		Sputtering 3	2	26 min	2
Stoker-6	Bay-6	Photo-etching	5	34 min	2
		Stripping	4	24 min	2
Stoker-7	Bay-7	Auto-appearance	3	114 min	2
		Repair (resection)	5	40 min	1
		Cleaning WCW	1	22 min	2
		Repair (connection)	2	10 min	1
Stoker-8	Bay-8	Dry etch DEC	1	46 min	2
		Dry etch DEI	1	41 min	2
		Dry etch DCH	1	53 min	2
		Auto-appearance	2	48 min	1
		Array inspect	2	91 min	2
Stoker-9	Bay-9	Auto-appearance	3	48 min	1
		Repair (connection)	3	10 min	2
		Appearance inspect	5	29 min	1
Stoker-10	Bay-10	Auto-appearance	3	114 min	2
		Repair (resection)	3	34 min	1
		Cleaning WCW	1	20 min	2
		Wet etch WEI	1	41 min	3
		Annealing	2	101 min	5
Stoker-11	Bay-11	Cleaning WCW	1	18 min	2
		Wet etch WEI	2	41 min	3

Table 6.8: (continued)

Stocker	Bay	Process	Number of critical equipment	Processing time	Capacity (lot/ cassette)
Stoker-12	Bay-12	Wet etch WEG	1	28 min	2
		Stripping	3	24 min	2
		Wet etch WED	1	34 min	2
		Dry etch DEI	1	41 min	2
Stoker-13	Bay-13	Wet etch WEG	2	28 min	2
		Wet etch WED	4	34 min	2
		Dry etch DEI	1	41 min	2
		Dry etch DCH	2	53 min	3
		Dry etch DEC	2	46 min	2
Stoker-14	Bay-14	Cleaning WCW	11	20 min	1
		Dry etch DEI	1	41 min	2
		Photo-etch	7	79 min	4
		Dry etch DEC	5	46 min	2

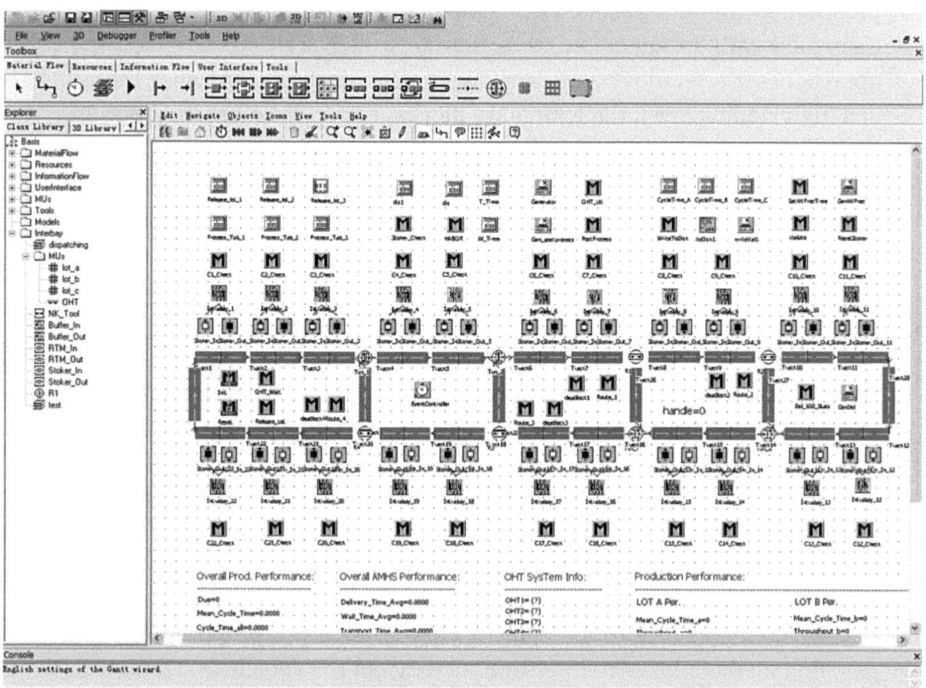

Figure 6.10: The simulation model built by eM-Plant.

The MHA and fuzzy logic control algorithm are realized by using VC++ and embedded into the simulation models as a dynamic link library (DLL). Simulation experiments were carried out to comprehensively compare the effectiveness of the proposed dispatching method and five traditional single-attribute and multi-attribute heuristic dispatching policies (Table 6.9). These traditional EDD, CLAB, FEFS, STD and RLWT dispatching rules and strategies have been adopted in vehicle dispatching of the Interbay system in SWFSs. The measurements of performance include delivery time, cassette waiting time, system throughput, total delivery amount, mean process cycle and vehicle utilization. Totally, four scenarios were designed for different combinations of two loading ratios (3 cassettes/2.5 h and 3 cassettes/2.75 h) and automated vehicles numbers (8 and 12). Each scenario was replicated in three simulation runs, and each run simulated a production period of 150 days with an aging time $T_{\max} = 2{,}000(s)$.

The simulation results are shown in Tables 6.9–6.12. In scenarios 1 and 2, the proposed MHAFLC has better performance in terms of mean delivery time, mean waiting time, system throughput and total delivery amount. As for the other two indexes, mean process cycle and vehicle utilization, those dispatching approaches have no significant difference. In scenarios 3 and 4, the proposed MHAFLC has better performance in terms of all measurements except the vehicle utilization. The conclusion can be obtained that when the system load is at a high level, the proposed MHAFLC approach may significantly improve the system performance. Compared with scenarios 3 and 4, the mean delivery times for all dispatching approaches in scenarios 1 and 2 are longer, which indicate that when the system is at a high loading level, the temporary blockages become more frequent.

For further analysis, an indicator of the comprehensive function D is introduced to realize the comparison of various dispatching rules. The comprehensive function D is defined as

$$D_p(x) = \left\{ \sum_{u \in U} w_u \left[\frac{f_u^* - f_u(x)}{f_u^* - f_u^-} \right]^p + \sum_{v \in V} w_v \left[\frac{f_v(x) - f_v^-}{f_v^* - f_v^-} \right]^p \right\}^{1/p} \tag{6.33}$$

Here, U and V denote the sets of performance measures corresponding to the maximum and minimum objectives, and w_u and w_v are weight coefficients for different measurements. $f_u^* = \max_{x \in X}(f_u(x))$ and $f_u^- = \min_{x \in X}(f_u(x))$ are the permitted maximal and minimal values for index u. f_v^* and f_v^- are defined in the same way. Figure 6.11 shows the D-function values of the proposed dispatching method and other rules $p = 1$. Table 6.12 reports the ANOVA results of different dispatching methods.

Furthermore, it is found that in all scenarios the proposed MHAFLC dispatching method has better performance except the vehicle utilization. In terms of D-function value, MHA-based dispatching methods have the best performance in comparison with the other rules. However, the fluctuation of its D-function

Table 6.9: Performance of different dispatching methods under scenario S1 (3 C/3 h, OHT = 12).

Dispatching strategy	Delivery time (s)		Waiting time (s)		Throughput (cassette)	Mean cycle time (s)	Delivery amount (cassette)	Vehicle utilization
	Mean	Deviation	Mean	Deviation				
MHAFLC	328.9	45,207	128.7	5,220	2,543	1,804,917	586,110	0.5247
MHA	321.2	45,273	121.4	5,109	2,415	2,053,857	587,250	0.5236
EDD	333.6	46,303	133.8	6,455	2,191	2,388,146	586,782	0.5114
CLAB	356.8	48,320	141.3	7,903	2,158	2,319,167	586,909	0.5267
FEFS	346.9	49,320	150.4	8,110	2,220	2,357,323	586,909	0.5182
STD	492.1	49,981	290.8	6,782	2,028	2,475,098	564,088	0.5100
RLWT	348.7	45,766	148.0	8,756	2,142	2,522,490	579,538	0.5190

Table 6.10: Performance of different dispatching methods under scenario S2 (3 C/2.5 h, OHT = 12).

Dispatching strategy	Delivery time (s)		Waiting time (s)		Throughput (cassette)	Mean cycle time (s)	Delivery amount (cassette)	Vehicle utilization
	Mean	Deviation	Mean	Deviation				
MHAFLC	328.1	45,414	128.5	5,128	2,593	978,251	589,625	0.6251
MHA	325.4	45,615	125.1	5,201	2,544	984,410	589,584	0.6366
EDD	332.6	46,334	132.2	6,725	2,489	113,694	577,329	0.5084
CLAB	362.1	48,271	156.5	7,741	2,528	1,063,915	582,040	0.5245
FEFS	351.5	49,330	196.8	7,910	2,589	1,000,430	589,341	0.5214
STD	494.0	49,181	292.0	6,397	2,369	1,649,776	564,088	0.5120
RLWT	348.8	45,966	147.5	8,636	2,505	1,133,740	579,538	0.5243

Table 6.11: Performance of different dispatching methods under scenario S3 (3 C/3 h, OHT = 8).

Dispatching strategy	Delivery time (s)		Waiting time (s)		Throughput (cassette)	Mean cycle time (s)	Delivery amount (cassette)	Vehicle utilization
	Mean	Deviation	Mean	Deviation				
MHAFLC	521.7	45,414	330.2	5,128	2,589	998,745	589,746	0.7258
MHA	521.2	45,615	328.7	5,201	2,587	1,006,923	589,036	0.7657
EDD	654.3	46,334	353.9	6,725	1,751	2,366,741	475,757	0.6019
CLAB	641.2	48,271	425.8	7,741	1,261	3,061,225	414,020	0.5319
FEFS	770.5	49,330	579.2	7,910	2,577	1,059,187	587,761	0.7596
STD	601.2	49,181	491.5	6,397	2,332	1,621,700	557,276	0.7215
RLWT	659.5	45,966	379.2	8,636	1,673	2,658,287	475,631	0.6167

Table 6.12: ANOVA results of different dispatching methods.

Proposed approach	Traditional approaches	Square deviation (SD) × 1,000	Freedom	Mean SD × 1,000	F-value
1	MHA	(221.5, 1112.5)	(1, 23)	(221.5, 48.4)	$4.58 > F_{0.05}(1, 23)$
2	RLWT	(189.1, 914.1)	(1, 23)	(189.1, 39.7)	$4.76 > F_{0.05}(1, 23)$
3	STD	(242.7, 1201.3)	(1, 23)	(242.7, 52.2)	$4.65 > F_{0.05}(1, 23)$
4	CLAB	(88.9, 612.1)	(1, 23)	(88.9, 26.6)	$3.34 > F_{0.05}(1, 23)$

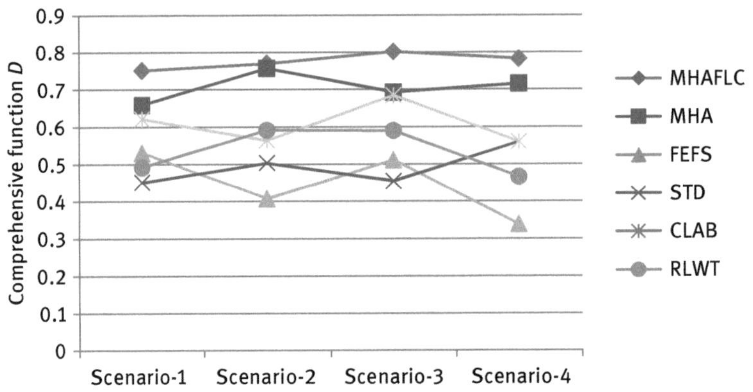

Figure 6.11: Comparison of *D*-function values of proposed dispatching method and other rules.

value indicates that the robustness needs to be strengthened. The proposed fuzzy-logic-based control is applied to dynamically adjust the weight and therefore improve the robustness.

6.3 Composite Rules-Based Interbay System Scheduling Method

Over the last few years, automated vehicle dispatching in material handling systems has received considerable attention. However, most of traditional vehicles dispatching approaches are usually utilized for single-objective optimization, which only consider the wafers' process time and waiting time, vehicles' location and travel time, and so on. In order to satisfy the demand of multiple-objective optimization, some other factors which radically influence the overall system performance and customer satisfaction, such as factors regarding wafer cassettes' priority, have to be considered. On the other hand, the Interbay material handling system is a dynamic and stochastic system in real-life semiconductor fabrication lines. Static attribute weights in traditional dispatching rules limit the ability for the material handling systems to manage unexpected environmental changes. Therefore, how to deal with the stochastic events such as the unexpected vehicle blockages needs to be answered. In this section, a multiple-objective scheduling model of Interbay AMHS is established [2, 3]. Aiming to meet the demands of dynamic adjusting and multiple-objective optimization, a genetic programming (GP)-based algorithm is proposed to generate composite dispatching rules (CDR).

6.3.1 Global Optimization Model of the Interbay System Scheduling Problem

The purpose of the scheduling for the Interbay material handling system is to allocate different vehicles to wafer cassettes so that the production and transportation efficiency can be optimized. Such a scheduling problem is actually a special dispatching problem where at any scheduling time t, there are n waiting wafer cassettes and m vehicles. In this section, the scheduling of Interbay material handling is accomplished in two steps. The first step is to prioritize the waiting cassettes by using CDR. The second step is to dispatch the vehicles with the status of "Idle" or "Retrieval" to cassettes according to specially designed transportation strategies and policies. The whole process of CDR-based scheduling method is shown in Figure 6.12.
Then, the mathematical model for the scheduling process in Interbay material handling systems is formulated.

1. Index and set
   - M set of waiting cassettes at time t, $|M| = m$;
   - K set of vehicles at time t, $|K| = n$;
   - Y set of rails and turntables;
   - S set of stockers;
   - PD^s set of loading/unloading events, $PD^s = \{(k, i) | \forall k, i \in M\}$;
   - h fixed loading/unloading time;

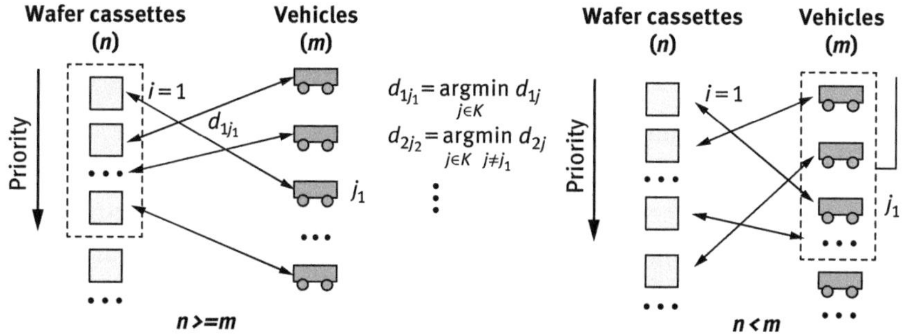

Figure 6.12: The material handling process of CDR-based scheduling method.

2. Attributes

   $L(i)$ wafer cassette of task $i \in M$;
   $DES(i)$ the target stocker of a wafer cassette for task $i \in M$;
   $WTS(i)$ the stocker of a wafer cassette for task $i \in M$;
   t_R^l the time when cassette $l \in L$ arrives at the current stocker;
   t_P^{ij} the time when the cassette of task $i \in M$ is loaded by vehicle $j \in K$;
   t_D^{ij} the time when the cassette of task $i \in M$ is unloaded by vehicle $j \in K$;
   t_Q^i the time when the cassette of task $i \in M$ requests a task;
   t_C^i the time when all operations of the cassette of task $i \in M$ are processed;
   t_S^i the time when the cassette of task $i \in M$ starts its next operation;
   E_j^t location of vehicle $j \in K$ at time t;
   e_{jy}^t is equal to 1, if vehicle $j \in K$ is located on rail $y \in Y$ at time t; otherwise e_{jy}^t is equal to 0;
   C_{ij}^t total handling cost of task $i \in M$ moved by vehicle j at time t;

3. Decision variables

   X_{ij}^t is equal to 1, if task $i \in M$ is processed by vehicle j at time t; otherwise X_{ij}^t is equal to 0.

4. Objective function

   For the scheduling of Interbay material handling, the main optimization objectives include the improvement of manufacturing performance and movement efficiency. The movement efficiency (cassette delivery time, cassette waiting time) and production performance (maximum lateness) are optimized.

$$\text{Min} \sum_{t=0}^{T} \sum_{i=1}^{m} \sum_{j=1}^{n} C_{ij}^t X_{ij}^t \tag{6.34}$$

where

$$C_{ij}^t = \frac{1}{m}\left\{ w_1 \cdot \frac{DT_{ij}^t - DT_i^-}{\overline{DT}_{\max} - \overline{DT}_{\min}} + w_2 \cdot \frac{WT_{ij}^t - WT_i^-}{\overline{WT}_{\max} - \overline{WT}_{\min}} + w_3 \cdot \frac{PDT_{ij}^t - PDT_i^-}{\overline{PDT}_{\max} - \overline{PDT}_{\min}} \right\}.$$

Among the above formulation, $DT_{ij}^t = (t_D^{ij} - t_Q^i)$ is the time when the cassette starts to be delivered, $WT_{ij}^t = (t_P^{ij} - t_R^{L(i)})$ is the waiting time of the cassette and $PDT_{ij}^t = [t_S^i - (t_D^{ij} + h)]$ is the lateness of the next operation of the cassette. DT_i^-, WT_i^- and PDT_i^- (i∈M) indicate the minimum of DTi, WTi and $PDTi$ at time t. $\overline{DT}_{\max}/\overline{DT}_{\min}$, $\overline{WT}_{\max}/\overline{WT}_{\min}$ and $\overline{PDT}_{\max}/\overline{PDT}_{\min}$ indicate the max/min of DT, WT and PDT from all delivery tasks in set M. wi is the weight.$\overline{DT}^t$, $\overline{WT}^t$ and $\overline{PDT}^t$ indicate the average value of DT, WT and PDT at time t.

$$\text{Min} \sum_{t=0}^{T} \left(\frac{\overline{DT}^t - \overline{DT}_{\min}}{\overline{DT}_{\max} - \overline{DT}_{\min}} + \frac{\overline{WT}^t - \overline{WT}_{\min}}{\overline{WT}_{\max} - \overline{WT}_{\min}} + \frac{\overline{PDT}^t - \overline{PDT}_{\min}}{\overline{PDT}_{\max} - \overline{PDT}_{\min}} \right) \tag{6.35}$$

5. Constraints

$$\sum_{i=1}^{m} X_{ij}^t = 1, X_{ij}^t \in \{0, 1\}, \forall t \in [0, T], \forall j \in K \tag{6.36}$$

$$\sum_{j=1}^{m} X_{ij}^t = 1, X_{ij}^t \in \{0, 1\}, \forall t \in [0, T], \forall j \in K \tag{6.37}$$

$$t_D^i X_{kj}^{t_D^i} + h \le t_D^i X_{ij}^{t_D^i}, \forall (k, i) \in H \subseteq PD^5, \forall j \in K \tag{6.38}$$

$$t_P^i X^{t_{ij}^i P} + t_D^i + h \le t_D^i X_{ij}^{t_D^i}, \forall (k, i), \subseteq PD^s, WTS(i) = DES(k), \forall j \in K \tag{6.39}$$

$$t_P^k + h \le t_P^i, \forall (i, k) \subseteq PD^5, (WT(i) = WT(k)) \wedge (t_R^k > t_R^i), \forall j \in K \tag{6.40}$$

$$t_D^i X_{kj}^{t_P^k} + CP_i \le t^i{}_P X^{t^i{}_D}{}_{ij}, \forall (k, i) \subseteq PD^s, L(i) = L(k) \wedge WTS(i) = WTS(k), \forall j \in K \tag{6.41}$$

$$t_R^{L(i)} + 1 \le t_P^i, \forall i \in M \tag{6.42}$$

$$E_1^t < E_2^t < \ldots < E_N^t, \forall t \in [0, T] \tag{6.43}$$

$$\sum_{j=1}^{M} e_{yj}^t \le 1, \forall t \in [0, T], \forall y \in Y \tag{6.44}$$

Among these constraints, formulation (6.36) ensures that a delivery request can be responded by only one vehicle; formulation (6.37) means that a vehicle can handle only one delivery task at one time; formulation (6.38) guarantees that the vehicle is not available until the current delivery task has been finished; formulation (6.39) indicates the constraint between the loading and unloading times of a cassette; formulation (6.40) indicates the constraint between the loading and unloading times of different cassettes in the same stocker; formulation (6.41) indicates the constraint between the unloading and loading times of a cassette in the same stocker; formulation (6.42) ensures that the loading of the cassette is later than the arrival of the cassette; formulation (6.43) describes the location constraints of vehicles and formulation (6.44) guarantees that there is only one vehicle on a rail at any time.

6.3.2 Architecture of Composite Rules-Based Interbay System Scheduling Method

The compound heuristic rule refers to a compound scheduling rule constructed by taking various heuristic information into consideration, such as various factors influencing the objective function and integrating human visual experience. It is possible to obtain the approximate optimal solution while satisfying the constraints of the optimization problems. To construct composite heuristic rules, two requirements should be satisfied: (1) be able to combine advantages of some single-heuristic rules, and does not require specific parameters related to the scene, with strong robustness and adaptability; (2) dynamic adjustment characteristics. It is able to make the appropriate adjustments according to changes of system optimization objectives at each stage.

Figure 6.13 shows the composite heuristic rules-based scheduling framework for an Interbay material handling system. Single-heuristic rules are combined together to obtain a variety of possible composite heuristic rules. With these composite heuristic rules, the scheduling process of an Interbay material handling system is carried out to obtain major transport performances and processing performances, which include wafer lot's average delivery time, due date satisfactory rate, average cycle time and wafer throughput. These indicators are used as the evaluation indexes of composite heuristic rules in order to analyze its effectiveness at different stages.

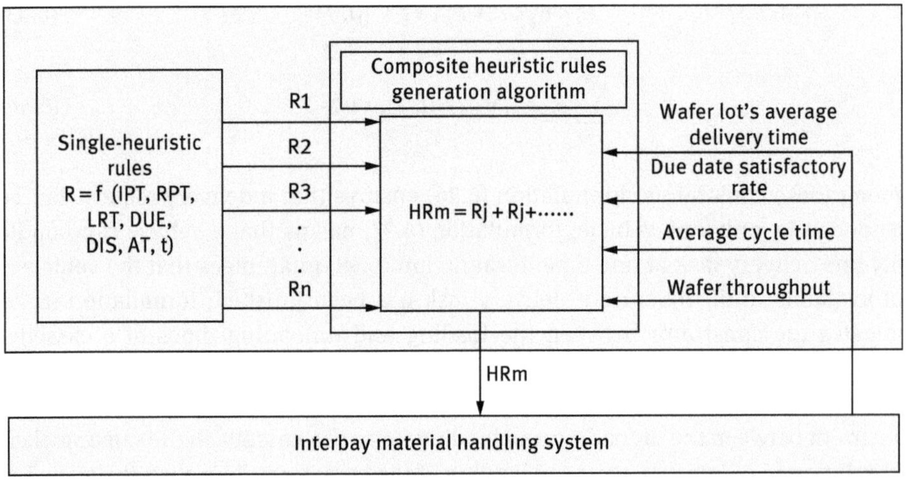

Figure 6.13: A composite heuristic rule-based scheduling framework of an Interbay material handling system.

6.3.3 Genetic Programming-Based Composite Dispatching Rule Algorithm

In this section, genetic programming-based composite rule generator (GPCRG) is introduced. In GPCRG, several system performance measures are grouped and the solutions with best fitness are obtained by searching from the tree-structure space of a feasible solution. These solutions correspond to different CDR. According to these rules, waiting cassettes are prioritized and vehicles are then assigned. The flowchart of the proposed GPCRG is depicted in Figure 6.14. The procedure includes encoding, initialization, fitness selection, cross and mutation, and stopping criterion. Details of the algorithm are presented in the following sub-sections.

1. Encoding scheme

   First of all, the set of terminals is defined in Table 6.13 and four fundamental operators are defined as Operator Set = { +, −, ×, \}. In order to obtain the CDR, a binary tree encoding scheme is used in the proposed GPCRG. The above terminals and operators are set as the nodes. It would be noticed that there are no child-nodes for all the terminals while all the operators are parent-nodes.

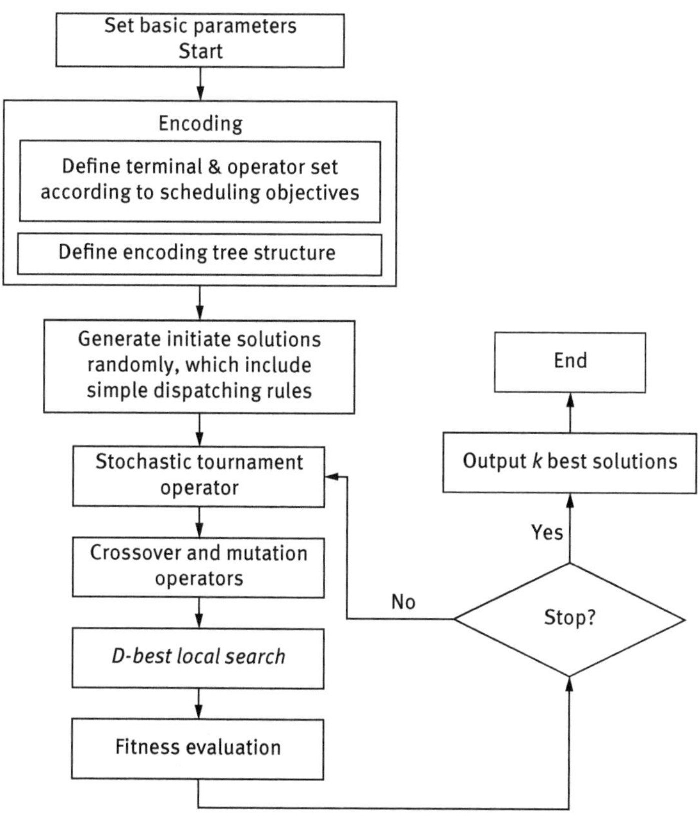

Figure 6.14: The flowchart of GPCRG.

Table 6.13: Terminal set of GPCRG.

Variable	Explanation
IPT	Start time of the next operation for the wafer cassette
RPT	Sum of processing time of all unfinished operations
LRT	Release time of the cassette
DUE	Due date of the wafer cassette
DIS	Distance between the cassette and the stocker
AT	Arrival time of cassette to stocker
T	Current time

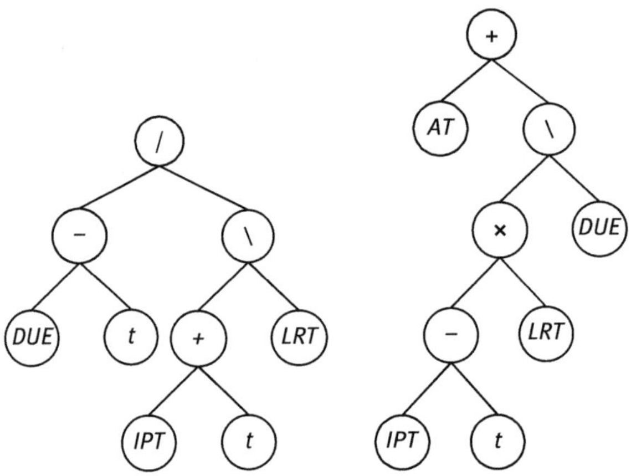

Figure 6.15: The diagram of coded binary tree of dispatching rules.

The diagrams in Figure 6.15 indicate the rules: $Rule\text{-}1 = (DUE - t)/[(t + IPT) * LRT]$ and $Rule\text{-}2 = [(IPT - t)*LRT]/DUE + AT$.

2. Evolution process
   (1) Population initialization
   The initial population of individuals is generated randomly. In order to keep the advantages of the simple dispatching rules, some basic rules such as MFEFS, STD, LWT and EDD are added to the initial population. The CDR with higher performance can be obtained as the coded binary tree evolves.

   (2) Genetic operators
   A. Stochastic tournament operator
   The procedure is as follows: choose two individuals rando mly and compare their fitness. The better one will be kept into next generation. Repeat this step for *M* times (suppose the population size is *M*); a new generation is then developed. This method will balance the computational efficiency and the keeping of good genes.

B. Crossover operator

In general, the crossover operator is regarded as a main genetic operator and the performance of the genetic algorithms depends, to a great extent, on the performance of the crossover operator used. Conceptually, the crossover operates on two chromosomes at a time and generates offspring by combining features of both chromosomes. In the proposed GP-based method, the procedure of crossover operator is shown in Figure 6.16.

C. Mutation operator

The aim of mutation is to introduce variability and diversity to the population so that the algorithm is able to escape from a local optimum. Two types of mutation exist in the proposed GA-based method: standard mutation and swap mutation. The standard mutation is to replace a child-tree randomly. The swap mutation is to exchange two child-trees in the individual. The procedure of the mutation is shown in Figure 6.17.

(3) Stopping criterion

Stopping criterion is a GP parameter which is used to control the termination of the genetic iterative process. Normally, a given number of generations are set as the stopping criterion.

3. Local search strategy

Traditional GP algorithms are easy to fall into the local optimal which results in premature convergence. To prevent this case, a local search strategy is introduced into the GPCRG. The efficiency of the local search algorithms depends on the generating mechanisms and search strategy. In this section, the λ-Interchange local search method is adopted.

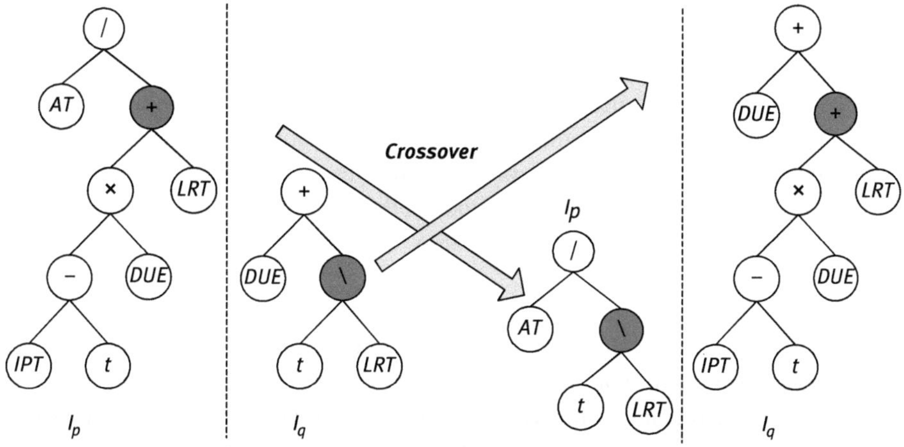

Figure 6.16: The crossover operator of GP-based CDR generating algorithm.

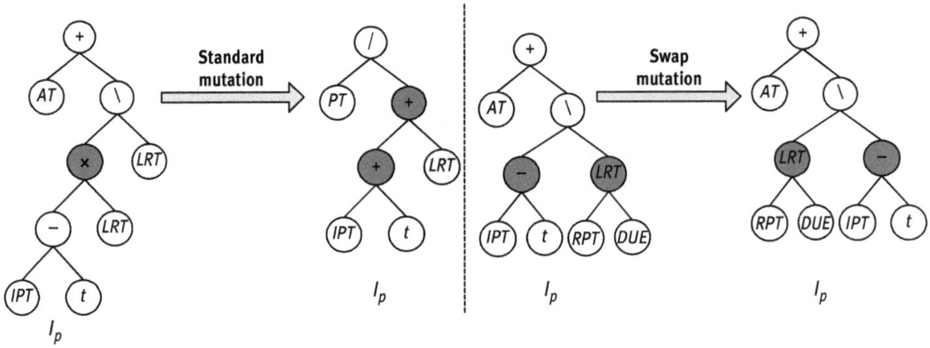

Figure 6.17: The mutation operator of GP-based CDR generating algorithm.

The λ-Interchange local search improves the quality of the solution by swapping child-trees from different individuals. In the proposed GP-based CDR generating algorithm, only $\lambda = 1$, $\lambda = 2$ and $\lambda = 3$ are considered, which means that between different individuals at most three child-trees can be swapped. According to the value of λ, three child-tree swap operators are defined: (1, 1), (2, 2), (3, 3). For example, the operator (2, 2) on an individual pair (I_p, I_q) indicates that choose two child-trees from individual I_p and add them into individual I_q; meanwhile choose two child-trees from individual I_q and add them into individual I_p. The procedure is shown in Figure 6.18.

During the procedure, only when the fitness of the individual is improved this swap is accepted. Since there are normally more than one child-tree in one individual, two strategies are applied to help choose the swap child-trees.

(1) First-Search (FS) strategy: choose the first child-tree which improves the fitness by searching from the neighborhood of current solution $N_\lambda(S)$.

(2) Global-Best-Search (GS) strategy: choose the child-tree which has the best performance for improving the fitness by searching from the neighborhood of current solution $N_\lambda(S)$.

Although GS strategy performs better in terms of obtaining the global optimal, the computational efficiency decreases. In this section, a compromised d-best strategy is proposed. d child-trees able to improve the fitness are searched from the neighborhood of the current solution and the best one will be chosen as the swap child-tree.

4. Fitness evaluation

Each individual can be decoded into corresponding CDR. In order to measure the performance of the individual, a method based on from-to-table is used. Suppose the CDR decoded from an individual x are named *Rule-x* and there are totally S stockers in the Interbay material handling system. Before the fitness evaluation, some assumptions are made as follows:

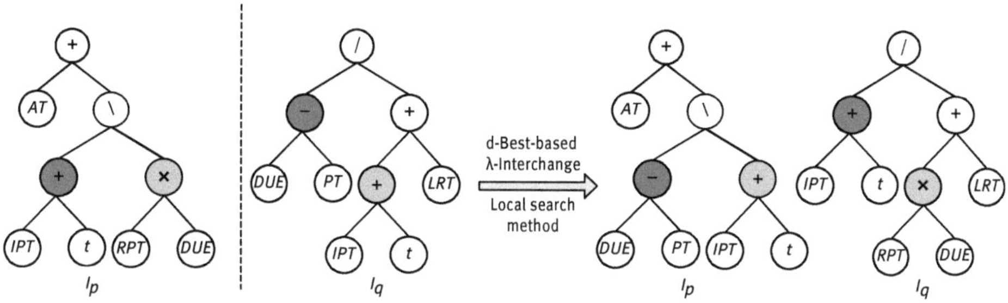

Figure 6.18: The local search operator (2, 2).

(1) In each time period T, the total number of requests from wafer cassettes is fixed. The arrival time of the requests is a Poisson process, which indicates that the time interval is exponentially distributed.
(2) Each vehicle can only deliver one wafer cassette at a time. The loading and unloading times obey the standard normal distribution.
(3) Delivery tasks are evenly assigned to vehicles.
(4) The movements of different vehicles are independent of each other.
(5) The possibility that the vehicle j is blocked by vehicle j' on the downstream transport rail at time t is proportional to the total quantity of vehicles and the material flow rate between stockers on basis of the from-to-table. It can be formulated as $\rho_j = a \cdot M + b \cdot FR_{(r_j, r_{j'})}$, where M indicates the total quantity of vehicles, $FR_{(r_j, r_{j'})}$ is the material flow rate between stockers, and a and b are the weights.
(6) There exist stochastic blockages of vehicles in the system and the block time obeys the Weibull distribution. This is because if the rail is blocked for a longer time, it is more possible that the rail becomes idle in the follow-up time. For any rail l, $\alpha_t^l < \alpha_{t'}^l, t < t'$, $\alpha_t^l = 1 - \exp(-((t - t_0)/\lambda)^k)$ indicates the possibility that the rail l is blocked before time t and becomes idle after time t, $t_c$is the time when the rail l becomes blocked, and k and λ are shape parameters of the Weibull function. Therefore, the expectation of the block time of rail l at time t_l when the vehicle enters rail l can be obtained by the following formulation:

$$BlockTime_t^l = \sum_{i=1}^{\infty} i(1 - \alpha_t^l)^{i-1} \cdot \alpha_t^l = 1/\alpha_t^l \tag{6.45}$$

The evaluation of individual fitness takes into account three performance measures: average delivery time, mean processing cycle time and waiting time of the cassette. The evaluation procedure consists of following steps:

Step 1. Calculate the arrival time t_R^i for each stocker according to the probability distribution, and sort all tasks based on their arrival times $t_R^1 \le \ldots \le t_R^m$.

Step 2. Calculate the starting times of all delivery tasks according to t_R^i. Suppose the set of arrival times of new tasks is $T_R = [t_R^1, t_R^2, ..., t_R^m]$, and the set of times when the vehicles finished the previous tasks is $T_D = [t_D^1, t_D^2, ..., t_D^{m-1}]$. At time $t \in T_R \cup T_D$ the dispatching is triggered, calculate the priorities of cassettes according to *Rule-x*.

Step 3. Calculate the delivery time for each vehicle. The delivery time is the sum of fixed moving time and block time. It is formulated as follows:

$$DeliveryTime_{ij} = dis(S_k^i, S_p^i)/v + \sum_{l \in R^j} \rho_j \, BlockTime^l{}_{t_l} \tag{6.46}$$

Here, $dis(S_k^i, S_p^i)$is the fixed distance between the location k of the current stocker and location p of the target stocker. v is the moving speed of vehicles. $BlockTime_{t_j}^l$ is the block time of vehicle j on downstream rail l.

In a real-life case, the rail constraints ensure that at any time there is only one vehicle on a rail. Therefore, the delivery time has to take into account the temporary block time due to the competence of vehicles. In order to obtain the average delivery time at any scheduling time t_{est}, a heuristic-based algorithm is proposed. Suppose there are $m = 3$ vehicles and $n = 4$ rails. The sequence of railways for all vehicles in current schedule is recorded as $l_1l_2l_3l_4 \,|l_2l_3l_4 \,|l_3l_4$, and a matrix $O[m][n]$ is defined as

$$O = \begin{bmatrix} (1) & (2) & (3) & (4) \\ (8) & (5) & (6) & (7) \\ (11) & (12) & (9) & (10) \end{bmatrix} \tag{6.47}$$

In this matrix O, each row represents a sequence of railways through which a vehicle will move in current schedule. Each railway is numbered in accordance with the sequence constraint. The cost of delivery time on railways for each vehicle is addressed in Table 6.14.

Sort all elements in the matrix O according to the time cost while the sequence constraints are still kept, and a vector $R = \{5, 1, 2, 9, 10, 3, 4, 6, 11, 12, 7, 8\}$ can be obtained which indicates the moving sequence of vehicles under the railway constraints. Define matrix $TM[n][nm]$ as follows: each row of TM represents a railway, and each column is corresponding to an element in matrix O. The value of the

Table 6.14: The cost of deliverytime on railways for each OHT.

Vehicle	l_1	l_2	l_3	l_4
1	(1)3	(2)3	(3)8	(4)3
2	(8)0	(5)2	(6)5	(7)3
3	(11)0	(12)0	(9)4	(10)5

elements in the matrix *TM* stands for the delivery time cost. If the element values 0, it means that the vehicle will not move through this railway in current schedule.

$$TM = \begin{bmatrix} 3 & 0 & 0 & 0 & 0 & 0 & 0 & 0 & 0 & 0 & 0 & 0 \\ 0 & 3 & 0 & 0 & 2 & 0 & 0 & 0 & 2 & 0 & 0 & 0 \\ 0 & 0 & 2 & 0 & 0 & 5 & 0 & 0 & 0 & 5 & 0 & 0 \\ 0 & 0 & 0 & 1 & 0 & 0 & 0 & 3 & 0 & 0 & 0 & 3 \\ (1) & (2) & (3) & (4) & (5) & (6) & (7) & (8) & (9) & (10) & (11) & (12) \end{bmatrix} \tag{6.48}$$

However, in the matrix *TM*, only the railway constraints can be reflected. The sequence constraints are still unknown from this matrix. In order to obtain the finish time of the delivery task for each vehicle, the matrix *TM* has to be mapped to another matrix $TT[n][nm]$ as follows:

$$TT = \begin{bmatrix} 3 & 0 & 0 & 0 & 0 & 0 & 0 & 0 & 0 & 0 & 0 & 0 \\ 0 & 0 & 0 & 0 & 0 & 0 & 0 & 0 & 0 & 0 & 0 & 0 \\ 0 & 0 & 0 & 0 & 0 & 0 & 0 & 0 & 0 & 0 & 0 & 0 \\ 0 & 0 & 0 & 0 & 0 & 0 & 0 & 0 & 0 & 0 & 0 & 0 \\ (1) & (2) & (3) & (4) & (5) & (6) & (7) & (8) & (9) & (10) & (11) & (12) \end{bmatrix} \tag{6.49}$$

In the matrix *TT*, each row indicates a railway. The element with a non-zero value in column $j \cdot n$ ($j = 1, 2, \ldots, m$) stands for the finish time of the delivery task processed by vehicle *j*.

When generating the matrix *TT*, precedence constraints have to be considered. If there is no precedence constraint, delivery time = cost time represented by current index $R(i)$ + max (all elements of row l in matrix *TM*) + stochastic block time. If there exist precedence constraints, delivery time = cost time represented by current index $R(i)$ + max (max (all elements of row l in matrix *TM*), max (all elements of row a in matrix *TM*)) + stochastic block time. l is the row index of the non-zero element in column $R(i)$ in matrix *TM*, a is the row index of the non-zero element in column $R(i) - 1$ in matrix *TM*. Based on the above algorithm, the matrix *TT* can be obtained as follows:

$$TT = \begin{bmatrix} 3 & 0 & 0 & 0 & 0 & 0 & 0 & 22 & 0 & 0 & 9 & 0 \\ 0 & 6 & 0 & 0 & 2 & 0 & 0 & 0 & 0 & 0 & 0 & 9 \\ 0 & 0 & 14 & 0 & 0 & 19 & 0 & 0 & 4 & 0 & 0 & 0 \\ 0 & 0 & 0 & 17 & 0 & 0 & 22 & 0 & 0 & 9 & 0 & 0 \\ (1) & (2) & (3) & (4) & (5) & (6) & (7) & (8) & (9) & (10) & (11) & (12) \end{bmatrix} \tag{6.50}$$

The average delivery time of all tasks at scheduling time t_{oss} can be calculated: $\bar{D} = (17 + 22 + 9)/3 = 16$.

The pseudo-code of the proposed algorithm for calculating the finish time of all tasks under the railway constraints and stochastic blockages is presented as follows:

Algorithm 6.1. Algorithm for calculating the finish time of all tasks under the railway constraints and stochastic blockages

TT = **0**; j = 0;
For i = 1 to n //First column in matrix TT
$TT[i][R(1)] = TM[i][R(1)]$;
For i = 2 to $\sum_{j=1}^{m} X_{ij} = 1$//Other columns in matrix TT
{ //row index of the non-zero element in column R(i) in matrix TM
l = Find NonZeroIndex(TM, i);
//row index of the non-zero element in column $R(i)$ – 1 in matrix TM
a = Find NonZeroIndex(TM, $R(i)$ - 1);
If (i%n == 1) //no precedence constraint
Then
{j ++;
$TT[l][R(i)]] = TM[l][R(i)]$ + MaxRowElem(TT, l) +p_j*$BlockTime$;
//MaxRowElem returns the maximum of all row elements
//p_j is the possibility that vehicle j meets block
}
Else //with precedence constraints
{$TT[a][R(i)]] = TM[a][R(i)]$ + max(MaxRowElem(TT, l), $TT(a, R(i) - 1)$)
+p_j*$BlockTime$;
}
}

At any scheduling time t_s, the average delivery time for all tasks is formulated as follows:

$$\overline{DT}^t(x) = \frac{1}{m}\sum_{i=1}^{m} TT[a][j \cdot n], j = 1, 2, ..., n \tag{6.51}$$

Here, a is the row index of the non-zero element in column$j \cdot n$ in the matrix TT.
Step 4. Prioritize all delivery tasks according to *Rule-x*, and calculate the average lateness of the next processing operation for current cassette.

For a delivery task$i \in M$:

$$PDT_t = \sum_{j=1}^{Q_{S_i}} PT_j / PC_{S_i} \tag{6.52}$$

Here, S_i is the location of the target stocker of task i. Q_{S_i} is the length of the queue in the buffer of the stocker. PT_j indicates the processing time of the current operation in the buffer. PC_{S_i} represents the maximum processing capacity of key equipment. The average lateness of the next operation for all cassettes can be formulated as

$$\overline{PDT}^t(x) = \sum_{i=1}^{|M|} \sum_{j=1}^{Q_{S_i}} PT_j / PC_{S_i} \tag{6.53}$$

Step 5. Calculate the average waiting time for all cassettes at scheduling time t.

$$\overline{WT}^t(x) = \sum_{i=1}^{|M|} WaitTime_i \tag{6.54}$$

In formulation (6.54), $j = \{j | j \in K \wedge d_{ij} = \arg\min_{j' \in K} d_{ij'}\}$ indicates the nearest vehi cle to the cassette of task $i \in M$, $WaitTime_i = (WT_{\max} - WT_i)/WT_{\max}$, $WT_i = t_P^{ij} - t_R^{ij}$, $WT_{\max} = const$ and $WaitTime_i$ is the waiting time for task $i \in M$.

Step 6. Calculate the fitness of *Rule-x* after the linear normalization and weighting for formulations (3.28), (3.30) and (3.32).

$$F_t(x) = F_t^1(x) + F_t^2(x) + F_t^3(x) \tag{6.55}$$

$F^i(x) \in [0, 1]$, $i = 1, 2, 3$, it is obtained by

$$F_t^1(x) = (\overline{DT}^t(x) - \overline{DT}_{\min}) / (\overline{DT}_{\max} - \overline{DT}_{\min}) \tag{6.56}$$

$$F_t^2(x) = (\overline{PDT}^t(x) - \overline{PDT}_{\min}) / (\overline{PDT}_{\max} - \overline{PDT}_{\min}) \tag{6.57}$$

$$F_t^3(x) = (\overline{WT}^t(x) - \overline{WT}_{\min}) / (\overline{WT}_{\max} - \overline{WT}_{\min}) \tag{6.58}$$

The fitness function for individual x in the whole scheduling period T can be formulated as

$$F_T(x) = \sum_{t=0}^{T} F_t(x) \tag{6.59}$$

The individual is better if the value of its fitness $F_T(x)$ is smaller, and has a relatively larger probability to be selected and kept to the next generation. The population is therefore evolved.

6.3.4 Simulation Experiments

To evaluate the CDR generated by proposed GPCRG, discrete event simulation models are constructed using the eM-Plant software. The GP-based CDR generating algorithm is realized by using VC++ and embedded into the simulation models as a DLL.

The established simulation models consist of three feed rates: 3 cassettes/2.5 h (high load), 3 cassettes/2.75 h (high load) and 3 cassettes/3 h (low load). Three cases of vehicle quantity are considered: 8, 10 and 12. The total simulation scenarios therefore count 3 × 3 = 9, and for each scenario the experiments repeat three times. The from-to-table of the current Interbay AMHS is obtained and presented in Table 6.15. Parameters of the proposed GP-based CDR generating algorithm are addressed in Table 6.16.

After 100 × 200 = 20,000 evaluations, 5 best individuals are obtained. The represented CDR are listed in Table 6.17. In scenario 1 where feed rate is 3 cassettes/2.5 h and vehicle quantity is 12, the scheduling results are shown in Table 6.18.

Table 6.15: The from-to-table of the current Interbay AMHS.

Stocker	STK1	STK2	STK3	STK4	STK5	...	STK14	Total
STK1	–	–	2.67	0.25	6.85	...	0.17	10.91
STK2	4.32	–	2.54	0.67	0.98	...	7.09	13.48
STK3	3.28	–	–	–	0.22	...	–	9.02
STK4	–	–	0.71	–	–	...	–	3.29
STK5	–	3.32	0.90	0.34	–	...	3.33	12.15
STK6	0.97	–	–	1.44	–	...	2.91	9.91
STK7	–	–	–	–	–	...	0.44	2.52
STK8	1.13	2.54	–	–	–	...	–	12.24
STK9	0.26	–	–	–	6.12	...	–	13.98
STK10	–	–	–	–	4.33	...	0.22	6.44
STK11	–	0.09	2.33	1.77	–	...	0.14	7.05
STK12	–	–	2.09	0.38	–	...	0.09	11.77
STK13	–	–	1.97	0.41	0.28	...	–	9.88
STK14	–	2.73	–	–	0.25	...	–	6.97
Total	9.96	8.78	12.31	5.26	19.03	...	14.29	127.93

Unit: Cassette per hour

Table 6.16: The parameter set for the GP-based CDR generating algorithm.

Parameter	Value	Parameter	Value
Population size	100	Possibility of standard mutation	0.05
Evolved generation	200	Possibility of swap mutation	0.05
Max depth of initial tree	10	Possibility of λ-interchange	0.1
Max depth of crossover tree	15	Value of d	2
Possibility of crossover	0.3	Weight (a, b)	(0.08, 0.05)

Table 6.17: The generated composite dispatching rules.

Composite dispatching rules	Formulations
GPCRG-R1	$\left(k.DIS+\frac{DUE.LRT(RPT+DUE)}{AT-LRT}\right).\left(\frac{DUE+RPT/DUE-RPT-AT}{t-AT}\right)--$ $k=\frac{AT(RPT+t-AT)((2RPT-t)+DUE)}{LRT}$
GPCRG -R2	$DUE-\frac{RPT.(DUE-t)}{(2LRT-DUE)+((DUE-t)/RPT).DIS}$
GPCRG -R3	$\left(DIS+\frac{AT}{IPT}\right)^2.\frac{(AT+RPT/(DUE-t))}{((AT^2+DUE.AT)(RPT+DUE))}$
GPCRG -R4	$\left(1-\frac{RPT}{LRT+IPT-t}\right)+DIS.(t-AT)$
GPCRG -R5	$LRT-t.\frac{DUE+r(RPT-AT)}{1/LRT+(t-DIS)/AT,\ r=\frac{(RPT+AT)(t-LRT)}{LRT}},\ e_l$

Table 6.18: Performance of the generated composite dispatching rules.

Composite dispatching rules	Mean cycle time (h)	Mean delivery time (s)	Due date satisfaction rate (%)	Throughput (c)
GPCRG-R1	507.62	325.01	87.91	2,539
GPCRG-R2	526.09	340.87	82.02	2,477
GPCRG-R3	516.44	323.14	91.22	2,514
GPCRG-R4	529.58	339.29	80.18	2,482
GPCRG-R5	511.99	324.90	89.14	2,511

Table 6.19: ANOVA results of *D*-value of generated composite dispatching rules.

GPCRG-R1	Square deviation (SD) × 1,000	Freedom	Mean SD × 1,000	*F*-value	Difference interval
R2	(401.1, 4972.5)	(1, 53)	(401.1, 93.8)	4.33	A
R3	(200.7, 9964.1)	(1, 53)	(200.7, 179.7)	1.07	B
R4	(325.5, 3910.3)	(1, 53)	(325.5, 73.8)	4.41	A
R5	(76.4, 2612.9)	(1, 53)	(76.4, 49.3)	1.55	B

The ANOVA results of *D*-value of the generated CDR are shown in Table 6.19. The benchmark is GPCRG-R1. "A" and "B" indicate the significant difference interval (confidence level less than 95%).

Figure 6.19 shows that GPCRG-R1, GPCRG-R3 and GPCRG-R5 have better comprehensive performance compared with the other two rules, and in addition are relatively more stable. The reason is that such three CDR contain more information of the

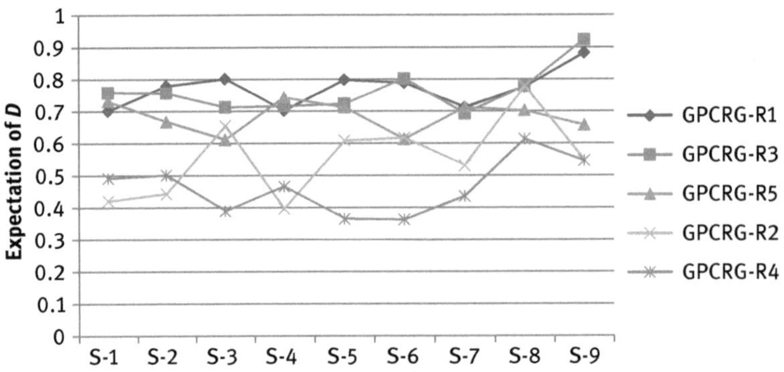

Figure 6.19: The *D*-value comparison among the proposed five composite rules in nine scenarios.

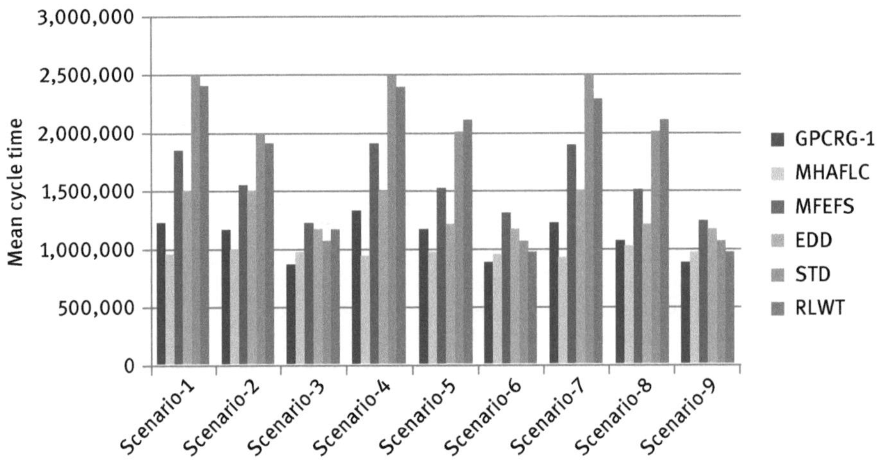

Figure 6.20: Comparison of mean cycle time among all dispatching methods in nine scenarios.

system. In contrast, GPCRG-R2 does not take into account the factor of due date *DUE* and GPCRG-R4 does not include the system time *t*.

Some other dispatching rules including MHA, EDD, STD, R-LWT and MFEFS are compared with generated CDR. These traditional dispatching rules and strategies have been adopted in vehicle dispatching of the Interbay system in SWFSs and been approved to be feasible and effective methods. The experimental results are depicted in Figures 6.20–6.22. An indicator of the comprehensive function *D* is introduced to realize the comparison of various dispatching rules. The variables in function *D* include average delivery time, wafer cassette throughput, mean processing cycle time and due date satisfaction rate.

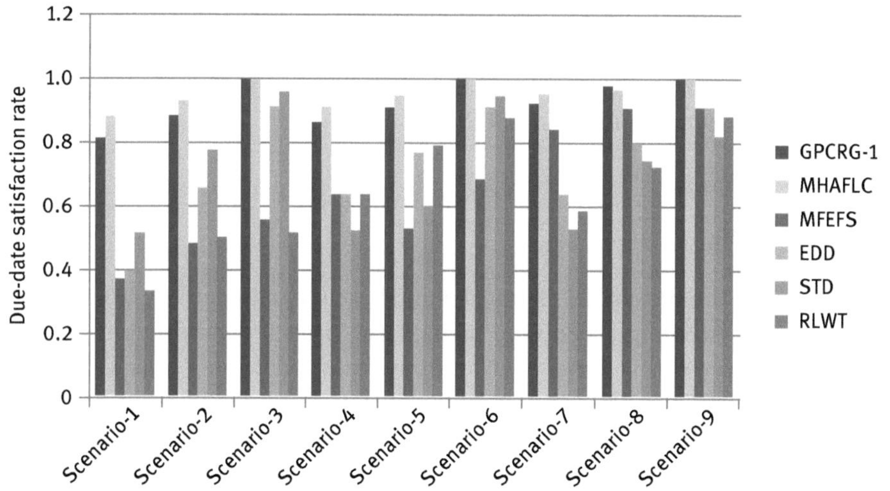

Figure 6.21: Comparison of due date satisfaction rate among all dispatching methods in nine scenarios.

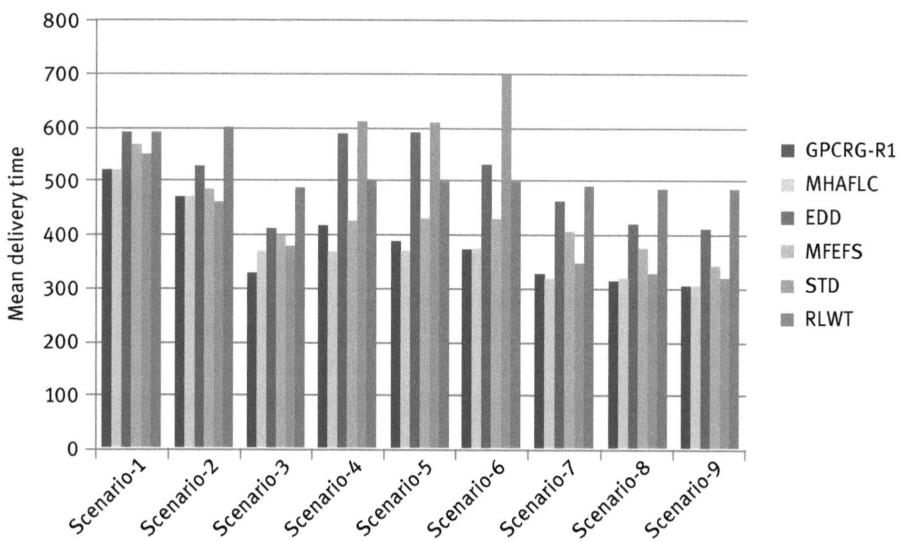

Figure 6.22: Comparison of mean delivery time in nine scenarios.

In terms of mean wafer processing cycle time and due date satisfaction rate, GPCRG-R1 and MHAFLC have best performance than the other methods in all nine scenarios – especially when the system is under a high load and vehicles are relatively not enough. It is also noticed that when the transport capacity becomes the bottleneck, the generated CDR are able to maximize the improvement of transport efficiency. As for the mean delivery time, when the system load is at a lower level, the performance of all scheduling methods improves. GPCRG-R1 and MHAFLC still have best performance than the other methods. However, when the system load is at a high level, MHAFLC performs better than GPCRG-R1.

6.4 Conclusion

In this chapter, an adaptive multi-parameter and hybrid intelligence (AMPHI)-based Interbay material handling scheduling method has been proposed based on four parameters: delivery times, wafer lots' waiting times, wafer lots' due dates and processing characteristics. A real-time material handling scheduling system has been optimized through a Hungarian algorithm and an MHA. The fuzzy logic control methods (including Mamdani fuzzy inference and Takagi-Sugeno fuzzy inference) have been used to adjust scheduling strategies dynamically according to the real-time system status. At the same time, this chapter has also studied the global optimization scheduling method based on composite heuristic rules, has proposed the global optimization model of an Interbay system scheduling problem and has designed a composite rule algorithm based on GP. To verify the effectiveness of these two proposed scheduling methods, the actual production data of a 300 mm semiconductor wafer fab has been used to establish a discrete time simulation model with eM-Plant software. By analyzing the experimental results, the AMPHI-based Interbay material handling scheduling method has taken into account randomness of the material handling environment and has met the requirements of the multi-objective optimization. However, it has showed inadequacy in real-time scheduling in solving a complex mathematical model. Therefore, this method is suitable for solving Interbay scheduling problems with higher demand of multi-objective optimization and lower real-time requirements. And it is better to adopt the composite rules-based Interbay system scheduling method to solve real-time scheduling problems. The composite rules can also guarantee the global performance to a certain extent. It is suitable for Interbay material handling scheduling problems with high real-time demand and global performance optimization.

References

[1] Qin, W., Zhang, J., Sun, Y.B. Dynamic dispatching for Interbay material handling by using modified Hungarian algorithm and fuzzy-logic-based control. The International Journal of Advanced Manufacturing Technology, 2013, 67(1 –4): 295–309.

[2] Wu, L. Research on Intelligent Scheduling Technologies of AMHS in Semiconductor Wafer Fabrication System. Shanghai Jiaotong University, 2011.

[3] Sun, Y. Research on Intelligent Scheduling Technology of Interbay System in Semiconductor Wafer Fabrication System [D]. Shanghai Jiaotong University, 2011.

[4] Lin, J.T., Wang, F.K., Yen, P.Y. Simulation analysis of dispatching rules for an automated Interbay material handling system in wafer fab. International Journal of Production Research, 2001, 39(6): 1221–1238.

[5] Qin, W., Zhang, J., Sun, Y.B. Multiple-objective scheduling for Interbay AMHS by using genetic-programming-based composite dispatching rules generator. Computers in Industry, 2013, 64 (6): 694–707.

7 Scheduling in Intrabay automatic material handling systems

The Intrabay material handling system is primarily responsible for wafer handling between processing machines and stockers within the processing bays. Since the Intrabay material handling systems are used to move material directly between processing machines, it is necessary for the Intrabay system to consider the status of each processing machine, which is obviously different from the Interbay system. In the production scheduling process in Intrabay material handling systems, it is first necessary to determine each wafer lot's priority according to their status. Second, the wafer lots with higher priority are assigned to vehicles according to dispatching rules. Timeliness constraints between processing operations should be considered seriously because Intrabay systems are directly connected with the processing areas. At the same time, the optimization objective in the Intrabay systems focuses on the performances of the processing machines, such as the wafer lot's average waiting time and average delivery time, which are closely related to the important performance indexes of semiconductor wafer fabrication systems (SWFSs), i.e., wafer lot's throughput and cycle time. In addition, similar to Interbay system scheduling problems, the Intrabay scheduling systems have a strong dynamic randomness, temporary congestions and deadlocks which should be taken into consideration.

7.1 Intrabay Automatic Material Handling Scheduling Problems

An Intrabay material handling system typically consists of transportation rails, load port for stockers and tools, and automatic vehicles (as shown in Figure 7.1). The automotive vehicles are guided by the transportation rail, which is typically a monorail system. Along the rail, automotive vehicles can transport wafer lots and interface with the load ports of stockers and tools. Automatic vehicles are adopted to move wafer lots. In this chapter, overhead hoist transports (OHTs) are selected as the automatic vehicle with which to investigate the vehicle dispatching problem in Intrabay material handling systems in a 300 mm SWFS.

When the Intrabay material handling system is running, wafers are fed into the stockers at random time and then transported to the processing machines. After all the processes are completed, the wafers are returned to the stockers leaving the Intrabay material handling system. The automatic vehicles are continuously transported in a one-way sequence on the transportation rails or respond to requests from load ports of stockers and processing machines.

https://doi.org/10.1515/9783110487473-007

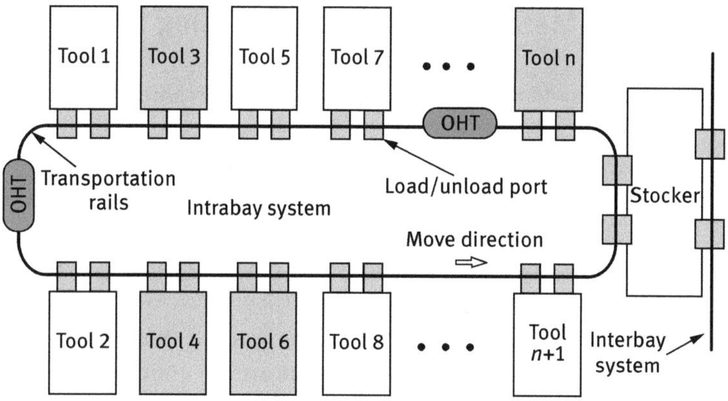

Figure. 7.1: The layout of an OHT-based Intrabay material handling system.

The wafer carriers waiting for the load port are transported to the next processing machine or stocker. Therefore, the Intrabay material handling scheduling problem can be described as the process of allocating the empty automatic vehicles to transport wafers in the load ports of stockers or processing machines. The Intrabay material handling scheduling problem has the following constraints:

1. Congestion constraints: In the Intrabay material handling system, automatic vehicles are continuously transported in a one-way sequence on the transportation rails. They cannot cross each other, and the downstream automatic vehicles parking for loading/unloading easily lead to the temporary blockage of upstream automatic vehicles.
2. Inter-process time constraints: Wafer waiting times between processing operations have timeliness restrictions (wafers in the last processing machine to complete the processing operation are required in the specified time to enter the next processing machine for processing, which is known as timeliness).
3. Deadlock constraints: A deadlock exists in the Intrabay material handling system (one wafer is jammed by another wafer waiting to be handled at the unloading port of its target processing machine, which causes the wafer to be unloaded after being transported by automatic vehicle to the target processing machine).

It is similar to the Interbay material handling scheduling problem. In this chapter, a greedy dynamic priority (GDP)-based Intrabay material handling scheduling method [1] and a pull and push strategy-based Intrabay material handling scheduling method [2, 3] are presented.

7.2 GDP-Based Intrabay Material Handling Scheduling Method

7.2.1 Formulation of the Intrabay Material Handling Scheduling Problem

In the Intrabay material handling system, the mathematical model is established with the objectives of minimizing normalized wafer lot's delivery time, wafer lot's total delivery time, cycle time, WIP quantity, and maximizing transport amount, vehicle utilization, wafer lot's throughput, wafer lot's due date satisfactory rates.

1. Notations

T_u – Smallest time unit in the material handling process (vehicle speed, loading/unloading time and wafer delivery request arrival time are integer multiples of T_u);

T – Cycle time period in the Intrabay system, $T \in \{T_u, 2T_u, 3T_u, ...\}$;

w – Within T, the quantity of wafers waiting for transportation in the Intrabay system;

q – Within T, the quantity of wafer lots entering the Intrabay system;

l – Within T, the quantity of wafer lots leaving the Intrabay system after finishing all processing operations;

n – Within T, the quantity of wafers delivered by vehicles;

m – The quantity of vehicles in the Intrabay system;

Y – Set of tracks and shortcuts in the Intrabay system;

S – Set of stockers in the Intrabay system;

e_{yj}^t – $e_{yj}^t = 1$ denotes that vehicle j is located at track or shortcuts y at time t, $y \in \{1, 2, ..., |Y|\}$;

α_u – Weighted coefficient of System performance indexes, $\alpha_u \in (0, 1]$, $u = 1, ..., 8$;

PD^s – Set of pick-up/drop-off events at stocker s, $PD^s = \{..., (k, i), ...\}$, $s \in S$;

$Pr(i)$ – Product type of wafer i;

$De(i)$ – Position of destination stocker or processing machine of wafer i;

$Re(i)$ – Position of stocker or processing machine of the waiting wafer i;

$T_{\max}^d$ – Maximum time in which a vehicle can transport a wafer lot;

$T_{\max}^t$ – Maximum value of wafer lots' total transportation time;

$T_{\max}^C$ – Maximum value of wafer lots' total cycle times;

$L_{\max}$ – Maximum value of wafer lots' throughput;

T_W^i – Maximum interval time between successive processing steps of wafer lot i;

T_D^i – The shortest delivery time for wafer lot i from the current stocker to the destination stocker;

E_j^t – Location of vehicle j at moment t;

F^i – Due date relaxation coefficient of wafer lot i;

T_p^i – Current processing time of wafer lot i;

h – The time that the vehicle spends in loading/unloading a wafer lot;

t_r^i – The moment that wafer lot i is released to the Intrabay system, t_r^i subject to certain probability distribution;

t_w^i– The moment that wafer lot i sends out a transportation request;

t_p^i – The moment that wafer lot i is loaded;

t_d^i – The moment that wafer lot i is unloaded;

t_{out}^i – The moment that wafer lot i leaves the Intrabay system after finishing all processing operations.

2. Decision variables

   X_{ij}^t – A 0–1 decision variable where X_{ij}^t is equal to 1, if wafer lot i is transported by vehicle j at moment t; otherwise, X_{ij}^t is equal to 0.

3. Objective function:

$$Obj = Min\left(\alpha_1 \frac{\sum_{i=1}^{n} (t_d^i - t_p^i)}{n \times T_{\max}^d} + \alpha_2 \frac{\sum_{i=1}^{n} (t_d^i - t_w^i)}{n \times T_{\max}^t} + \alpha_3 \frac{(w-n)}{w} + \alpha_4 \frac{\sum_{i=1}^{n} (t_d^i - t_p^i + 2h)}{m \times T} \right.$$
$$\left. + \alpha_5 \frac{(L_{\max} - l)}{L_{\max}} + \alpha_6 \frac{\sum_{v=1}^{l} (t_{out}^v - t_r^v)}{l{\star}T_{\max}^C} + \alpha_7 \frac{(q-p)}{q} + \frac{\alpha_8}{l} \sum_{v=1}^{l} \frac{(t_{out}^v - t_r^v)}{F^v \times T_p^v} \right) \tag{7.1}$$

4. Constraints:

$$\sum_{j=1}^{m} X_{ij}^t = 1, \quad X_{ij}^t \in \{0,\ 1\},\ \forall t \in [0, T], \quad i = 1, \ldots, n \tag{7.2}$$

$$\sum_{i=1}^{n} X_{ij}^t = 1, \qquad X_{ij}^t \in \{0,\ 1\},\ \forall t \in [0, T],\ j = 1, \ldots, m \tag{7.3}$$

$$t_d^k X_{kj}^{t_d^k} + h + 1 \le t_p^i X_{ij}^{t_p^i} \tag{7.4}$$

$$t_p^i X_{ij}^{t_p^i} + T_D^i + h \le t_d^i X_{ij}^{t_d^i} \tag{7.5}$$

$$t_d^i - t_w^i + h \le T_W^i \tag{7.6}$$

$$t_p^k X_{kj}^{t_p^k} + h \le t_d^i X_{ir}^{t_d^i}\ ,\ \forall\ Re(k) = De(i), \quad (k, i) \in PD^s \tag{7.7}$$

$$t_d^k X_{kj}^{t_d^k} + T_p^k \le t_p^i X_{ir}^{t_p^i},\ \forall\ Pr(k) = Pr(i),\quad (k,i) \in DP^s \tag{7.8}$$

$$t_r^i < t_w^i < t_p^i \tag{7.9}$$

$$E_1^t < E_2^t < \cdots < E_m^t \tag{7.10}$$

$$\sum_{j=1}^{m} e_{yj}^t \le 1\ ,\ \forall\ t \in [0, T],\ \forall\ y \in Y \tag{7.11}$$

where formula (7.1) is the objective function considering normalized wafer lot's delivery time, wafer lot's total delivery time, transport amount, vehicle utilization, wafer's throughput, cycle time, WIP quantity and wafer lot's due date satisfactory rate; formula (7.2) ensures that each wafer lot can be transported by only one vehicle; formula (7.3) makes sure that a vehicle can only transport a wafer lot at a time; formula (7.4) ensures that the vehicle can start next transportation task unless it has completed current task; formula (7.5) makes sure that the loading and unloading times of the same wafer lot must satisfy certain constraints; formula (7.6) makes sure that a wafer lot needs to meet the timeliness constraints between processes; formula (7.7) ensures that the loading and unloading times of a wafer lot at a processing machine or stocker must satisfy certain constraints; formula (7.9) makes sure that loading times of different wafer lots in the same processing machine must meet chronological constraints; formula (7.10) ensures the location constraints between vehicles; formula (7.11) makes sure that each track, shortcut or turntable can only carry one vehicle at any time.

7.2.2 Architecture of the GDP-Based Intrabay Material Handling Scheduling Method

The architecture of the proposed GDP-based Intrabay material handling scheduling method is shown in Figure 7.2. First, with the advantages of rapid calculating and dynamic identifying of fuzzy set theory, the wafer lots' dynamic priority decision-making model is developed. Second, a simplified Hungarian method is constructed to assign OHT vehicles to the wafer lots with higher priority in order to reduce the temporary blocking of automatic vehicles during wafer lot

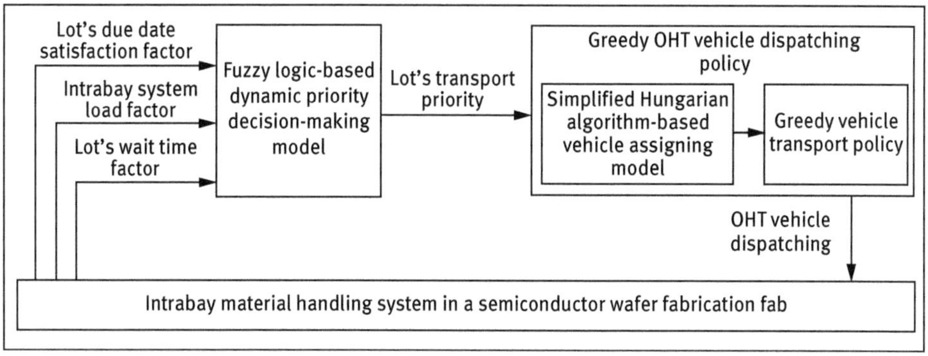

Figure 7.2: Architecture of the GDP-based Intrabay material handling scheduling method.

handling. Finally, according to the material handling optimization assignment results, a greedy vehicle transport policy is proposed to use automatic vehicles to transport wafer lots quickly and efficiently in order to avoid the deadlock in the Intrabay material handling process.

7.2.3 Fuzzy Logic-Based Dynamic Priority Decision-Making Model

In this section, the membership functions of the input and output variables are defined by using a triangular membership function. The fuzzy inference policy is determined by the Mamdani inference. And the aggregation policy is determined by the Mini-max rule. The maximal membership function approach is taken as the defuzzified policy. The architecture of the proposed fuzzy logic-based [4–6] dynamic priority decision-making model [7] is shown in Figure 7.3.

1. Input variables

   The input variables of the fuzzy logic-based dynamic priority decision-making model consist of the Intrabay system load factor, the lot's due date satisfaction factor and the lot's wait time factor. Detailed descriptions of these input variables are presented as follows.

   (1) Lot's due date satisfaction factor (DDSF)

   The DDSF is used to measure the current lot's de date satisfaction. The DDSF is calculated by eq. (7.12):

$$DDSF = \frac{RT_i}{RP_i \times Fac} \tag{7.12}$$

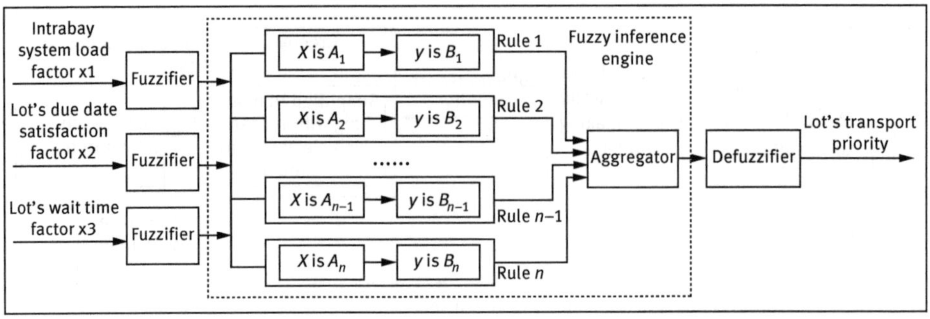

Figure 7.3: The architecture of the fuzzy logic-based dynamic priority decision-making model.

$$Fac = \sum_{i=1}^{n} \frac{(Cycle\ \ time)_i}{PT_i} \times \frac{1}{n} \tag{7.13}$$

where Fac is the average value of the historical records for each type lot, and Fac is computed as in eq. (7.13); RT_i is the remaining time to the due date of lot i; PT_i is the overall processing time of lot i; RP_i is the remaining processing time of lot i; and n is the number of historical records for the current type of lot.

(2) Lot's waiting time factor (WTF)
The WTF is adopted to measure the current wafer lot's waiting time. WTF is defined as

$$WTF = \frac{CW_i}{AW} \tag{7.14}$$

where AW is the average waiting time of the current waiting lot, and AW is calculated as $AW = \sum_{i=1}^{n} CW_i \times \frac{1}{n}$. CW_i is the current waiting time for lot i, and n is the number of historical records for the current type of lot.

(3) System load factor (SLF)
The SLF is used to measure the current Intrabay system load ratio. A higher SLF value of the Intrabay system indicates a heavy system load. The SLF is calculated by

$$SLF = \frac{NC}{NV} \tag{7.15}$$

where NC is the current number of lots waiting for moving in the Intrabay material handling system and NV is the number of OHTs in the Intrabay material handling system.

2. Membership functions and the rule table

   The input variables include DDSF, WTF and SLF. The DDSF can be categorized into bad satisfaction (BS), normal satisfaction (NS) and good satisfaction (GS). The membership function of DDSF is shown in Figure 7.4(a). The WTF can be categorized into not long (NL), medium long (ML) and very long (VL). The membership function of the WTF is shown in Figure 7.4(b). The SLF can be categorized into not heavy (NH), medium heavy (MH) and very heavy (VH). The membership function of the WTF is shown in Figure 7.4(c).

   The lot's transport priority is the output variable of the fuzzy logic-based dynamic priority decision-making model.

   The transport priority can be classified as most high priority (MHP), very high priority (VHP), high priority (HP), low priority (LP) and very low priority (VLP), which indicate levels 9, 8, 7, 6 and 5, respectively. The membership function of the output variable is shown in Figure 7.4(d). Table 7.1 lists the membership functions of three input variables and one output variable.

   The fuzzy rule table is developed as shown in Table 7.2 in accordance with the simulation experiences of OHT dispatching and the definitions of the variables DDSF, WTF and SLF.

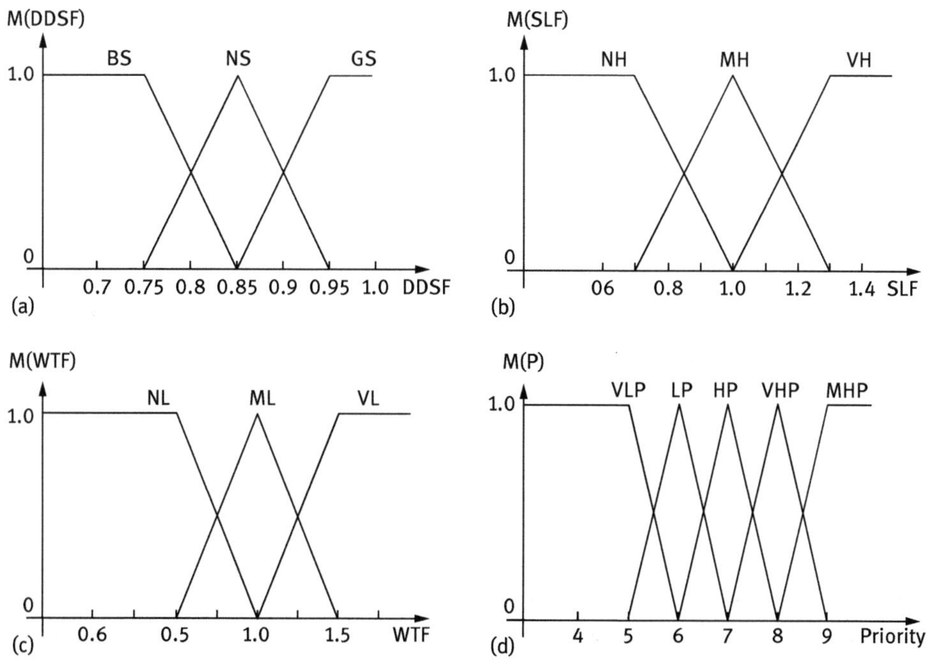

Figure 7.4: (**a**) Membership function of the DDSF. (**b**) Membership function of the SLF. (**c**) Membership function of the WTF. (**d**) Membership function of the lot's transport priority.

Table 7.1: The membership function of the input variables of weight modification model.

Variables	Fuzzy sets (linguistic descriptions)	Fuzzy set membership functions
Lot's due date satisfaction factor (DDSF)	Bad satisfaction (BS)	$(-\infty, 0.0, 0.75, 0.85)$
	Normal satisfaction (NS)	$(0.75, 0.85, 0.95)$
	Good satisfaction (GS)	$(0.85, 0.95, 1.0, +\infty)$
System load factor (SLF)	Not heavy (NH)	$(-\infty, 0.0, 0.5, 1.0)$
	Medium heavy (MH)	$(0.5, 1.0, 1.5)$
	Very heavy (VH)	$(1.0, 1.5, 2.0, +\infty)$
Lot's wait time factor (WTF)	Long (NL)	$(-\infty, 0.0, 0.7, 1.0)$
	Medium long (ML)	$(0.7, 1.0, 1.3)$
	Very long (VL)	$(1.0, 1.3, 2.0, +\infty)$
The lot's transport priority	Very low priority (VLP)	$(-\infty, 0.0, 5.0, 6.0)$
	Low priority (LP) and	$(5.0, 6.0, 7.0)$
	High priority (HP),	$(6.0, 7.0, 8.0)$
	Very high priority (VHP)	$(7.0, 8.0, 9.0)$
	Most high priority (MHP)	$(8.0, 9.0, 10.0, +\infty)$

Table 7.2: Fuzzy rule table.

Order	Rule ID				
	SLF	DDSF	WTF		
			NL	ML	VL
1	VH	BS	VLP	VLP	VLP
2		NS	VLP	VLP	VLP
3		GS	VLP	VLP	MHP
4	MH	BS	VHP	VHP	MHP
5		NS	HP	VHP	MHP
6		GS	LP	LP	MHP
7	NH	BS	MHP	MHP	MHP
8		NS	HP	HP	MHP
9		GS	LP	LP	MHP

7.2.4 Hungarian Algorithm-Based Vehicle Dispatching Approach

In the previous section, the dynamic priorities of the wafer lots waiting for transportation are computed. In this section, a Hungarian algorithm-based vehicle dispatching approach [8] is developed to allocate wafer lots to automatic vehicles. First, all wafer lots waiting for transportation are sorted by priority in descending order and a certain number of wafer lots with higher priority are selected as the candidate jobs. If the number of wafer lots waiting for transportation *Nw* is more than the number of available vehicles Nv, then the first Nv wafer lots with higher priority are selected as the candidate wafer lots. Otherwise, if $Nw \leq Nv$, then all Nw wafer lots are selected as candidate wafer lots. Second, the Hungarian algorithm is used to assign the available vehicles to the candidate wafer lots to find the global optimal vehicle dispatching solutions. Third, after the optimal dispatching solutions are found, the available OHT vehicles begin to execute the dispatching solutions according to the greedy vehicle transport policy, which can reduce OHT vehicles' waste movement effectively.

In the Intrabay material handling system, the problem of dispatching available vehicles can be formulated as the following mathematical problem. Assume that there are m wafer lots to be assigned to n available vehicles at any one time. The available vehicles dispatching problem can be formulated as follows:

$$\text{Minimise} \sum_{i=1}^{n}\sum_{j=1}^{m} C_{ij}X_{ij} \tag{7.16}$$

subject to

$$\sum_{i=1}^{n} X_{ij} = 1, \text{ for } i = 1, 2, \ldots, n, \qquad \sum_{j=1}^{m} X_{ij} = 1, \text{ for } j = 1, 2, \ldots, m \tag{7.17}$$

where $X_{ij} \in \{0, \ 1\}$, and for all i and j, C_{ij} is the travel time of available vehicle j to wafer lot i. The cost matrix C_0 is described as

$$C_0 = \begin{bmatrix} C_{11} & C_{12} & \ldots & C_{1n} \\ C_{21} & C_{22} & \ldots & C_{2m} \\ \ldots & \ldots & \ldots & \ldots \\ C_{m1} & C_{m2} & \ldots & C_{mn} \end{bmatrix} \tag{7.18}$$

The solution procedure using the simplified Hungarian method is described as follows.

Step1: For the original cost matrix C_0, if $m < n$, then dummy candidate jobs are added to make the same number. Then the average solution μ is measured as $\mu = \frac{1}{n}\sum_{i=1}^{n} C_{ii}$.

Step2: The variance σ_{ij}^2 of each cell in the original cost matrix C_0 is calculated to form a new cost matrix C_1. The variance σ_{ij}^2 is measured as $\sigma_{ij}^2 = (C_{ij} - \mu)^2$.

Step 3: For each row in the original cost matrix C_1, the minimum number is subtracted from all numbers. Then a reduced cost matrix C_2 is formed.

Step 4: For each column in the reduced cost matrix C_2, the minimum number is subtracted from all numbers. Then a reduced cost matrix C_3 is formed.

Step 5: The minimum number of lines is drawn to cover all zeros in the cost matrix C_3. If this number is equal to n (size of the cost matrix), then stop the process: the optimal solution is obtained. Otherwise, go to Step 4.

Step 6: The minimum uncovered number d is determined. Then it is calculated as follows: (1) d is subtracted from the uncovered numbers; (2) d is added to the numbers covered by two lines and (3) the numbers covered by one line are the same as before. As a result, a new reduced cost matrix C_3 is formed. Then go to Step 5.

7.2.5 The Greedy Optimization-Based Vehicle Scheduling Strategy

The greedy vehicle scheduling strategy is developed on the basis of the push–pull approach, in which the vehicle is selected according to the states of the current wafer processing machine and the next process machine (i.e., the target processing machine). The procedure of the greedy vehicle scheduling strategy is given in Figure 7.5.

When the optimal vehicle dispatching solution is yielded based on the Hungarian algorithm, the OHT vehicle begins to execute the solution. First, the OHT vehicle moves

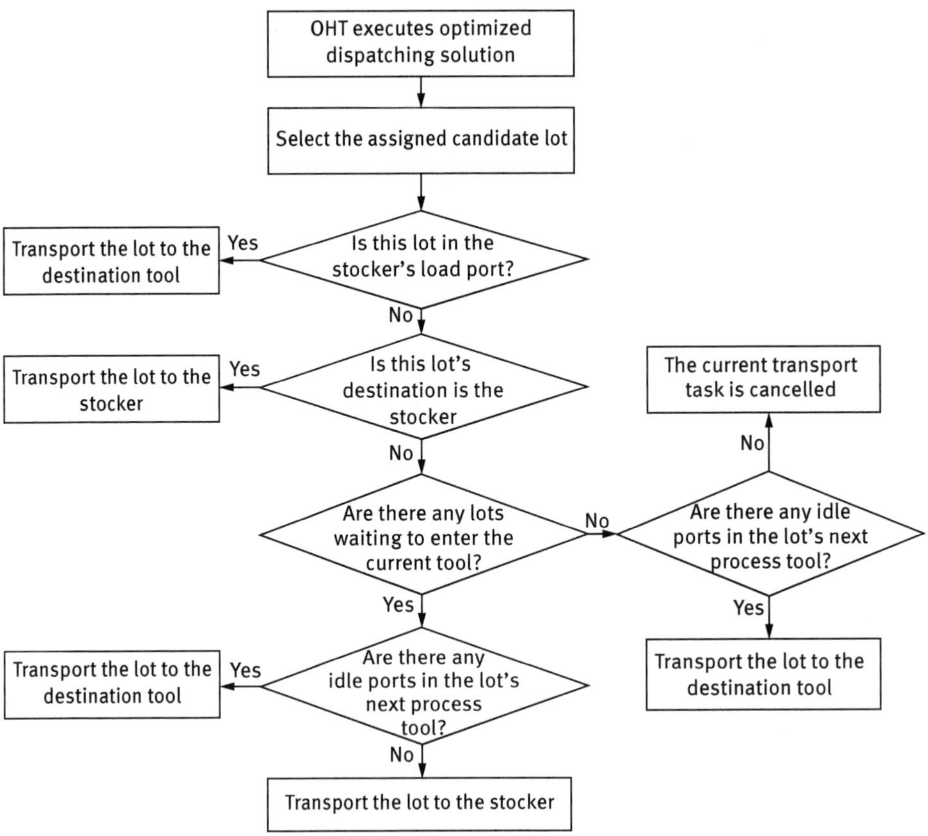

Figure 7.5: The procedure of the greedy OHT vehicle scheduling strategy in the Intrabay material handling system.

to the original tool where the assigned candidate lot is located. Then a look-ahead mechanism is used to check whether there is any lot waiting to enter the current original tool. If there are some, then a push mechanism-based vehicle transport policy is executed. Otherwise, a pull mechanism-based vehicle transport policy is executed. In the push mechanism-based vehicle transport policy, the state of the next processing tool of the current assigned candidate lot is evaluated. If there is any idle port in the next processing tool, then the current assigned candidate lot is transported to the destination tool's load port. Otherwise, if there is no idle port in the next processing tool, then the current assigned candidate lot is transported to the stocker. On the other hand, in the pull mechanism-based vehicle transport policy, the state of the next processing tool of the current assigned candidate lot is also evaluated. If there is any idle port in the next processing tool, then the current assigned candidate lot is transported to the destination tool's load port. Otherwise, if there is no idle port in the next processing tool, the current assigned candidate lot is held in the original tool. Finally, if the optimal vehicle dispatching solution is accomplished or a new empty OHT vehicle arrives, the GDP dispatching policy is reassigned again.

7.2.6 Simulation Experiments

In this section, the effectiveness of the proposed approach is demonstrated by using simulation experiments on eM-Plant. The Intrabay data are collected from a 300 mm SWFS. In this study, there are 2 stockers, 28 tools and a 108 m Intrabay rail loop. The average loading/unloading operation time of the OHT vehicles is 16 seconds. The speed of the OHT vehicles is 1.8 m/s. The number of OHTs can be 5, 6 and 7 in the Intrabay system. The system loading ratios can be 90%, 100% and 110% of the Intrabay system. Each Intrabay experiment was run for 15 days with a warm-up of two days.

The proposed approach is compared with the traditional first encounter first served (FEFS) rule and the highest priority lot first rule (HP) in order to evaluate its effectiveness. The experimental results of the GDP approach are summarized in Tables 7.3 and 7.4. The results demonstrate that the GDP approach outperforms the FEFS rule- and HP rule-based dispatching approaches.

A desirability function $D(X)$ is used to compute the comprehensive performance indexes in order to further verify the effectiveness of the proposed approach. The desirability function is a geometric mean of all performance indexes:

$$D = \left(\prod_{i=1}^{n} d_i \right)^{1/n} \tag{7.19}$$

where d_i is the individual desirability function of the performance index and n is the number of performance indexes of the automatic material handling system and the SWFS. In this study, all performance indexes are treated as having the same weight.

Table 7.3: Comparison of the GDP approach and FEFS rule.

Scenario		Dispatching policy	Movement (lot)		Delivery time (s)		Transport time (s)		OHT utilization (%)	
System loading	OHT number		Mean value	Improved (%)	Mean value	Improved (%)	Mean value	Improved (%)	Mean value	Improved (%)
90%	5	FEFS	64,218	8.9	117.0	4.0	50.5	2.2	49.51	10.7
		HP	59,479	1.7	125.5	10.5	49.4	0.0	44.74	1.2
		GDP	58,478	/	112.3	/	49.4	/	44.20	/
	6	FEFS	63,310	7.6	106.7	4.4	51.1	1.6	41.15	9.2
		HP	59,481	1.6	105.4	3.3	50.4	0.2	38.48	2.9
		GDP	58,478	/	101.9	/	50.3	/	37.37	/
	7	FEFS	62,926	7.0	102.0	4.5	51.6	1.7	35.45	9.1
		HP	58,482	0	99.3	1.9	51.0	0.6	32.51	0.9
		GDP	58,481	/	97.4	/	50.7	/	32.23	/
100%	5	FEFS	64,868	0	120.4	5.9	51.3	1.6	50.74	1.3
		HP	64,868	0	121.7	6.9	51.0	1.0	50.43	0.7
		GDP	64,849	/	113.3	/	50.5	/	50.08	/
	6	FEFS	64,868	0	109.5	2.5	51.9	1.7	42.79	2.9
		HP	64,871	0	108.6	1.7	51.3	0.6	42.37	3.9
		GDP	64,857	/	106.8	/	51.0	/	42.22	/
	7	FEFS	64,866	0	104.4	2.3	52.8	4.0	37.05	0.3
		HP	64,871	0	105.0	2.9	52.5	3.4	37.42	0.4
		GDP	64,853	/	102.0	/	50.7	/	35.96	/

110%	5	FEFS	69,905	0.6	142.6	5.3	53.1	0.2	56.66	0.7
		HP	70,280	1.1	151.0	10.5	53.0	0	56.95	1.2
		GDP	69,529	/	135.0	/	53.0	/	56.25	/
	6	FEFS	69,729	0.2	125.9	1.5	53.3	0.4	47.26	0.2
		HP	69,953	0.6	133.3	7.0	53.4	0.6	47.36	0.4
		GDP	69,598	/	124.0	/	53.1	/	47.17	/
	7	FEFS	70,031	0.6	127.6	11.9	55.1	0.2	41.44	0.5
		HP	69,627	0.1	123.5	9.0	54.7	1.1	41.29	0.1
		GDP	69,579	/	112.4	/	54.1	/	41.23	/
Total average		FEFS	66,080	2.7	117.3	4.8	52.3	1.7	44.67	3.8
		HP	64,657	0.6	119.3	6.4	51.9	1.0	43.51	1.2
		GDP	64,300	/	111.7	/	51.4	/	42.97	/

Table 7.4: Comparison of the GDP approach and FEFS rule.

Scenario		Dispatching policy	Cycle time (s)		Throughput (lot)		Due date satisfaction (%)		WIP in Intrabay (lot)	
System loading (%)	OHT number		Mean value	Improved (%)	Mean value	Improved (%)	Mean value	Improved (%)	Mean value	Improved (%)
90	5	FEFS	2,657.8	3.9	13,658	0	98.63	1.4	30	3.3
		HP	2,622.6	2.6	13,657	0	98.72	1.3	29	0
		GDP	2,553.7	/	13,658	/	100.0	/	29	/
	6	FEFS	2,561.8	2.8	13,658	0	98.63	1.4	29	3.4
		HP	2,503.0	0.6	13,659	0	98.72	1.3	28	0
		GDP	2,488.4	/	13,659	/	100.0	/	28	/
	7	FEFS	2,523.7	2.6	13,657	0	98.63	1.4	28	0
		HP	2,467.7	0.4	13,659	0	98.72	1.3	28	0
		GDP	2,458.9	/	13,659	/	100.0	/	28	/
100	5	FEFS	3,178.4	2.2	15,152	0	98.42	1.6	39	2.6
		HP	3,165.0	1.8	15,152	0	98.52	1.5	39	2.6
		GDP	3,108.4	/	15,152	/	100.0	/	38	/
	6	FEFS	3,116.9	1.0	15,152	0	98.42	1.6	38	0
		HP	3,096.5	0.4	15,152	0	98.52	1.5	38	0
		GDP	3,084.2	/	15,152	/	100.0	/	38	/
	7	FEFS	3,081.5	0.7	15,152	0	98.42	1.6	38	0
		HP	3,079.2	0.6	15,152	0	98.52	1.5	38	0
		GDP	3,058.8	/	15,152	/	100.0	/	38	/

110	5	FEFS	25,496.0	25.5	16,100	1.0	94.34	5.2	340	23.2
		HP	25,405.0	25.2	16,206	0.4	94.67	4.9	290	10
		GDP	18,988.0	/	16,263	/	99.29	/	261	/
	6	FEFS	24,824.0	25.6	16,124	0.9	94.29	3.0	331	23.3
		HP	24,651.0	25.1	16,100	1.1	94.58	2.7	342	25.6
		GDP	18,458.7	/	16,276	/	97.15	/	254	/
	7	FEFS	24,381.0	24.9	16,108	1.1	94.27	5.6	339	25.6
		HP	24,423.0	25.0	16,100	1.1	94.63	5.2	339	25.7
		GDP	18,296.0	/	16,278	/	99.58	/	252	/
Total Average		FEFS	10,202.3	21.0	14,973	0.4	97.12	8.8	135	20.7
		HP	10,157.0	20.7	14,982	0.3	97.29	9.0	130	17.7
		GDP	8,055.0	/	15,028	/	88.56	/	107	/

Using the desirability function $D(X)$, the desirability values of the FEFS rule, HP rule and proposed GDP approach in different scenarios are calculated, as shown in Figure 7.6. A statistical test of desirability values using analysis of variance (ANOVA) is executed, as shown in Table 7.5. The F value of 26.72 in Tab. 7.5 indicates that there are significant differences between the FEFS rule and the proposed GDP approach. The F value of 16.22 in Table 7.5 means that there are significant differences between the HP rule and the proposed approach. The comparison of the desirability values of the proposed GDP approach, FEFS rule and HP rule in Figure 7.6 illustrates that the proposed GDP approach clearly outperforms the other two rules in terms of comprehensive performance. (In Figure 7.6, L indicates system loading ratios; e.g., L90 means 90% loadings in the Intrabay system. V indicates the OHT vehicle number; e.g., V5 means five OHT vehicles in the Intrabay system.)

In all experimental scenarios, the computation time of the proposed GDP approach is more than the FEFS rule- and HP rule-based OHT vehicle dispatching

Table 7.5: ANOVA of the FEFS rule and the HP rule with GDP dispatching approach.

Compared rules	Source	Sum of squares (SS)	Degrees of freedom (DF)	Mean squares (MS)	*F*-value	Prob.>*F*
FEFS and GDP	Columns	0.20895	1	0.20895	26.72	<0.001
	Error	0.12512	16	0.00782	–	–
	Total	0.33408	17	–	–	–
HP and GDP	Columns	0.12366	1	0.12366	16.22	<0.001
	Error	0.122	16	0.00762	–	–
	Total	0.24566	17	–	–	–

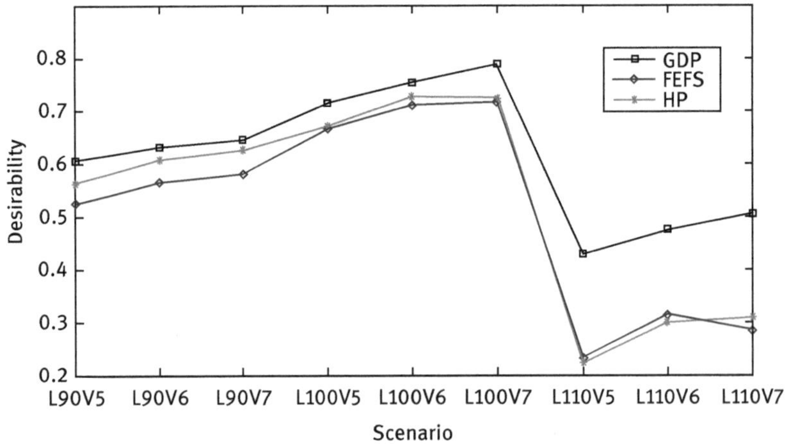

Figure 7.6: Comparison of the desirability values of the GDP policy, FEFS rule and HP rule.

methods. However, the maximal computation times in all experimental scenarios are less than 1.0 seconds. This means that the proposed GDP approach is efficient enough for real-time OHT vehicle dispatching.

The reasons why the proposed approach has better comprehensive performance are briefly analyzed as follows. (1) The main performance indices of the Intrabay material handling system are thoroughly considered and used as the input variables in the dynamic priority decision-making model. Those wafer lots with worse performances can be assigned higher transport priority and transported first. (2) The Hungarian algorithm is a rapid and effective approach to find the optimal or suboptimal OHT vehicle dispatching solution. The vehicle blockage phenomenon in the Intrabay material handling system can be eliminated effectively. (3) The transportation deadlock phenomenon can be solved and the OHT vehicle's waste movement can be reduced based on the greedy vehicle transport policy. Therefore, the comprehensive performance indices of both the Intrabay material handling system and the wafer manufacturing system could be improved effectively.

7.3 Pull and Push Strategy-Based Intrabay Material Handling Scheduling Method

Figure 7.7 depicts the layout of an Intrabay material handling system in a 300 mm wafer fab, which contains a stocker, multiple bypasses, a single-loop Intrabay material handling system and multiple shortcuts. In general, the vehicles are the OHTs, and the tools are steppers and scanners. The information of these tools is summarized in Table 7.6.

In this study, assume that the input wafers in this system are 20,000 per month; the mix products are 70% Logic IC (products X and L) and 30% DRAM (products D and F). The inter-arrival times for each product in this Intrabay material handling system are summarized in Tables 7.7 and 7.8.

7.3.1 Architecture of a Pull and Push Strategy-Based Intrabay Material Handling Scheduling Method

The push strategy in the Intrabay material handling scheduling process is that vehicles' transportation strategy is adjusted according to wafer lots' status (i.e., vehicles search lots (VSL)). On the contrary, the pull strategy is that wafer lots dynamically select vehicles according to their status (i.e., lots' search vehicles (LSV)). Neither push strategy nor pull strategy can maximize the performance of the Intrabay material handling system. Using either of the two strategies alone, there is a situation in which vehicles or wafer lots are idle or waiting. Therefore, a pull/push strategy is introduced so as to improve the efficiency of the Intrabay material handling system. Two strategies are

Table 7.6: The tools' information.

	Type	Max batch size	Load ports	Quantity	Average lot processing time (s)	Average throughput per hour (wafers)
Scanner	KxxN	1	4	11	1,343	67
	PxxK	1	4	1	2,903	31
	BxxK	1	4	2	2,000	45
Stepper	AxxA	1	4	6	400	225
	AxxI	1	2	4	514	175
	AxxD	1	2	4	400	225
	MxxO	1	2	1	600	150

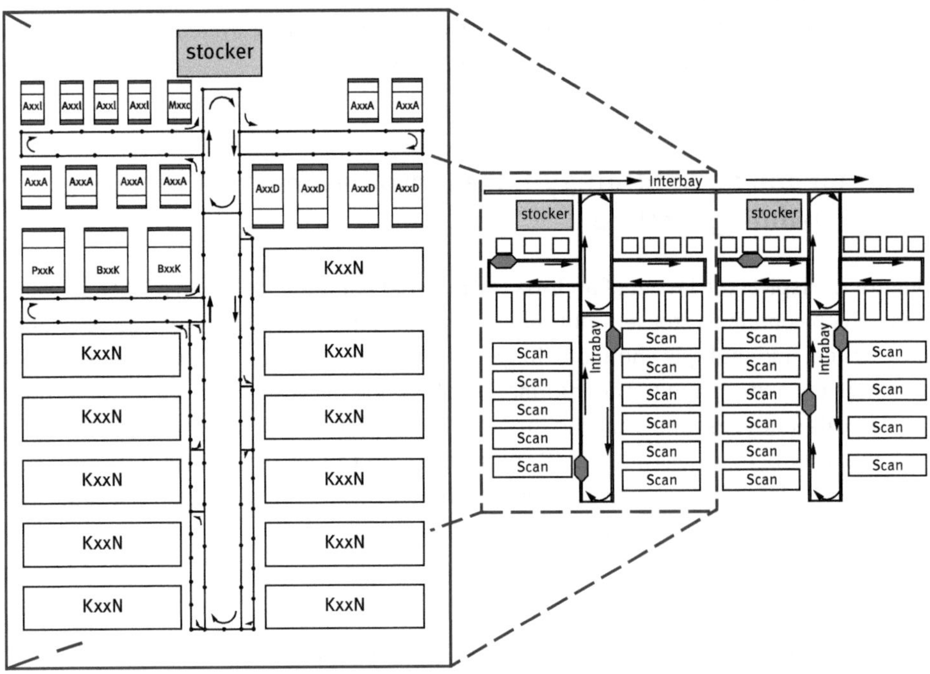

Figure 7.7: The layout of an Intrabay material handling system in a 300 mm wafer fab.

switched according to the real-time status of the Intrabay material handling system, as shown in Figure 7.8. Three types of indicators of the Intrabay system are used to verify the efficiency of the pull/push strategy in different scenarios. These indicators include production performance indicators (throughput, wafer lots' cycle times, WIP in the system and peak of WIP quantity), wafer lot performance indicators (due date, 95% due date and delivery time) and vehicle performance indicators (transportation amount and average vehicle utilization).

Table 7.7: The routings for logic IC in the Intrabay material handling system.

Product	Step 1	Step 2	Step 3	Step 4	Step 5	Flow count	Interarrival time (h/lot)
X	KxxN	AxxD	AxxA	AxxI		20	0.1
	BxxK	KxxN	AxxD	AxxA	AxxI	1	2.6
	AxxI					3	0.9
	PxxK	AxxI				1	2.6
	MxxO					5	0.5
L	KxxN	AxxD	AxxA	AxxI		26	0.1
	AxxD	AxxA	AxxI			6	0.4
	BxxK	AxxI				1	2.6
	PxxK	AxxI				1	2.6
	MxxO					7	0.4

Table 7.8: The routing for DRAM in the Intrabay material handling system.

Product	Step 1	Step 2	Step 3	Step 4	Step 5	Flow count	Interarrival time (h/lot)
D	MxxO					4	1.5
	BxxK	AxxI				1	6.0
	KxxN	AxxD	AxxA	AxxI		7	0.9
	AxxD	AxxA	AxxI			1	6.0
	AxxI					5	1.2
	KxxN	AxxI				8	0.8
	KxxN	AxxD	AxxA	AxxI		1	6.0
	BxxK	KxxN	AxxD	AxxA	AxxI	1	6.0
	KxxN	AxxD	AxxA	BxxK	AxxI	1	6.0
	MxxO	KxxN	AxxD	AxxA	AxxI	1	6.0
F	MxxO					1	6.0
	MxxO					2	3.0
	KxxN	AxxD	AxxA	AxxI		20	0.3
	AxxD	AxxA	AxxI			3	2.0
	BxxK	KxxN	AxxD	AxxA	AxxI	4	1.5
	PxxK	AxxI				1	6.0

7.3.2 VSL and LSV Dispatching Rules Based on Pull and Push Strategy

In general, the push strategy provides a production command according to the forecast of the production system and the inventory system. The pull strategy works together with Kanbans. The dispatching decisions for the AMHS are either the VSL or the LSV as shown in Figures 7.9 and 7.10, respectively. Using this hybrid

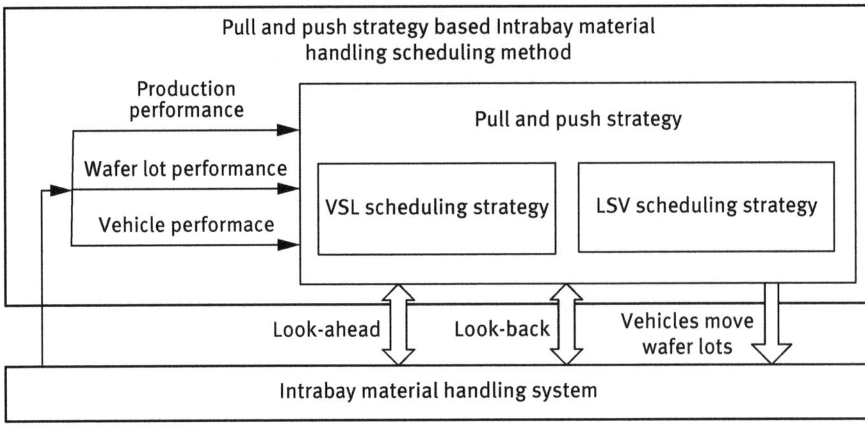

Figure 7.8: Architecture of the pull/push strategy-based Intrabay material handling scheduling method.

pull and push dispatch rule, the tool's utilization can be improved, and the stocker's WIP and waste movement can be reduced.

7.3.3 Simulation Experiments

7.3.3.1 Establishment of the Simulation Model

The system performance was evaluated by using a discrete-event simulation model in order to investigate the above vehicle scheduling mechanism. As shown in Figure 7.11, the input data of this AMHS system is the actual moving rate between tools. In this study, stocker's crane cycle times are generated by normal distribution $N(18, 5^2)$. The mix products are 70% Logic IC and 30% DRAM. The average load/unload times of the OHT is 20. The number of vehicles can be 2, 4 and 6. Each simulation ran 30 days with a warm-up period of two days. The arrival rate for the input wafers per month can be 17,000 or 26,000.

7.3.3.2 Analysis of the Simulation Results

The simulation results of the PP-NV-FEFS rule and the NV-FEFS rule are shown in Table 7.9. According to the simulation results, we can summarize and draw the following conclusions:

1. In scenario H-2, the PP-NV-FEFS rule increases the mean amount of the empty vehicle utilization and the mean amount of throughput, but decreases other performance measures. In scenarios H-4 and H-6, all performance measures excluding the throughput are improved by the PP-NV-FEFS rule.
2. For scenarios (L-2, L-4, L-6), all performance measures are not affected by the PP-NV-FEFS rule.

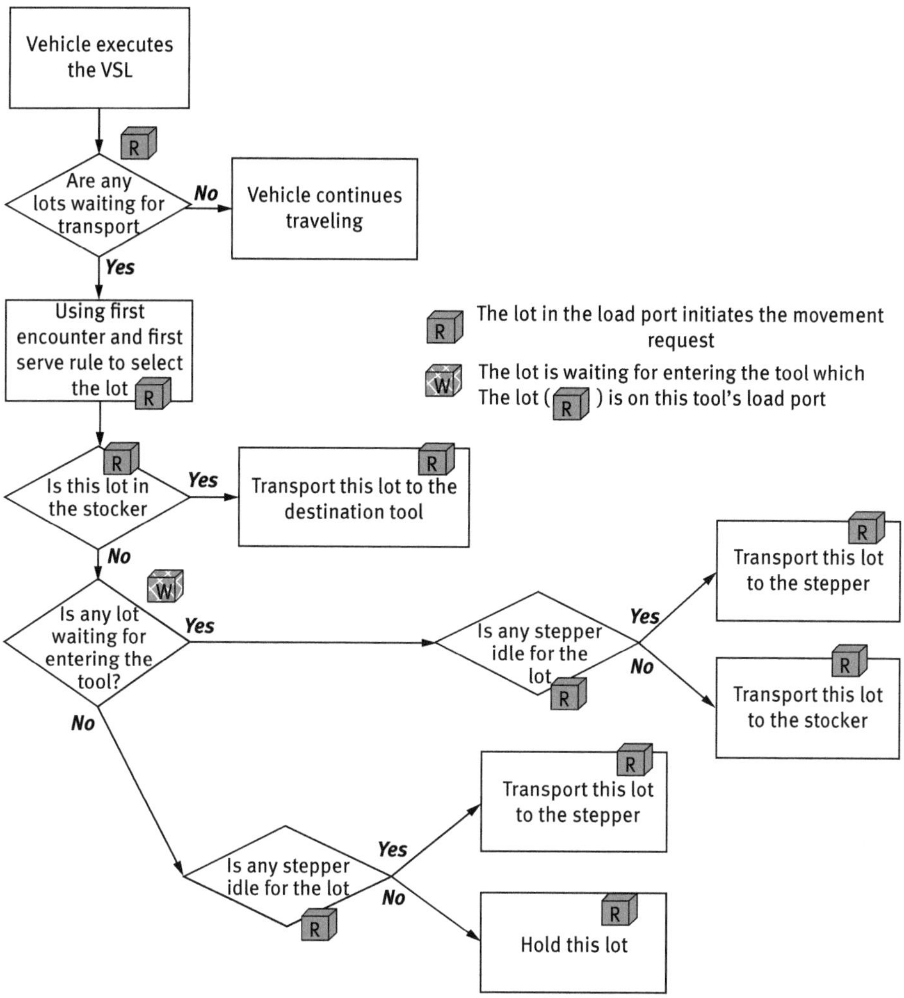

Figure 7.9: The procedure of the push/pull (PP) rule for the VSL.

7.4 Conclusion

The Intrabay material handling system is complex due to complicated mixes of products, dynamic transportation demands, vehicle blockages and transportation deadlocks. In this chapter, a GDP dispatching policy and a pull- and push-based Intrabay system scheduling method have been introduced. With experimental data from a 300 mm semiconductor wafer fabrication line, a simulation model of an Intrabay material handling system has been built. In comparison with the FEFS and HP rules, these two dispatching methods have shown good performance in obtaining a better OHT dispatching control solution for the Intrabay material handling system.

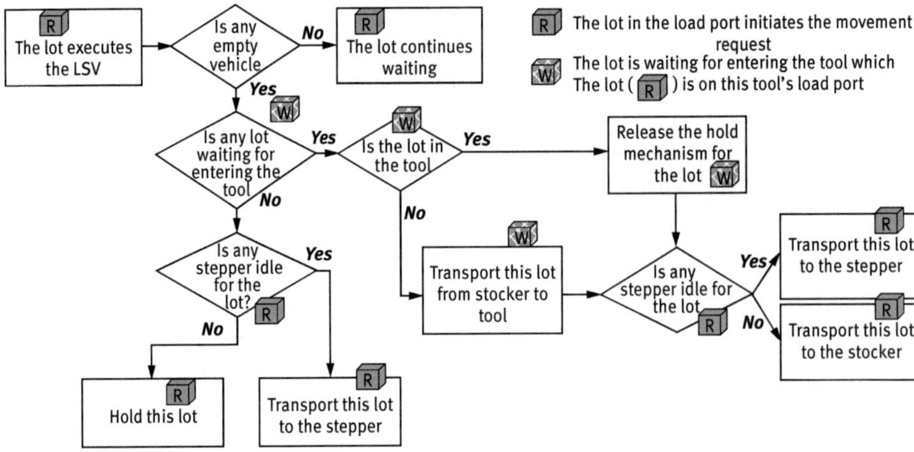

Figure 7.10: The procedure of the push/pull (PP) rule for the LSV.

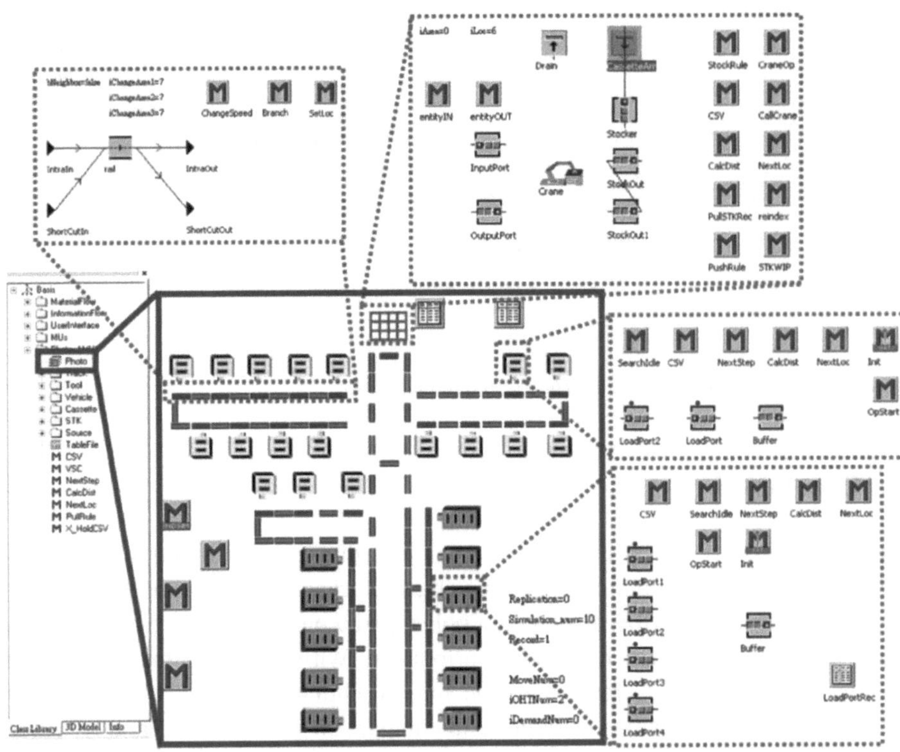

Figure 7.11: The simulation model of the Intrabay.

Table 7.9: The comparison of the PP rule and no PP rule.

Scenario		Throughput	Improved (%)	Wafer cycle time	Improved (%)	System WIP	Improved (%)	Stocker WIP	Improved (%)	Stocker peak WIP	Improved (%)	95% delivery time	Improved (%)	Delivery time	Improved (%)	Transport time	Improved (%)	Movement throughput	Improved (%)	Empty vehicle utilization	Improved (%)
H-2	pp	21,599	0.3	3,623.6	3.8	30.6	32.4	2	93.2	17.3	89.1	161.5	58	65.5	42	23.7	2.3	89,098	11.1	29.4	6.9
		(142)		(62)		(1.2)		(0.3)		(2.8)		(3.6)		(0.8)		(0.1)		(652)		(0.2)	
	xpp	21,484		3,764.9		67.3		38.6		159.4		385.5		112.3		24.2		100,220		27.5	
		(111)		(64)		(4.5)		(8.8)		(29.6)		(73.6)		(14.2)		(0.1)		(2781)		(0.2)	
H-4	pp	21,549	0	3,286.8	2.4	27.2	1.8	1.7	45.2	16.7	43.4	62.7	1.5	39.2	2.2	24.1	2.1	88,992	6.7	13.6	vsp: 1.0
		(129)		(65)		(0.5)		(0.2)		(1.8)		(1.1)		(0.1)		(0.1)		(359)		(0.1)	9.7
	xpp	21,511		3,367.8		27.8		3.2		29.5		63.6		40.1		24.6		95,414		12.4	
		(103)		(72)		(0.7)		(0.5)		(7.7)		(1.3)		(0.2)		(0.1)		(1004)		(0.1)	
H-6	pp	21,511	0	3,223.9	4.3	26.6	5.0	1.7	38.0	16.6	39.2	61.6	2.2	37.8	2.7	24.5	2.7	88,992	6.1	8.3	9.2
		(104)		(55)		(0.4)		(0.2)		(1.8)		(0.2)		(0.1)		(0.1)		(362)		(0)	
	xpp	21,511		3,363.1		27.9		2.8		27.3		63		38.9		25.2		94,794		7.6	
		(104)		(44)		(0.9)		(0.3)		(5.7)		(0.2)		(0.1)		(0.1)		(740)		(0)	
L-2	pp	13,984	0	2,621.5	0.3	14.2	0.7	0.1	0	6.2	7.5	81.1	0.5	44.8	2.2	22.9	0	57,832	0.1	17	0
		(151)		(20)		(0.3)		(0)		(0.9)		(1.1)		(0.3)		(0.1)		(541)		(0.2)	
	xpp	13,984		2,629		14.3		0.2		6.7		81.5		44.9		22.9		57,918		17	
		(151)		(17)		(0.5)		(0)		(1.1)		(1.1)		(0.3)		(0.1)		(551)		(0.2)	

(continued)

Table 7.9: (continued)

L-4	pp	13,945	0	2,557.8	0	13.7	0	0.1	0	6.3	3.1	57.3	0	36.2	0	23.1	0	57,553	0.1	7.2	0
		(149)		(18)		(0.3)		(0)		(0.8)		(0.1)		(0.1)		(0)		(672)		(0)	
	xpp	13,945		2,557.9		13.7		0.2		6,5		57.3		36.2		23.2		57,626		7.2	
		(149)		(17)		(0.3)		(0)		(0.7)		(0.1)		(0.1)		(0)		(680)		(0)	
L-6	pp	13,900	0	2,554.4	0	13.8	0	0.2	0	6.1	0	57.6	0	35.6	0	23.5	0	57,387	0.1	4.5	0
		(106)		(14)		(0.3)		(0)		(0.7)		(0.1)		(0)		(0)		(486)		(0)	
	xpp	13,900		2,552.7		13.7		0.2		6.1		57.6		35.6		23.5		57,387		4.5	
		(106)		(13)		(0.3)		(0)		(0.7)		(0.1)		(0)		(0)		(487)		(0)	

Note: H shows that the arrival rate of the input wafer is 26,000; L shows that the arrival rate of the input wafers is 17,000; xPP shows that no push rule or no pull rule.

References

[1] Wu, L.H. Research on Intelligent Scheduling Technologies of AMHS in Semiconductor Wafer Fabrication System. Shanghai Jiaotong University, 2011.

[2] Li, B., Yang, C.J. Production logistics system based on a hybrid push/pull control strategy in make-to-order environments. International Journal of Innovative Computing Information and Control, 2009, 5(5): 1343–1350.

[3] Lin, J.T., Wang, F.K., Chang, Y.M. A hybrid push/pull-dispatching rule for a photobay in a 300 mm wafer fab. Robotics and Computer-integrated Manufacturing, 2006, 22(1): 47–55.

[4] Dou, Z. Fuzzy Logic Control Technology and Application. Beijing?Beijing University of Aeronautics and Astronautics Press, 1995.

[5] Kecman, V. Learning and Soft Computing: Support Vector Machines, Neural Networks, and Fuzzy Logic Models. Cambridge: MIT Press, 2001.

[6] Mok, P.Y., Kwong, C.K., Wong, W.K. Optimsation of fault-tolerant Fabric-cutting schedules using genetic algorithms and fuzzy set theory. European Journal of Operational Research, 2007, 177(26): 1876–1893.

[7] Wu, L., Zhang, J, Sun, Y., et al. A greedy dynamic priority dispatching policy for intrabay in semiconductor wafer fabrication system Proceedings of the 6th CIRP-Sponsored International Conference on Digital Enterprise Technology. Springer Berlin Heidelberg, 2010, pp. 927–938.

[8] Wu, L.H., Mok, P.Y., Zhang, J. An adaptive multi-parameter based dispatching strategy for single-loop interbay material handling systems. Computers in Industry, 2011, 62(2): 175–186.

8 Integrated scheduling in AMHSs

Material handling scheduling problems in automated material handling systems (AMHSs) mainly focus on scheduling automated vehicles. Although numerous approaches have solved global optimization scheduling problems in material handling, which have shown good performances, there are some problems in the existing scheduling methods:

1. A single method is adopted by present studies for material handling scheduling problems in AMHSs. However, intelligent methods, operations research methods, mathematical programming methods and scheduling rules cannot effectively combine wafer scheduling, OHT dispatching and OHT routing process scheduling.
2. The automatic vehicle handling process in an AMHS consists of two portions: OHT dispatching and OHT routing. In view of the vehicle path optimization, a scheduling method is designed so as to reduce large amount of temporary congestions in the material handling process.
3. Although the current global optimization method for AMHSs can meet the requirements of the multi-objective optimization process, the methods including composite scheduling rules, rule switching and dynamic parameter adjustment are linear weighting of multiple sub-objectives, which is no breakthrough in the weighting process. In the real-time scheduling process, the priority decision-making process of each sub-objective is largely dependent on preset parameters, which is difficult to accurately control [1, 2].

In this chapter, two scheduling approaches are developed for integrated scheduling problems in AMHSs. One is integrated scheduling method based on parallel multi-objective genetic algorithm (PMOGA), and another is complex heuristic scheduling method based on genetic algorithms and route library (GARL) [3].

8.1 Description of Integrated Scheduling Problems in AMHSs

The integrated scheduling process in AMHSs is to combine the wafer processing process and the vehicle handling process. The integrated scheduling process mainly consists of three portions: lot scheduling, OHT dispatching and OHT routing.

1. Lot scheduling

   In Intrabay material handling systems and Interbay material handling systems, there is a large amount of work-in-process wafer inventories. In scheduling methods based on scheduling rules, the lot scheduling process is to sort parts by their priorities and to allocate resources according to the priority sorting. However, it only considers the scheduling process before the part is loaded, and it does not

https://doi.org/10.1515/9783110487473-008

consider the scheduling process after reaching the objective process technology area. In fact, there are a number of processing bays to produce the same process in Interbay material handling systems. Therefore, the lot scheduling process is combined with the OHT routing process in order to effectively reduce the load imbalance and reduce the average processing cycle time of workpieces in Interbay material handling systems.

2. OHT dispatching
   The OHT dispatching problems in AMHSs have been presented in detail in the previous chapter, and would not be discussed here.

3. OHT routing
   The OHT routing process is a handling process, in which parts are transported through a particular route to the objective processing bay after they are loaded on the vehicle at the starting stocker. The time spent in this process is also part of the overall machining cycle. Therefore, it is important to optimize the OHT routing process.

The framework of the integrated scheduling process for wafer processing and vehicle handling is illustrated in Figure 8.1. The integrated scheduling process mainly consists of three portions: lot scheduling, OHT dispatching and OHT

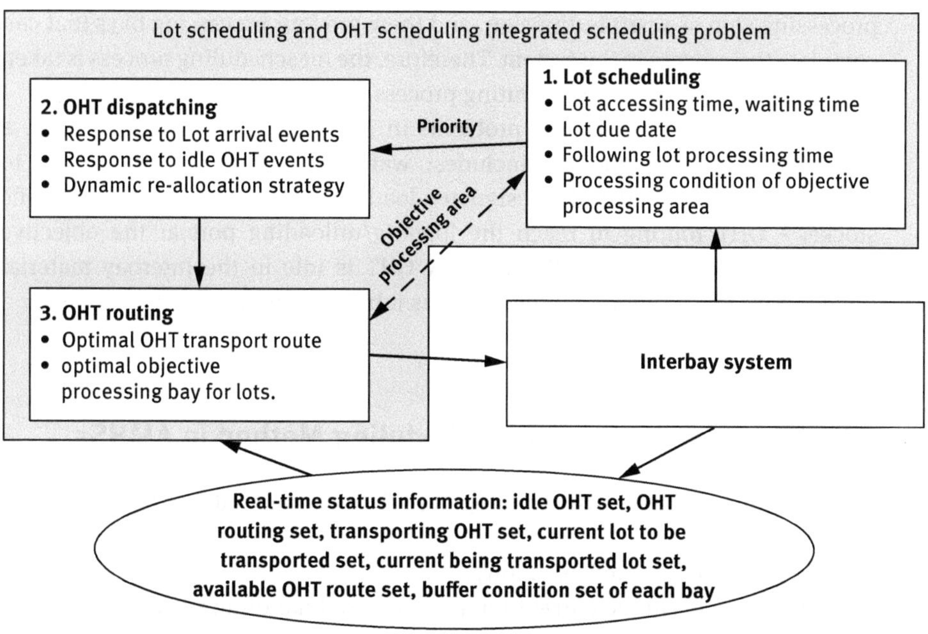

Figure 8.1: The framework of the integrated scheduling model for wafer processing and vehicle handling.

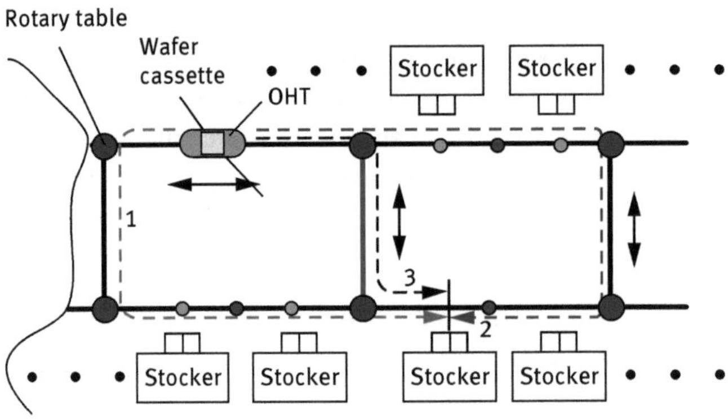

Figure 8.2: Temporary congestion problems in material handling systems.

routing. All the scheduling and calculation procedures are developed on the basis of the real-time information of the material handling process in AMHSs.

The complex heuristic rules GPCRG-1 presented in Chapter 6 are adopted to prioritize parts, that is to say, the part with the highest priority will be moved at first.

The OHT dispatching process and the OHT routing process focus on the OHT with the optimal path in order to avoid the temporary congestion, as shown in Figure 8.2 (red lines indicate a temporary blockage path). The objective processing bay selection is considered in order to improve the load balance in Intrabay material handling systems. The selection process is shown in Figure 8.3. The next processing step of a part is diffusion, and there are four processing bays that can complete the process in the system. Therefore, the lot scheduling process is taken into consideration in the OHT routing process.

In the integrated scheduling problems in AMHS of wafer manufacturing, a complete scheduling process includes: wafer scheduling (prioritization) to accept the job →arrived at the designated loading/unloading port at the specific stocker→ OHT routing to reach the loading/unloading port at the objective stocker→ unload the part. When a new OHT is idle in the Interbay material handling system (completing the previous job) or a part arrives, it will trigger a complete rescheduling process.

8.2 PMOGA-Based Integrated Scheduling Method in AMHSs

8.2.1 Formulation of PMOGA-Based Integrated Scheduling Model in AMHSs

Index and definition are listed as follows:

K – Set of vehicles in the current Interbay material handling system;

K_1 – Set of vehicles with the status of "*Delivery*";

K_2 – Set of vehicles with the status of "*Idle*";

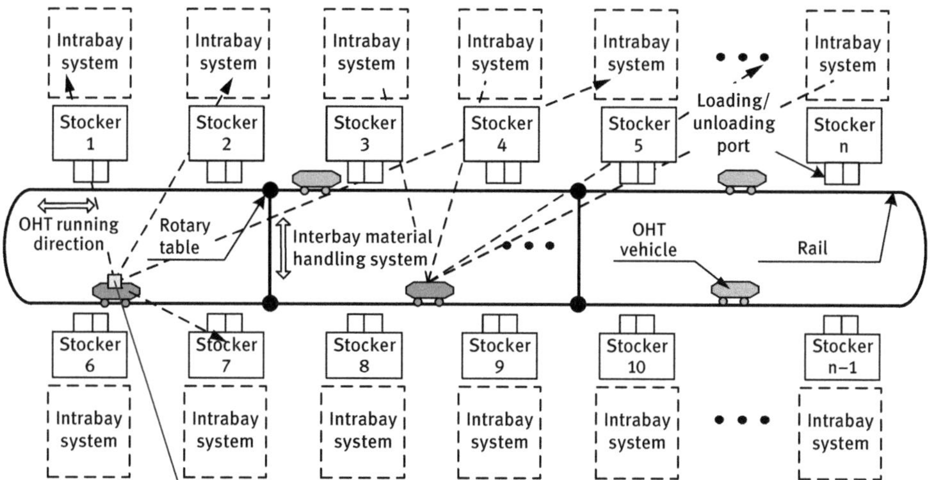

9	30:00.0000	30:00.0000	C10	1	1
10	1:30:00.0000	0.0000	C7	5	3
11	30:00.0000	30:00.0000	C3	1	1
12	30:00.0000	1:00:00.0000	C22	1	1
13	30:00.0000	30:00.0000	C4	1	1
14	1:00:00.0000	2:00.00.000	C9	1	2
15	30:00.0000	30:00.0000	C13	1	1
16	30:00.0000	1:00:00.0000	C3	1	1
17	1:00:00.0000	1:30:00.0000	C14	1	2
18	30:00.0000	30:00.0000	C20	1	1
19	30:00.0000	30:00.0000	C12	1	1
20	30:00.0000	30:00.0000	C17	1	1

Diffusion Process

Bay-1
Bay-2
Bay-5
Bay-7

Figure 8.3: Sketch map of several objective processing bays for wafer lots.

L – Set of wafers in the current Interbay material handling system;
L_1 – Set of wafers being delivered;
L_2 – Set of wafers waiting for delivering;
P_l – Set of available bays for wafer $l \subset L$;
N – Set of nodes, $N = \{n_0, n_1, \ldots\ldots, n_r\}$;
E – Set of directed links, $E \subset (N \times N)$, $\forall (n, n') \in E$ indicates the link from node n to n';
E^r – Set of links included in routine $r \in R^k$;
$SUS(n)$ – Successor node set for node $n \in N$;
R^k – Set of feasible rails of OHT vehicle $k \in (K \cup K_2 \cup K_3)$;

Intermediate variables are defined as follow:
C_k^r– Time consumption of routine $r \in R^k$, in which $k \in (K_1 \cup K_2)$;
CB_k^r – Clogging time consumption of routine $r \in (K_1 \cup K_2)$, in which $j \in Y - T$;
$CM_{l_2}^{k_2}$ – Time consumption of the matching of wafer $l_2 \in T_2$ and $k_2 \in K_2$;

x$C_{l_1}^p$ – ime consumption for the water $l_1 \in T_1$ choosing processing bay $p \in P_l$.

The decision variables are:
$X_{l_1}^p$ – Binary number 0/1, 1 if wafer $l_1 \in T_1$ chooses processing bay $p \in P_l$, or 0;
Y_{k1}^r – Binary number 0/1, 1 if vehicle $k_1 \in K_1$ chooses processing bay $r \in R^{k1}$, or 0;
$Z_{k_2}^r$ – Binary number 0/1, 1 if vehicle $k_2 \in K_2$ chooses processing bay $r \in R^{k2}$, or 0
$\alpha_{k_1}^r$ – Binary number 0/1, 1 if vehicle $k_1 \in K_1$ unloads the wafer at the stocker in routiner $\in R^{k1}$, or 0;
$\beta_{k_2}^r$ – Binary number 0/1, 1 if vehicle $k_2 \in K_2$ unloads the wafer at the stocker in routine $r \in R^{k1}$, or 0;
$\gamma_{l_2}^r$ – Binary number 0/1, 1 if the objective wafer is $l_2 \in T_2$ for the routine $r \in R^k$, or 0;
λ_e^{rt} – Binary number 0/1, $\forall t \in T$ the value is 1 if there is a vehicle in any link $e \in E^r$ in routine $r \in R^k$, or 0.

According to the definition of the problem description, to minimize handling costs and maximize manufacturing performance objectives, if two portions of the objective are unified converted into the time-consuming, the integrated scheduling objective function of wafer manufacturing and OHT handling is developed as follow:

$$Min \sum_{k_1 \in K_1} \sum_{r \in R^{k_1}} \{\sum_{l_1 \in L_1} \sum_{p \in P_l} \cdot ((X_{l_1}^p \cdot C_{l_1}^p + h) \cdot \alpha_{k_1}^r + Y_{k_1}^r \cdot (C_{k_1}^r + B_{k_1}^r))\} + \sum_{k_2 \in K_2} \sum_{r \in R^{k_2}} \{\sum_{l_2 \in L_2} (Z_{k_2}^r \cdot (C_{k_2}^r + \beta_{k_2}^r \cdot h + B_{k_2}^r))\} \tag{8.1}$$

In the objective function, the idle OHTs and the handling OHTs are considered separately in the system. The optimization objective of the idle OHTs is the optimal assignment, while the optimization objective of the handling OHTs is the optimal routing and wafer manufacturing performances. The constraints are listed as follow:

$$\sum_{p \in P_l} X_{l_1}^p = 1, \forall l_1 \in T_1 \tag{8.2}$$

$$\sum_{r \in R^{k_1}} Y_{k_1}^r = 1, \forall k_1 \in K_1, \sum_{r \in R^{k_2}} Z_{k_2}^r = 1, \forall k_2 \in K_2 \tag{8.3}$$

$$\sum_{r \in R^{k_1}} \alpha_{k_1}^r = 1, \forall k_1 \in K_1, \sum_{r \in R^{k_2}} \beta_{k_2}^r = 1, \forall k_2 \in K_2 \tag{8.4}$$

$$\sum_{l_2 \in T_2} \gamma_{l_2}^r \leq 1, \forall r \in R^k, \forall k \in K \tag{8.5}$$

$$\sum_{r \in R^k} \gamma_{l_2}^r \leq 1, \forall l_2 \in T_2 \tag{8.6}$$

$$(n_i, n_{i+1}) \in E, \forall n_i \in r, r \in R^k, \forall k \in K, i = 0, 1, \ldots, m-1 \tag{8.7}$$

$$\sum_{r \in R^k} \lambda_e^{rt} \leq 1, \forall e \in E^r, r \in R^k, \forall k \in (K_1 \cup K_2), \forall t \in T \tag{8.8}$$

Constraints (8.2) ensure that each wafer can only select one processing area every time. Constraints (8.3) represent that each wafer can only be loaded on one OHT and an OHT can only load a wafer. Constraints (8.4) make sure that an OHT can only be loaded in a stocker. Constraints (8.5) ensure that each waiting job can only be completed by one OHT. Constraints (8.6) make sure that each path only corresponds to a job. Constraints (8.7) ensure that there is only one OHT on any rail at any time, which also said that there were one OHT to unload wafer in the AS/RS port at any time. If the current rail is occupied, then the subsequent OHT will wait until the current rail is idle. Constraints (8.8) make sure that the OHT only unload at an objective stocker (a wafer can only enter a workstation).

8.2.2 Architecture of PMOGA-Based Integrated Scheduling Method in AMHSs

In this section, a PMOGA is proposed to solve the above problem. A methodology which combines a PMOGA and a local search strategy is developed to improve the quality of solutions in order to optimize simultaneously multiple sub-objectives.

8.2.3 Description of PMOGA

8.2.3.1 Parallel Process of PMOGA

1. Parallel genetic algorithm
   With the development of science and technology, the scales of many problems are expanded and the complexities of many problems are increased. It is necessary for GAs to improve their calculation speeds and the quality of solutions. The genetic algorithm has a natural parallelism, which is very suitable for parallel implementation. In a parallel genetic algorithm (PGA), the natural parallelism of GAs is combined with the modern parallel computing method. Hence, the PGA not only improves the calculation speed, but also increases the population size and isolates

sub-populations, which can preserve the diversity and distribution of the population so as to reduce the possibility of premature convergence and improve the solution quality. The PGA is defined by using a 6-tuple equation:

$$PGA = \{\mathrm{DMM}, \mathrm{X}, \mathrm{Z}, \Delta, \varphi, \mathrm{SGA}\} \tag{8.9}$$

where
DMM – a set to achieve the parallel process or processor pi in LAN;
X – A set of information exchange objects for each processor;
Z – A set of content information exchange;
Δ – The time interval or frequency at which information is exchanged;
φ – Individual replacement operator for information exchange;
SGA – The simple GA runs on each processor.

The PGAs can be briefly classified into four categories: master-slave, coarse-grained, fine-grained and mixed-model [4].

(1) Master-slave model

It is a direct parallel scheme of GAs, which does not change the basic structure of GAs, and has only an evolutionary population. The selection operation, the crossover operation, the mutation operation and other basic operations are serially completed by the master node machine (or host processor). The fitness evaluation process and decoding process are parallelly performed by nodes (or sub-processors). It is relatively easy to implement this model. If the calculation section focused on the fitness evaluation, the master-slave model is very effective in parallel programs. However, it is necessary to consider the load balancing of master and slave nodes or processors load balancing and the problem of communication bottlenecks.

(2) Coarse-grained model

It is also known as a distributed model or an island model, which is the most adaptable and the most widely used parallel model in GAs. It divides the population into several sub-groups according to the number of nodes (or processors). Each sub-group runs a GA on its own node (or processor). After a certain number of iterations, each sub-population will exchange several individuals. On the one hand, more excellent individuals are introduced, and on the other hand, the diversity of the population is enriched greatly so as to prevent premature convergence.

(3) Fine-grained model

The difference between the fine-grained model and the coarse-grained model is that the fine-grained model has only one population in the whole evolutionary process, but its sub-population is very small. The ideal state is that each sub-node has an individual, and there is a strong communication capability between sub-nodes. For each chromosome, selection and crossover operations are done in the node and its neighborhood nodes. Since the entire evolutionary process rarely requires global operation, the parallel nature of GAs plays an important role.

(4) Mixed-model
The hybrid model is a model in which the three basic models are mixed to form a hierarchical structure. It can be further classified into three major classes: a coarse-grained–fine-grained model, a coarse-grained–coarse-grained model and a coarse-grained–master-slave model.

2. The parallel model in PMOGAs
The PMOGA proposed in this chapter is a coarse-grained-based PGA. The migration topology, the migration scale and the migration strategy must be determined in this model. The fully interconnected migration topology is adopted, that is, individuals in any sub-group exchange information with individuals in a randomly selected sub-group to ensure the diversity and distribution of the population, as shown in Figure 8.4. The migration period of any sub-group is denoted as Tmig = (1/10) * Gen, in which Gen is the maximum number of iteration, and the migration ratio is 10%.
In migrating groups, the non-dominated Pareto set with the highest rank is migrated out to be replaced by the non-dominated Pareto set with the lowest rank. This migration strategy is adopted by the PMOGA. There are two cases: the number of individuals contained in the migrated optimal non-dominated Pareto set is greater than or less than the optimal individuals contained in the replaced non-dominated Pareto set in order to ensure that the total number of individuals moved into the sub-group would be kept stationary throughout the process.

8.2.3.2 Multi-Objective Evolutionary Process of a PMOGA

1. Multi-Objective Genetic Algorithms (MOGAs)
There are essential differences between multi-objective problems (MOP) and single-objective problems. In general, a MOP consists of n decision variables, k

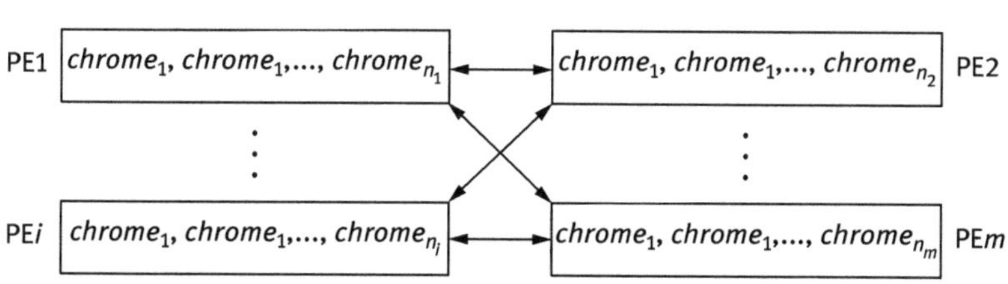

Figure 8.4: The interconnected topology migration in PMOGAs.

Algorithm 8.4.1: Pseudocode of sub-group migration strategy in PMOGAs

---

if *Popsize*(P_{mig_in} <= *Popsize*(P_{mig_out}) **then**
//Select individuals witd *Popsize*(P_{mig_in}) from P_{mig_out}
P_{mig_in} = *RandIndSel*(P_{mig_out}, P_{mig_in})
else if (*Popsize*(P_{mig_out}) ≤ *Popsize*(P_{mig_in}) **then**
//Select all individuals from P_{mig_out}, together with randomly generated individuals as the migration set
P_{mig_in} = P_{mig_in} ∪ *RandPopCreate*(*Popsize*(P_{mig_out} − P_{mig_in}))
end
P_{mig_in}: non-dominated Pareto set with the lowest rank needs to be replaced in moved-in sub-groups
P_{mig_out}: non-dominated Pareto set with the highest rank in moved-out sub-groups
RandIndSel(*Pop_src, Pop_dest*): Individual selection function
RandPopCreate(*popsize*): Random individual generation function

---

objective functions and m constraint conditions. There is a functional relationship among the objective function, the constraint condition and the decision variable.

The processing objective of a multi-objective evolutionary algorithm (MOEA) is a multi-objective optimization problem. MOEAs have been widely applied in many fields. A Pareto-based MOEA is the most popular technique. The main feature of this approach is to integrate the Pareto-optimal concept in the selection mechanism. The basic process of a Pareto-based MOEA is illustrated in Figure 8.5.

The common framework of a GA is adopted by a MOGA. The Pareto sorting strategy is used by MOGAs to solve multi-objective optimization problems. Its evolution process is consistent with the MOEA. In this chapter, a PMOGA is proposed on the basis of MOGAs. A rapid non-dominated Pareto set constructor is used, and the solution set is further divided based on an aggregation method to improve the solution quality and computational efficiency.

2. Non-dominated Pareto set
Current methods for constructing optimal Pareto solution set includes non-dominated sorting method, exclusion, Challenge Cup method and divide and conquer recursive method. We use a fast non-dominated sorting method which has higher computational efficiency than the above-mentioned methods.

The basic steps are: Select an individual x in the population Pop (select the first individual) as the comparison object (pivot). Compare according to the partial order, which is defined as:

$$\succ_d : \forall x, y \in Pop, x \succ_d y \text{ if } (x \succ y \text{ or x and y do not dominate each other})$$

All the individuals in the Pop are divided into two parts by using x as the middle boundary after sorting. One part is x-dominated individuals (would not be

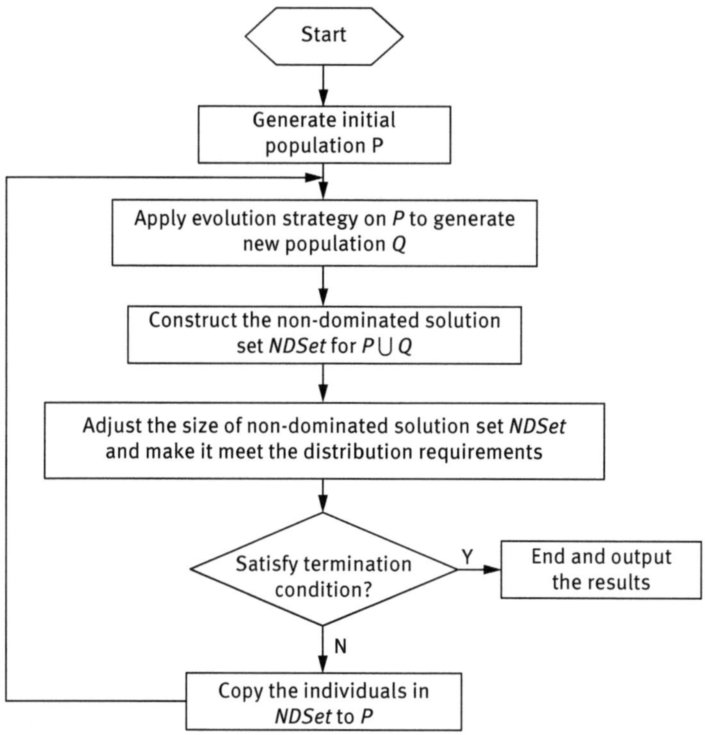

Figure 8.5: Basic optimization process of a Pareto-based MOEA.

considered in the next round of sorting); the other part is an individual that dominates x or an individual that is not related to x. If x is not dominated by all of these individuals in the second part, then x can be incorporated into a non-dominant set. But as long as any of them dominate x, then x is still dominated by the individual, and continues to participate in the next round of sorting until there is only one individual in the second part.

Suppose that the population is $Pop = \{y_1, y_2, ..., y_m\}$, $\forall x_k \in Pop$, Pop is divided as two parts, $\{x_1, x_2, ..., x_{k-1}\}$ and $\{x_{k+1}, x_{k+2}, ..., x_m\}$. $y \succ_d x_k$ and $x_k \succ z$ if $\forall y \in \{x_1, x_2, ..., x_{k-1}\}$, $\forall z \in \{x_{k+1}, x_{k+2}, ..., x_m\}$. It is easy to prove that the non-dominated set constructed by this algorithm satisfies the following conditions:

(1) Suppose that the Pop is $\{y_1, y_2, ..., y_m\}$ after the sequencing r objectives according to $\succ_d$, then $\exists k$, $\{y_1, y_2, ..., y_k\}$ is the non-dominated set of Pop. $y' \succ y$ if $\forall y \in \{y_{k+1}, y_{k+2}, ..., y_m\}$, $\exists y' \in \{y_1, y_2, ..., y_k\}$ if.
(2) After sequencing according to $\succ_d$, if $k_1 < k_2 < ... < k_n$, $P_1 = \{y_1, y_2, ..., y_{k_1}\}$, $P_2 = \{y_{k_1+1}, y_{k_1+2}, ..., y_{k_2}\}$, ..., $P_n = \{y_{k_{n-1}+1}, y_{k_{n-1}+2}, ..., y_{k_n}\}$ satisfying $\cup P_{i \in \{1, 2, ..., n\}} = Pop$ and $\forall i, j \in \{1, 2, ..., n\}, i \neq j, P_i \cap P_j = \emptyset$, $P_1 \succ P_2 \succ ... \succ P_n$.

The above two conditions are the basic conditions for constructing the optimal Pareto solution set.

3. Solution distribution based on the aggregation degree method
An improved clustering density method is adopted to maintain the distribution of the proposed PMOGA. When a new population is generated, individuals with a high degree of clustering density will be kept to the next generation as much as possible. Suppose that $d[i]$ is the gathering distance of individual i, $d[i]\ f_m$ is the objective function.

$$d[i] = \sum_{j=1}^{k}\sum_{m=1}^{r}(d[i+j].f_m - d[i-j].f_m)/k \tag{8.10}$$

Since there are individuals with aggregated distance ∞, the aggregation distances of all individuals are normalized:

$$Maxd = \max_{i\in PF^t,\, d[i]\neq\infty} d[i],\ Mind = \min_{i\in PF^t,\, d[i]\neq 0} d[i] \tag{8.11}$$

$$\hat{d}[i] = \begin{cases} \dfrac{(Maxd + \text{Mind})}{\text{Mind}} & if d[i] = \infty \\ \dfrac{d[i]}{\text{Mind}} & \text{otherwise} \end{cases} \tag{8.12}$$

8.2.3.3 Genetic operators in PMOGAs

1. Path encoding
A real-coded manner is adopted by the path encoding process.
Denote $G = (N, E)$, in which $N = \{n_0, n_1, ..., n_r\}$ is the node set; $E \subset (N \times N)$ represents a set of directional links between nodes; $n, n' \in L$ indicates the link from node n to n' Cp_m, $Adj(n) = \{m : (n, m) \in L\}$ represents the adjacent node set of node n. First, all the links of consecutive nodes in the system are numbered. Assume that there are r nodes, m links, then the serial numbers are assigned in accordance with the following manner:

$$(n_1, n_2) = 1, (n_2, n_3) = 2, ..., (n_{r-1}, n_r) = m \tag{8.13}$$

Denote path $p = \{n_0, n_1, ..., n_k\}$ as a sequence from node n_0 to n_k. This node sequence corresponds to a sequence of links, and each path can be represented by a string of serial numbers from 1 to m. Since the walking route of a vehicle is different, the variable length chromosome is used to represent individuals.

If there are k OHTs, then a multi-parameter cascade coding is used to represent a chromosome, each vehicle occupies x_i digit (i = 1, 2, ……, k), as follow:

$$\text{Chromosome } = b_{11}b_{12}...b_{1l_1}|b_{21}b_{21}...b_{2l_2}|...|b_{k1}b_{k2}...b_{kl_k} \tag{8.14}$$

The total length of the string is $\sum_{i=1}^{k} l_i$.

2. Population initialization
The initial population has a greater impact on the results. Artificial selection method is not applicable to the present problem, and randomly generated rule will generate a lot of unavailable paths. Solomon [5] proposed a heuristic algorithm for VPR to generate a good individual, and then generate some individuals in the neighborhood of the individual, which account for a certain proportion and the rest of the individuals are randomly generated. Eppstein proposed a K shortest path algorithm that can be used to generate an initialization path. In addition, there are network flow methods and some point-to-point shortest path generation algorithms. Since this part of the calculation is off-line calculation, the time complexity of the algorithm would not be taken into consideration. Dijkastra-Floyd algorithm [7] is used to generate the initial optimal path, and part of solutions is selected as the initial population. In the population initialization procedure, the shortest path from any vertex in graph G = (N, L) to all other vertices is found to calculate the initial optimal path. The weighted adjacency matrix of G is presented as follow:

$$D = \begin{pmatrix} 0 & d_{12} & \infty & \cdots & \infty \\ d_{21} & 0 & d_{23} & \cdots & \infty \\ \infty & d_{32} & 0 & \cdots & \infty \\ \cdots & \cdots & \cdots & \cdots & \cdots \\ \infty & \infty & \cdots & d_{n,n-1} & 0 \end{pmatrix} \tag{8.15}$$

In addition to the subscript elements and diagonal elements in matrix D, the remaining elements are infinite. The shortest distance and optimal path set *DFSet* between all vertex pairs in the Interbay material handling system is obtained by using Dijkastra–Floyd algorithm. The remaining paths are generated by using a random algorithm. In each real-time scheduling trigger, the appropriate path is selected from *DFSet* as the initial path according to the OHT starting position corresponding to the node.

3. Select operator
Random tournament selection operator [8] is adopted in this algorithm. This selection method has both random and deterministic characteristics, and its calculation efficiency is very high. Since PMOGA is a multi-objective algorithm, the steps of its selection operator are presented as follows:

(1) n ($n = 2$) individuals are randomly selected from the population for comparison, and the individuals with the highest rank are inherited to the next generation. If the sort is the same, the individuals with smaller density are selected.
(2) The above process is repeated N times, and N individuals are generated for the next generation.

4. Crossover operator
   Due to the special nature of OHT movement, the single point or multipoint crossover operator will produce discontinuous infeasible path, which is not conducive to the evolutionary convergence of the algorithm. Therefore, a hybrid node crossover operator is used in PMOGA. For two randomly selected individuals, only the points with the exactly same serial number are selected for crossover operation. When there is more than one coincidence node, a node is randomly selected. If there is less than one node, no crossover operation will be performed. This ensures that no intermittent paths are generated. The procedure is illustrated in Figure 8.6.
   The basic steps of the crossover operator are presented as follows:
   (1) Two paths R1 and R2 with a common node are randomly selected from the population in accordance with crossover probability Pc;
   (2) A node from all the common nodes is selected as a crossover point;
   (3) All sub-path (node sequence) positions are exchanged after the intersection is crossed

5. mutation operator, insert operator, and delete operator.
   The mutation operation of a PMOGA consists of three operations. (1) Randomly remove a serial number from the chromosome (excluding the start and end numbers); (2) Randomly select a node from the chromosome to insert a new serial number; (3) Randomly select a serial number in the individual and replace it with a new serial number. Since these three ways are likely to produce

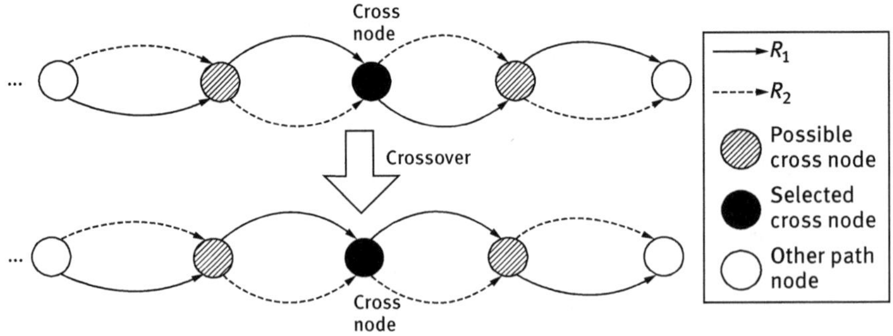

Figure 8.6: Schematic diagram of the path crossover.

intermittent path, two special operators (an insert operator and a delete operator) are added. For example, for path R, the mutation operation, insertion operation and deletion operation are presented as follow:

- $R = (13, 14, 15, 7, 8, 9, 17)$
- Mutation: $R = (13, 14, 15, 12, 8, 9, 17)$
- Insertion: $R = (13, 14, 15, [16, 17, 21], 12, [13, 18], 8, 9, 17)$
- Deletion: $R = (13, 14, 15, 16, 17)$

8.2.3.4 Fitness evaluation

As the previous presentation, if $E \subset (N \times N)$ contains m links, then a vector is used to represent the status of the rail at scheduling time t:

$$s = \{S^1(t), S^1(t), ..., S^m(t)\}$$

Suppose that

$$S^l = \begin{cases} 1 & \text{if link } l \text{ is blocked} \\ 0 & \text{otherwise} \end{cases} \tag{8.16}$$

If there is at least one busy OHT on the link l, it is called blocked. For any link l, the status vector $\{S^l(t), t = 0, 1, ...\}$ related to status $S^l(t)$ at time t = 0, 1, ... is a Markov chain, and any two status vectors $\{S^{l_1}(t), t = 0, 1, ...\}$ and $\{S^{l_2}(t), t = 0, 1, ...\}$ are Independent [9], in which l_1 l_2. The status transition matrix is:

$$P^l_{t,t+1} = \begin{pmatrix} 1-\alpha^l_t & \alpha^l_t \\ 0 & 1 \end{pmatrix} \tag{8.17}$$

wherein α^l_t represents the probability that the directed link l is blocked before time t, and is idle after time t, e.g., $S^l(0) = 1$, $S^l(1) = 1$, $\ldots, S^l(t) = 1, S^l(t+1) = 0$. The status transition matrix shows probability α^l_t that any directed link l changes from blocked to idle at time t, as shown in Figure 8.7.

Three factors including the average OHT delivery time, longest OHT delivery time and average processing cycle of a workpiece are considered in the fitness evaluation process. The smaller the fitness value, the better the individual.

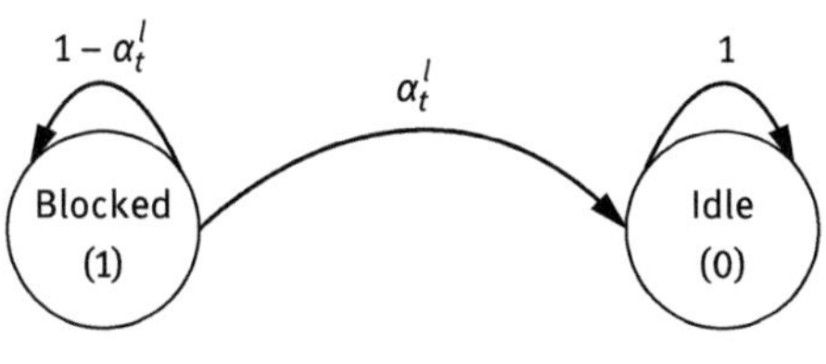

Figure 8.7: Sketch map of the state transition characteristics of link l.

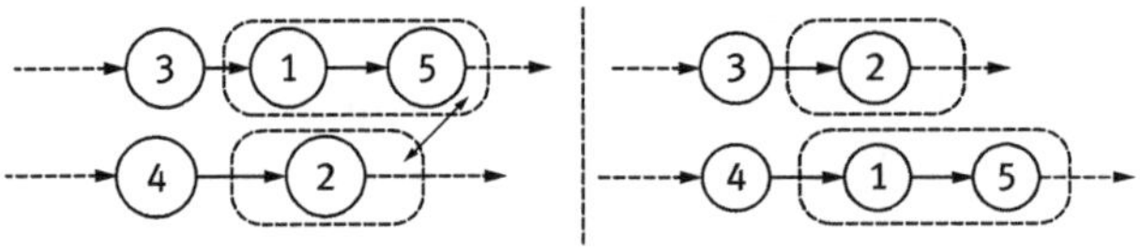

Figure 8.8: Exchange operator in the local search.

8.2.3.5 Local search strategy

A λ-interchange local search method is adopted in the local search strategy. Since the processed individuals in PMOGAs are paths, the local search process improves the quality of the solution by exchanging nodes in different paths. For example, for a given path (Rp, Rq), the exchange operator (1, 2) represents the replacement of one node from path Rp to path Rq, and one node is changed from path Rq to path Rp., as shown in Figure 8.8.

The FS strategy is used to select switching nodes. Due to the difference between multi-objective and single-objective optimization problems, the solution *S'* that leads to the cost reduction is an optimal solution to dominate the current solution.

The mixed local search strategy can significantly improve the quality of the genetic algorithm, but on the other hand, it will increase the computational complexity. Therefore, it only assigns a higher local search probability to the non-dominated individuals at the forefront of Pareto (the highest rank). In the experiment, the initial setting probability of the local search process is 10% and the local search probability of remaining individuals is 1%.

8.2.3.6 PMOGA execution process

The overall execution process of a PMOGA is illustrated in Figure 8.9. The initial population is divided into several sub-populations according to the number of processors, and each sub-population is concurrently running on the respective processors. After a certain evolutionary iteration, each sub-population will exchange a number of individuals in accordance with the migration topology and the migration strategy. In the process of merging sub-populations, in order to maintain the new final population size N (the size of the sub-population Popi), when $Pop^t{}_i + F_i > N$, select $N\text{-}Pop^t{}_i + F_i$ individuals with the smallest aggregation distance from F_i to enter the new population. When the algorithm finally converges, an optimal Pareto solution set is obtained, and the individual with the largest clustering distance is chosen as the final optimal solution from the optimal solution set.

8.2.4 Simulation Experiments

8.2.4.1 Computational Efficiency Analysis

The PMOGA parallelization scheme proposed in this chapter is developed on the basis of C ++ and OpenMP, and the basic parameters of GA and parallel parameter

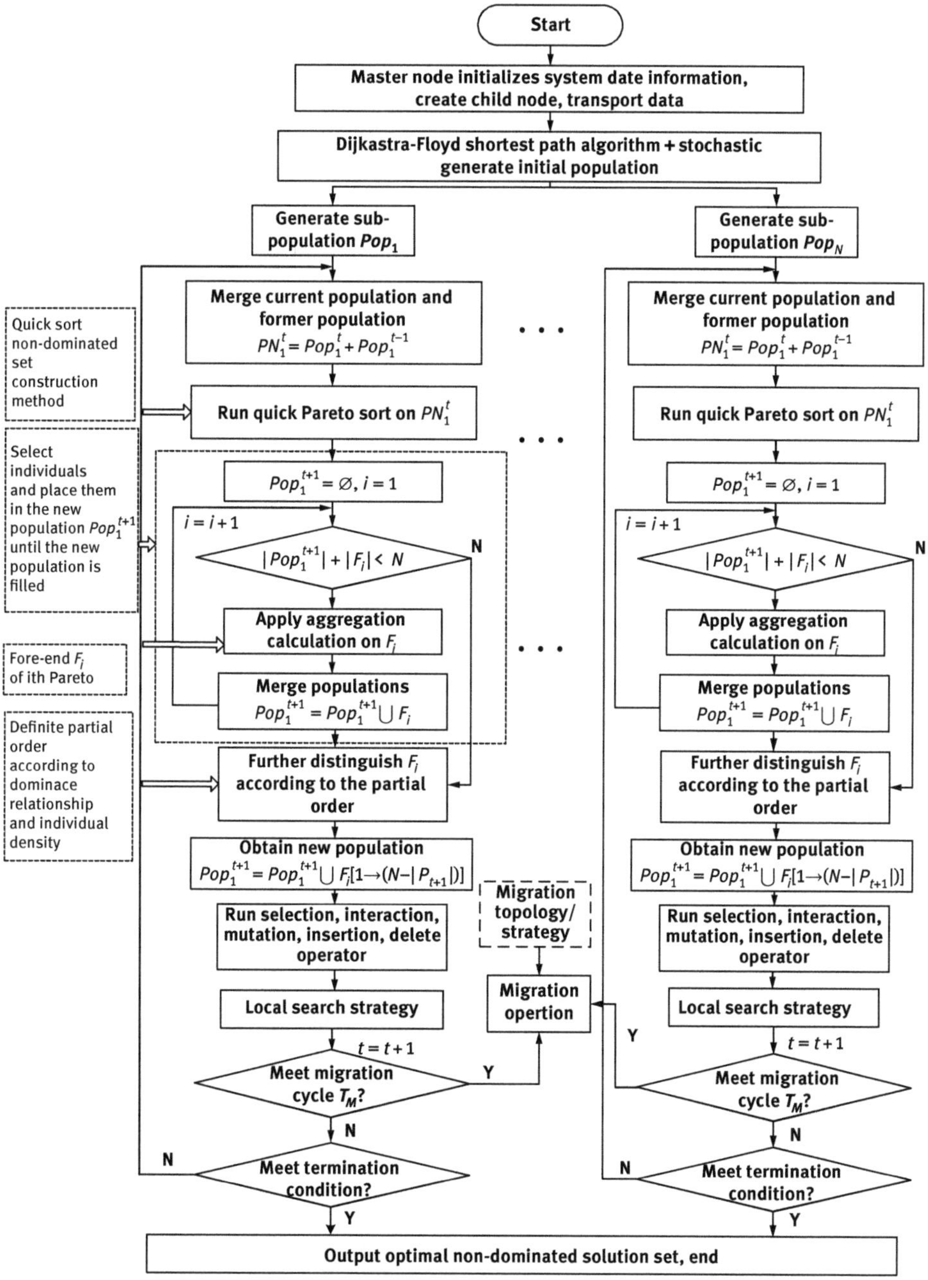

Figure 8.9: General flowchart of PMOGA.

Table 8.1: The basic parameters of a GA and parallel parameter settings.

Name	Value	Name	Value
The maximum length of chromosome	200	Crossover probability	0.3
Iteration number	200–500	Mutation probability	0.1
Population size	200–400	λ-Interchange local	0.1–0.2
Objective number	3	Search probability	2
CPU number	4	*d*-Value	

Table 8.2: The basic parameters of a GA and parallel parameter settings.

Index	Objective number	Population size	Iteration number	λ	CPU number	CPU time (ms)
1	3	200	200	0.1	1	1,700
2	3	200	200	0.1	4	540
3	3	200	500	0.1	4	540
4	3	400	200	0.1	4	911
5	3	400	500	0.1	4	2,139
6	3	400	500	0.2	8	1,218

settings are shown in Tables 8.1 and 8.2. (The same experiments are repeated three times to obtain the average values.)

8.2.4.2 Convergence Analysis

Since there is no known optimal solution set PF_{true} for the multi-objective problem currently studied, the approaching degree evaluation method [10] is used, in which the approximation is measured by calculating the minimum distance of the solution set to the reference set. The smaller the degree of deviation, the higher the quality of solution.

The optimal Pareto set of solutions and the convergence curves of the three objectives and the two objectives (only considering objectives 1 and 2) are shown in Figures 8.10 and 8.11, respectively.

8.2.4.3 Comprehensive analysis

The integrated scheduling method based on PMOGA proposed in this chapter is compared with the previous scheduling method in three indexes (average delivery time, maximum delivery time and average workpiece processing cycle). The basic parameters and process parameters are the same as given in Section 8.2. Three release rates (3 cards/2.5 h, 3 cards/2.75 h and 3 cards/3 h), and three OHT numbers (8, 10, 12) are implemented. The total number of simulation scenes is $3 \times 3 = 9$, and the experiment is repeated once for each scene. The results obtained are shown in Figures 8.12 to 8.15.

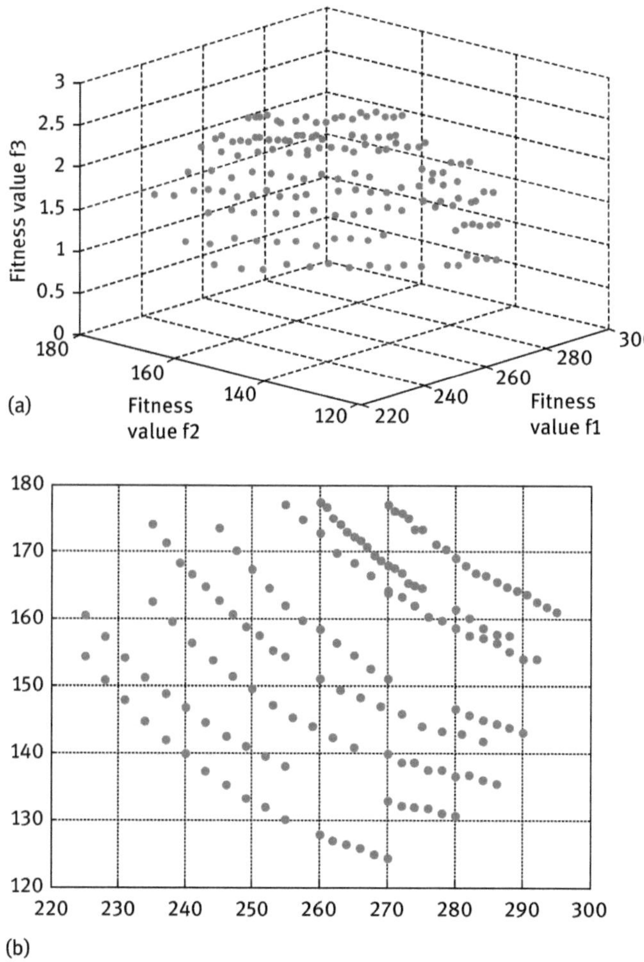

Figure 8.10: Optimal Pareto solution sets: (a) optimal Pareto solution distribution when there are three objectives and (b) optimal Pareto solution distribution when there are two objectives.

There are no significant differences for these methods in the average delivery time. However, the PMOGA-based scheduling method can significantly reduce the maximum delivery time of scheduling tasks, which is important for an Interbay material handling scheduling problem with time constraints. For the average processing cycle, the PMOGA-based method has no significant difference compared with the GPCRG-R1 and MHAFLC methods. For the distribution of processing time (Figure 8.16), since the multi-objective optimization algorithm proposed in this chapter balances multiple optimization objectives, the average processing time is in the vicinity of the mean, which has a small mean difference.

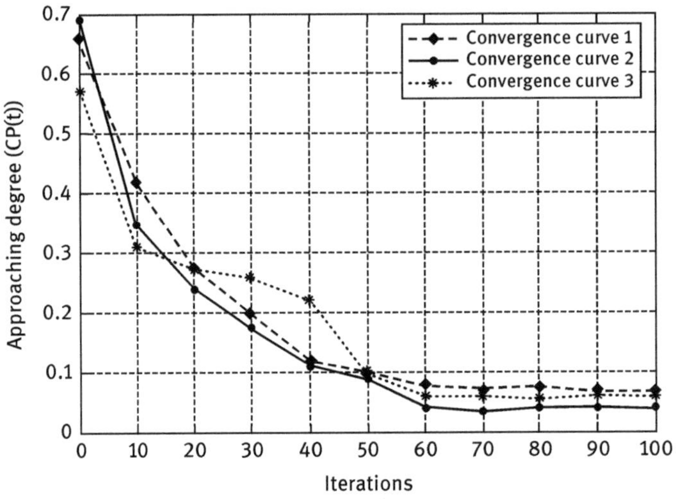

Figure 8.11: Convergence process of optimal Pareto solution sets.

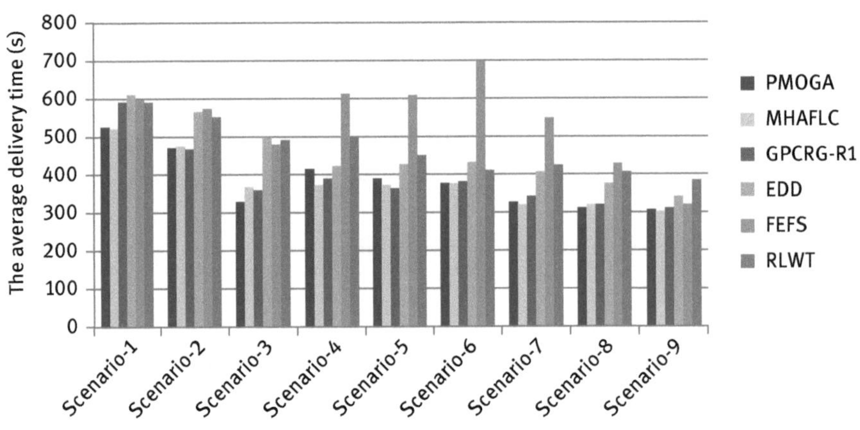

Figure 8.12: Comparison on average delivery time.

8.3 GARL-Based Complex Heuristic Integrated Scheduling Method in AMHSs

In this chapter, on the basis of GP-based complex heuristic rule constructing method presented in Chapter 6, a GARL-based complex heuristic integrated scheduling method is proposed to complete the integrated scheduling process for the wafer processing, vehicle assignment and path planning by using path planning methods and specific dispatch rules. The integrated scheduling mathematical model

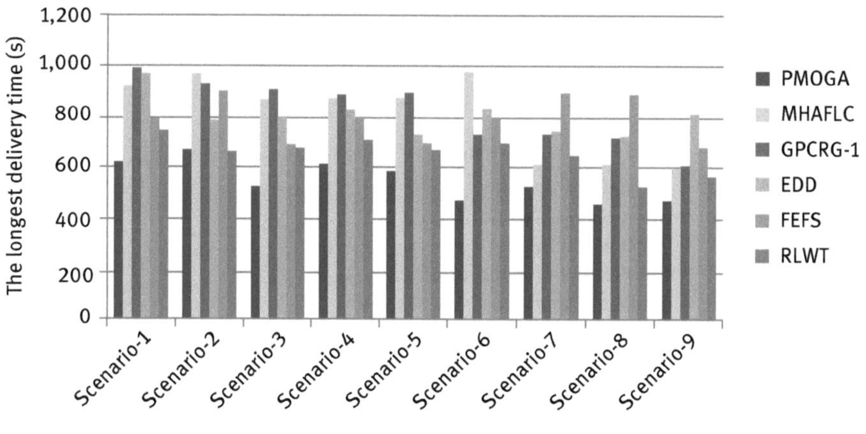

Figure 8.13: Comparison on longest delivery time.

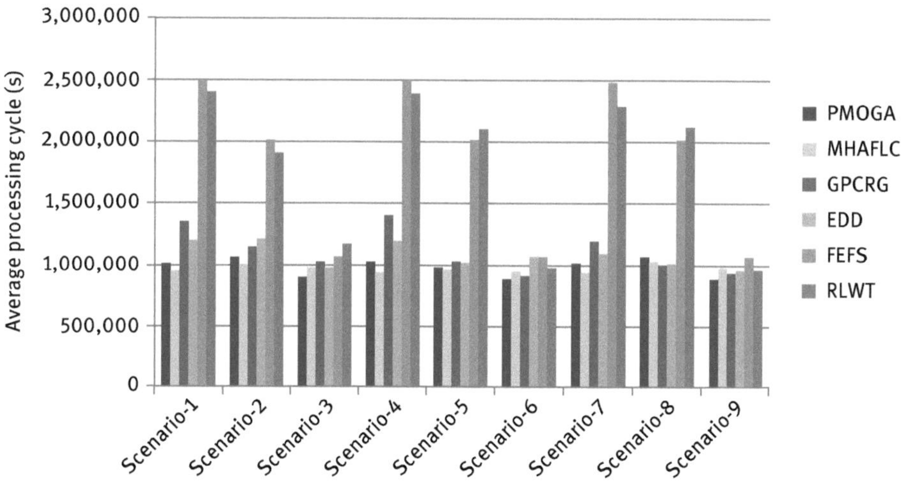

Figure 8.14: Comparison on average processing cycle.

established in Section 8.2.1 is adopted to obtain the optimal Pareto solution for minimizing the average vehicle delivery time and minimizing the production cost by searching for reasonable vehicle dispatch rules and vehicle paths.

8.3.1 Basic Rules for Automated Material Handling System Scheduling

The GARL-based integrated scheduling method consists of two parts: vehicle assignment composite heuristic rule generation and vehicle path planning. The vehicle assignment composite rules are developed on the basis of the assignment rules shown in Table

Table 8.3: Basic assignment rules in AMHSs.

Categories	Index	Description
S1	S11	First in first out
	S12	Shortest remaining time priority
	S13	Critical ratio priority
	S14	The earliest plan priority
S2	S21	First in first serve
	S22	Longest waiting part priority
	S23	Shortest handling distance priority
	S24	Longest handling distance priority
S3	S31	Nearest idle vehicle priority
	S32	Farthest idle vehicle priority
	S33	Longest idle time vehicle priority
	S34	Lowest utilized vehicle priority

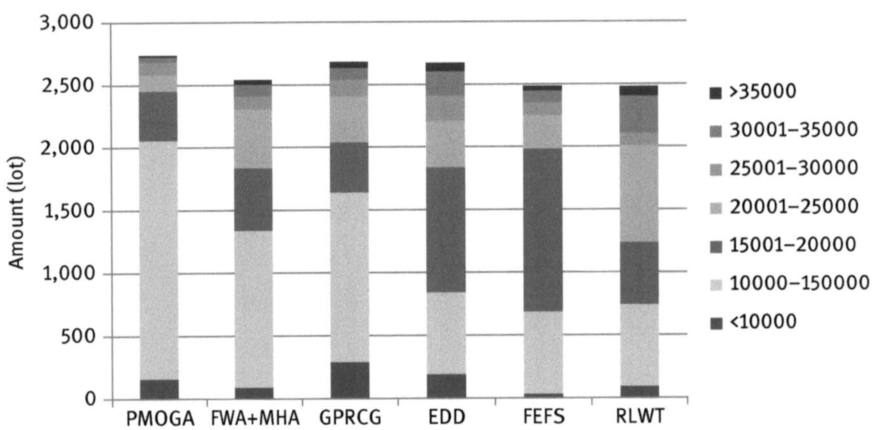

Figure 8.15: Comparison on average processing cycle distribution.

8.3. Rule S1 is used to make decisions on the parts entering the free processing area from the warehouse when there is an idle processing area. Rule S2 is applied to make decisions on the next handling task of a vehicle when a free vehicle faces multiple workpieces to be transported. Rule S3 is used to make decisions on which a free vehicle is assigned to the newly arrived handling task when multiple vehicles are idle.

8.3.2 Construction of the Vehicle's Path Library

The requirements for real-time vehicle path planning in AMHSs are investigated in this section. The vehicle dispatching and path planning are implemented simultaneously by real-time selecting the specific paths. The steps are presented as follows:

1. First, the handling model of the AMHS is developed. Suppose that there are N loading and unloading points, a link set $E \subset (N \times N)$ is constructed, $\forall (n_0, n_k) \in E$ indicates the link from node n_0 to node n_k. When a vehicle moves from node n_0 to node n_k, the path $p \equiv \{n_0, n_1, \ldots, n_{k-1}, n_k\}$is found. $\tau(p) = \sum_{i=0}^{k-1} d(n_i, n_{i+1})/v$ represents the time that the vehicle goes through path p, where $d(n_i, n_{i+1})$ indicates the physical distance of the link (n_i, n_{i+1}), and v represents the speed of the vehicle.
2. Second, an offline vehicle path library is developed. When a vehicle moves from node n_0 to node n_k, the shortest path and the near shortest path are considered. Z shortest paths are extracted to build up the off-line vehicle path library $r = \{p_1, p_2, \cdots, p_Z\}$. At the same time, the far paths are considered, a vehicle needs to choose a far path in the actual situation to sacrifice the shortest distance in order to reduce the vehicle block. Therefore, Z paths are randomly chosen to expand the off-line vehicle path library $R = \{p_1, p_2, \ldots, p_{2Z}\}$. In the actual operation, according to the handling tasks allocated to each vehicle, the path is selected in its corresponding off-line vehicle path library to complete the path planning process.

8.3.3 Genetic Algorithm-Based Intelligent Routing Selection Approach

This section describes how to use genetic algorithms to select intelligent paths. As shown in Figure 8.16, the chromosomes are coded as {S1|S2|S3|R} corresponding to three different situations to generate composite rules, and select the vehicle path after the handling task is generated. Its corresponding decoding process is shown in Figure 8.17. The optimal Pareto solution for minimizing the transportation cost and minimizing the production cost is obtained by using the chromosomal fitness evaluation based on the Pareto dominance relationship.

8.3.4 Simulation Experiments

In this section, the proposed scheduling method-based integrated composite heuristic rules and path library is implemented by using C ++ and OpenMP, and the basic GA parameters are shown in Table 8.4.

The proposed integrated scheduling method is compared with rules such as first come first serve (EDD), first empty first-served (FEFS) and long waiting time rule (RLWT). The processing parameters are summarized in Table 8.5.

Three release rates (3 cards/2.5 h, 3 cards/2.75 h and 3 cards/3 h), and three OHT numbers (8, 10, 12) are implemented. The total number of simulation scenes

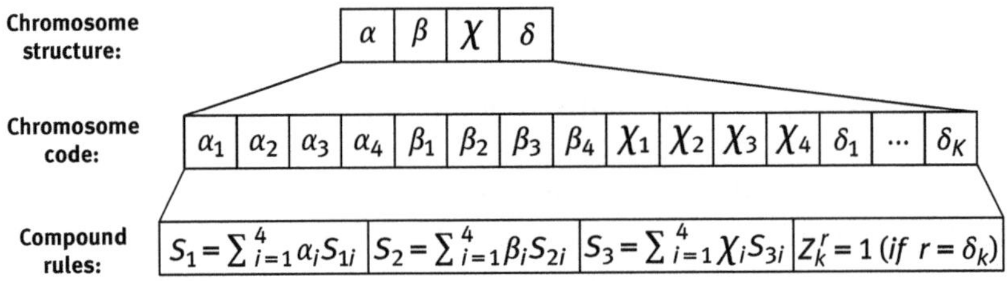

Figure 8.16: Chromosome encoding.

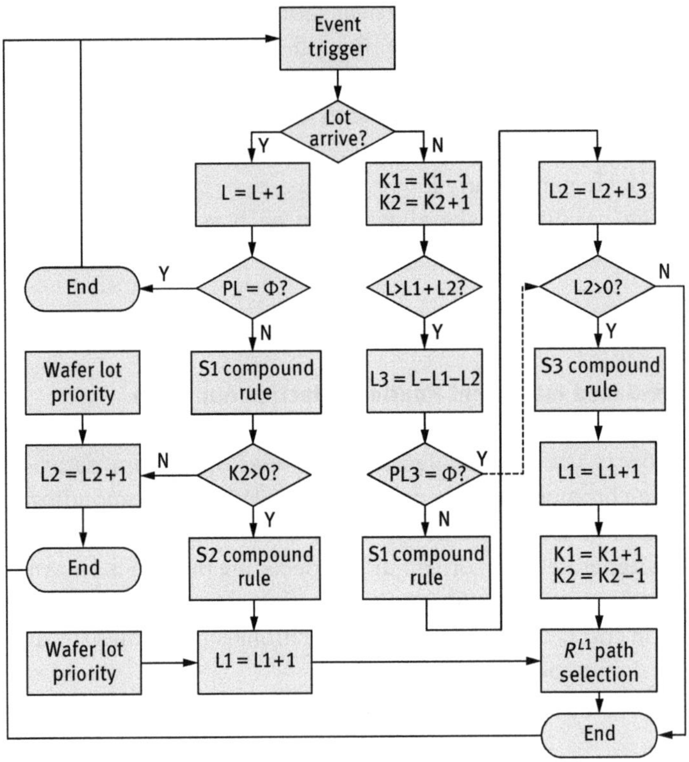

Figure 8.17: Chromosome decoding of OHT dispatching and path planning.

is $3 \times 3 = 9$, and the experiment is repeated once for each scene. The results obtained are shown in Figures 8.18 to 8.19.

The algorithm proposed in this chapter can significantly reduce the handling cost and the average response time during operation in AMHSs. On one hand, these results are benefited from the proposed integrated scheduling method of the vehicle assignment compound rule generation and vehicle path selection. On the other hand,

Table 8.4: Basic GA parameters and parallel setting parameters.

Parameter name	Parameter value	Parameter name	Parameter value
Maximum chromosome length	200	Crossover probability	0.3
Evolutionary iteration number	200	Mutation probability	0.1
Population size	400	λ-Interchange local search probability	0.1–0.2

Table 8.5: Processing technology and key equipment data.

Warehouse number	Bay number	Corresponding process	Number of critical equipment	Processing time (min)	Equipment processing capacity (lot)
Stocker-1	Bay-1	Wash WIS	3	20	2
		Lithography	4	73	3
		Sputtering 1	3	35	2
Stocker-2	Bay-2	Electrical Testing	3	44	1
		Sputtering 2	1	28	2
		Sputtering 3	1	26	2
		Chemical vapor deposition	5	43	2
Stocker-3	Bay-3	Wash WCW	3	19	2
		Chemical vapor deposition	5	41	3
Stocker-4	Bay-4	Automatic appearance	2	48	1
		Wash WCW	4	21	3
Stocker-5	Bay-5	Automatic appearance	1	48	1
		Sputtering 2	1	28	2
		Sputtering 3	2	26	2
Stocker-6	Bay-6	Lithography	5	34	2
		Stripping	4	24	2
Stocker-7	Bay-7	Automatic appearance	3	114	2
		Repair (excision)	5	40	1
		Wash WCW	1	22	2
		Repair (connection)	2	10	1

(continued)

Table 8.5: (continued)

Warehouse number	Bay number	Corresponding process	Number of critical equipment	Processing time (min)	Equipment processing capacity (lot)
Stocker-8	Bay-8	Dry etching DEC	1	46	2
		Dry etching DEI	1	41	2
		Dry etching DCH	1	53	2
		Automatic appearance	2	48	1
		Array check	2	91	2
Stocker-9	Bay-9	Automatic appearance	3	48	1
		Repair (connection)	3	10	2
		Appearance inspection	5	29	1
Stocker-10	Bay-10	Automatic appearance	3	114	2
		Repair (excision)	3	34	1
		Wash WCW	1	20	2
		Wet etching WEI	1	41	3
		Annealing	2	101	5
Stocker-11	Bay-11	Wash WCW	1	18	2
		Wet etching WEI	2	41	3
Stocker-12	Bay-12	Wet etching WEG	1	28	2
		Stripping	3	24	2
		Wet etching WED	1	34	2
		Dry etching DEI	1	41	2
Stocker-13	Bay-13	Wet etching WEG	2	28	2
		Wet etching WED	4	34	2
		Dry etching DEI	1	41	2
		Dry etching DCH	2	53	3
		Dry etching DEC	2	46	2
Stocker-14	Bay-14	Wash WCW	11	20	1
		Dry etching DEI	1	41	2
		Lithography	7	79	4
		Dry etching DEC	5	46	2

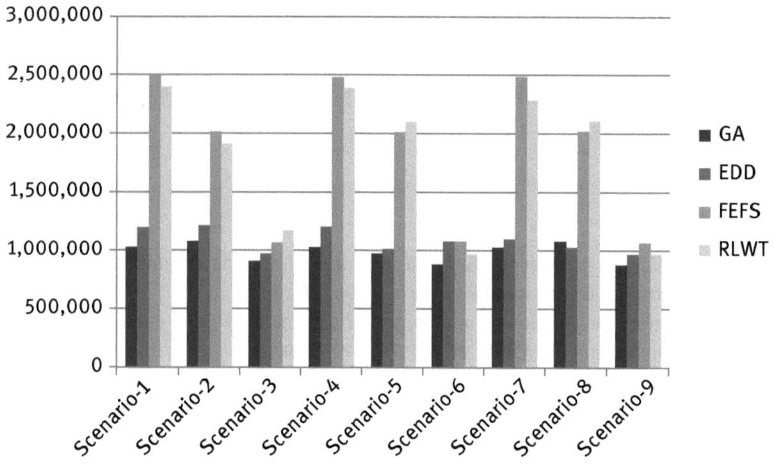

Figure 8.18: Comparison on average handling cost.

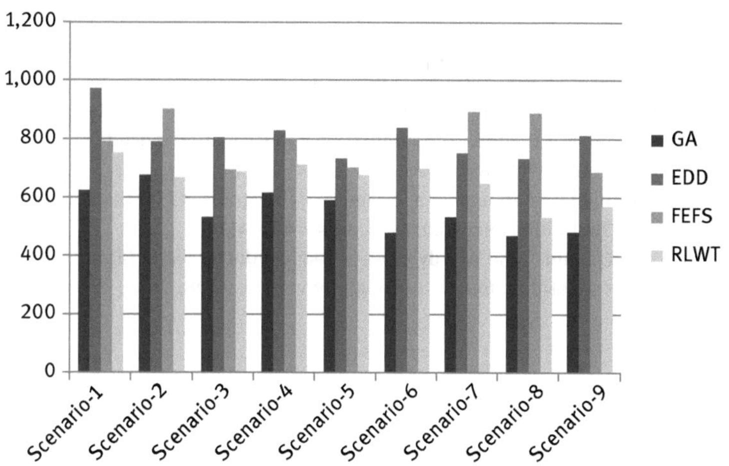

Figure 8.19: Comparison on average response time.

they are also benefited from the Pareto domination for multi-objective problems to find the optimal Pareto solution.

8.4 Conclusion

In this chapter, the integrated scheduling problems of Interbay and Intrabay material handling processes in wafer fabrication systems have been investigated and their corresponding scheduling methods have been introduced. The integrated scheduling method based on parallel multi-objective genetic algorithms and the integrated scheduling method based on multi-agent cooperation have been presented in detail. The methods and algorithms have been verified by using the data from a Shanghai wafer fabrication manufacturer.

References

[1] Le-Anh, T., De-Koster, M.B.M. A review of design and control of automated guided vehicle systems. European Journal of Operational Research, 2006, 171: 1–23.

[2] Vis, I.F.A. Survey of research in the design and control of automated guided vehicles systems. European Journal of Operational Research, 2006, 170: 677–709.

[3] Yin-Bin, S. Intelligent Scheduling Wafer Fabrication Interbay Material Handling System. Shanghai Jiaotong University, 2011.

[4] Jensen, M.T. Reducing the run-time complexity of multiobjective EAs: The NSGA-II and other algorithms. IEEE Transactions on Evolutionary Computation, 2003, 7(5): 503–515.

[5] Solomon, M.M. Algorithms for vehicle routing and scheduling problems with time window constraints. Operations Research, 2010, 35(35): 254–266.

[6] Eppstein, D. Finding the k shortest Paths in a network. Management Science, 1971, 19(17):712-716.

[7] Cormen, T.H., Leiserson, C.E., Rivest, R.L., et al. Introduction to Algorithms (Second Edition). Cambridge: The MIT Press, Mcgraw-Hin, 2001, pp. 1297–1305.

[8] Liao, D.Y., Wang, C.N. Differentiated preemptive dispatching for automatic materials handling services in 300 mm semiconductor foundry. The International Journal of Advanced Manufacturing Technology, 2006, 29(9): 890–896.

[9] Osman, I.H., Metastrategy simulated annealing and tabu search algorithms for the vehicle routing problem. Annals of Operations Research, 1993, 41(4): 421–451.

[10] Zhen-Cai, C., Qi-Di, W., Fei, Q., Zun-Tong, W. Advances in semiconductor wafer fabrication scheduling based on genetic algorithm. Journal of Tongji University, 2008, 36(1): 97–102.

9 Scheduling Performance Evaluation of Automated Material Handling Systems in SWFSs

Over the last few years, performance evaluation and analysis of the AMHS have received considerable attention. This chapter investigates the modeling process of the AMHS in order to evaluate the performances of various scheduling methods. Since the production scheduling is within a real-time environment, the evaluation model is required to be not only effective but also efficient. An agent-oriented and knowledge-based colored timed Petri net (AOKCTPN)-based approach is therefore proposed.

9.1 Modeling Requirements of Scheduling Performance Evaluation in the AMHS

Traditional modeling approaches for manufacturing systems include queuing networks, perturbation analysis, Markov chain, Petri net (PN) and so on. Due to the special characteristics of a SWFS, the PN as an analytical tool for discrete event dynamic systems has become the most widely adopted method for modeling semiconductor manufacturing systems and analyzing the schedule performances. In recent years, there have been many reports of PN and extended PN applications in modeling SWFSs.

Object-oriented Petri net (OOPN) methodologies were widely investigated in order to reduce the model scale and improve its re-configurability and re-usability. Zhang et al. [1] proposed an extended OOPN modeling approach to build up a simulation platform for real-time scheduling of SWFS. Liu et al. [2, 3] proposed an extended object-oriented Petri nets (EOPNs) for effectively modeling SWFSs. The resulting model validates that the EOPNs may cope well with complex SWFSs modeling. Wang and Wu [4] proposed an object-oriented hybrid PN model (HOPN) approach for semiconductor manufacturing lines, which provides a good foundation for studying scheduling policies. Zhou et al. [5] proposed an object-oriented and extended hybrid PNs to describe a semiconductor wafer fabrication flow. The modeling results indicated that the proposed modeling approach can deal with the dynamic modeling of a complex photolithography process effectively. However, most of the approaches proposed in above researches are applied for SWFSs; meanwhile few studies can be found for modeling AMHSs. On the other hand, the OOPN approach can only support for the predefined control flow, not for distributed dynamic route control in AMHSs.

For the distributed environment, agent-oriented PN methodologies get increasing attention in semiconductor manufacturing systems. Zhang et al. [6] and Zhai et al. [7] constructed an AOCTPN for SWFSs, which reduced the complexity and

https://doi.org/10.1515/9783110487473-009

improved the reusability of the model, and could be used to describe the coordination scheduling and controlling between agents. However, in these agent-oriented PN approaches, the information and knowledge of scheduling and controlling process of SWFSs and AMHSs cannot be easily described. Therefore, the established models have poor performances in supporting the decision-making.

Based on the above brief literature review, some conclusions are obtained and listed as follows:

1. Although the PN has been widely used in modeling and performance analyzing for SWFSs and AMHSs, most of these models are not suitable for large-scale and complex SWFSs and AMHSs, especially for the 300 mm wafer fab. In addition, such PN-based models are highly system dependent and lack re-configurability and re-usability. Therefore, the adaptability of the model for changing market and dynamic environment cannot be guaranteed.
2. The OOPN-based modeling approach can deal with the dynamic modeling for a complex manufacturing system and has been used to model and analyse SWFSs. However, few researches in this field can be found for AMHSs. At the same time, traditional OOPN approaches can only support the predefined control flow, which may not be applied for distributed dynamic route control in AMHSs.
3. In order to support decision-making such as scheduling and controlling, the model for AMHSs should be capable of describing the status information of AGV and furthermore the knowledge of scheduling and controlling process of AMHSs. However most of the PN-based approaches cannot provide such information and knowledge and therefore are not easy to be integrated with the decision-making process.

In order to fill the above research gaps, the chapter introduces a modeling approach based on an AOKCTPN to evaluate the performances of various scheduling methods. An AOKCTPN is an object-oriented approach, which can achieve great reusability, and adopts knowledge-based decision-making methods for places and transitions in PNs, which can improve the decision-making ability and adaptability.

9.2 A Modeling Approach Based on an Agent-Oriented Knowledgeable Colored and Timed Petri Net (AOKCTPN)

The method based on the simulation model and the method based on the PN model are less reusable and inefficient in evaluating the scheduling performances of an AMHS. In this chapter, the advantages of the PN model in dealing with dynamic discrete events and complex parallel systems are adopted. At the same time, the advantages of the high-level OOPN model in reducing the size and reusability of the problem are combined. Therefore, this chapter proposes a reusable and modular scheduling performance evaluation modeling method: agent-oriented knowledgeable colored and timed Petri net (AOKCTPN).

9.2.1 Definition of an AOKCTPN Model

1. An AOKCTPN method
An AOKCTPN is developed on the basis of an agent-oriented PN modeling theory. By improving the agent-oriented PN modeling theory and expanding the knowledge description of the place and transition based on the object model, the complexity of large-scale SWFS modeling is effectively solved. According to the real SWFS hierarchical architecture, the AOKCTPN model is built up in a hierarchical form in order to make the established system model more concise and accurate. An AOKCTPN can be defined as

$$AOKCTPN = \{APS,\ TC,\ C,\ D,\ A,\ KB,\ M_i\}$$

where

1. APS = $APS = \{AP_1, \ldots, AP_i, \ldots AP_n\}$, which is the set of agents in the model. APS includes basic agent place set and composite agent place set, which is denoted by $APS = \{LAPS\}U\{LCAPS\}$.
$LAPS = \{LAP_1,\ LAP_2, \ldots,\ LAP_i, \ldots,\ LAP_w\}$, which corresponds to a set of basic agent places in the model, where LAP_i is the ith agent place in the set. LAP_i represent the physical resource agents (such as processing machine, transportation rail, shortcut, turntable and stocker) and logical resource agents (such as releasing, dispatching, routing) in an AMHS.
$LCAPS = \{LCAP_1,\ LCAP_2, \ldots,\ LCAP_i, \ldots,\ LCAP_w\}$, which corresponds to a set of composite agent places in the model, where $LCAP_i$ is the ith composite agent place in the set, which is denoted by $LCAP_i = \{LAPS,\ TC,\ C,\ D,\ I,\ O,\ KB,\ M_0\}$.
2. $TC = \left\{TC_1,\ TC_2, \ldots TC_j, \ldots TC_m\right\}$, which corresponds to a limited set of communication transitions, where TC_j is the jth communication transition that provides communication among basic agent places and composite agent places. In addition, $LAP_i \cup TC_j \neq \Phi, LCAP_i \cup TC_j \neq \Phi, LAP_i \cap TC_j = \Phi$, *and* $LCAP_i \cap TC_j = \Phi$.
3. $C = \{C(AP_1), \ldots, C(AP_i), \ldots, C(AP_n)\}U\{C(TC_1), \ldots, C(TC_j), \ldots,\ C(TC_m)\}$, which corresponds to color set of places and communication transitions. $C(AP_i) = \left\{a_{i1},\ a_{i2}, \ldots a_{i\mu_i},\right\}$, which is a color set of agent places AP_i, where $\mu_i = |C(AP_i)|(i = 1,\ 2, \ldots, n)$. $C(TC_j) = \left\{b_{j1},\ b_{j2}, \ldots b_{j\mu_j},\right\}$, which is a color set of agent places TC_j, where $\mu_j = |C(AP_j)|(j = 1,\ 2, \ldots, m)$.

4. D corresponds to time delay set of places and transitions. $D(AP_i) = \{d_{il}, \ldots, d_{jh} \ldots, d_{iu}\}$, $\mu_i = |D(AP_i)| = |C(AP_i)|$ $(i = 1, 2, ..., n)$, where d_{ih} is the time delay required by the agent place AP_i with respect to the hth color a_{ih}, denoted by $d_{ih} = D(a_{ih})$.
5. $A \subseteq I \cup O$, which is the set of directed arc functions of agent place AP and communication transition TC, where $I(AP_i, TC_j)(a_{ix}, b_{jy})$: $C(LAP_i) \times C(TC_j) \to N$ is the input function, and $O(AP_i, TC_j)$ (a_{ix}, b_{jy}): $C(LAP_i) \times C(TI_j) \to N$ is the output function.
6. KB is the set of descriptive and control/decision-making knowledge with respect to agent place AP. $KB : KB(AP_i) - > S_{APi}$ is the mapping from agent place AP_i to the descriptive and control/decision-making knowledge set S_{APi}.
7. $M_0 = [M(AP_i)]_{n*1}$, which is the initial mark of the AOKCTPN model, where $M(AP_i) : APS \to C(AP_i)$ is the mark of AP_i, and $i = 1, 2, ..., n$.

The interior behavior of the basic agent place is encapsulated in an AOKCTPN model. A basic agent place LAP_j is defined by a 7-tuple:

$$LAP_j = \{P^j, T^j, C^j, A^j{}_d, A^j{}_{inh}, K_p, K_T\}$$

1. $P^j = \{LP^j\} U \{MP^j\} U \{SP^j\}$, which corresponds to the set of places in agent place LAP_j, where LP^j is a set of resource places to describe the interior state of the agent; $MP^j = \{IM^j, OM^j\}$ is a set of message places to communicate between agent places, where IM^j corresponds to an input message place from other agent to agent place LAP_j and OM^j corresponds to an output message place; SP^j is a set of information place to describe the real-time information of an agent place.
2. $T^j = \{TI^j, TS^j, TE^j\}$, which is a set of delay time transitions, where TI^j is the set of immediate transitions, TS^j is a set of the stochastic transitions and TE^j is a set of the deterministically timed transitions. In addition, $(LP^j or\ MP^j or\ SP^j) \cup (TI^j or\ TS^j or\ TE^j) \neq \Phi$, $(LP^j or\ MP^j or\ SP^j) \cup (TI^j or\ TS^j or\ TE^j) = \Phi$.
3. C^j corresponds to a color set of places and transitions in agent place LAP_j. $C^j(LP^j)$, $C^j(MP^j)$, $C^j(SP^j)$, $C^j(TI^j)\ and\ C^j(TE^j)$ corresponds to the respective color set of LP^j, MP^j, SP^j, TI^j, $TS^j and TE^j$.
4. $A^j{}_d$ is a set of connecting arc functions between places and transitions, in which $\{LP^j or\ IM^j or\ SP^j\} \times \{TI^j or\ TS^j or\ TE^j\} \to \{0, 1\}$ is the input function and $\{TI^j or\ TS^j or\ TE^j\} \times \{LP^j or\ OM^j or\ SP^j\} \to \{0, 1\}$ is the output function.
5. $A^j{}_{inh}$ is a set of inhibited arc functions, and $Inh(p_l, t_m) : p_l \times t_m \to \{0, 1\}$ is a inhibited arc function between place p_l and transition t_m.
6. K_p is the mapping from place set P^j to place knowledge set S_{pj}. K_T is a mapping from transition set T^j to transition knowledge set S_{Tj}.

Table 9.1: The knowledge classification.

Knowledge subset name	**Detail description of the subset**
Basic descriptive knowledge subset of places and transitions	1) State information of places and transitions. 2) The agent information that places and transitions belong to. 3) Token-type information corresponding to places and transitions. 4) Places' capacity information. 5) Other information.
Statistical descriptive knowledge subset of places and transitions	1) Token's quantity statistical information in places. 2) Fired time statistical information of delay transitions and stochastic transitions. 3) Transitions' fired frequency statistical information. 4) Other information.
Control and decision-making knowledge subset of places and transitions	1) The control and decision-making rules used to select tokens in places. 2) The control and decision-making rules used to solve conflicts among multi-transitions. 3) Other information.

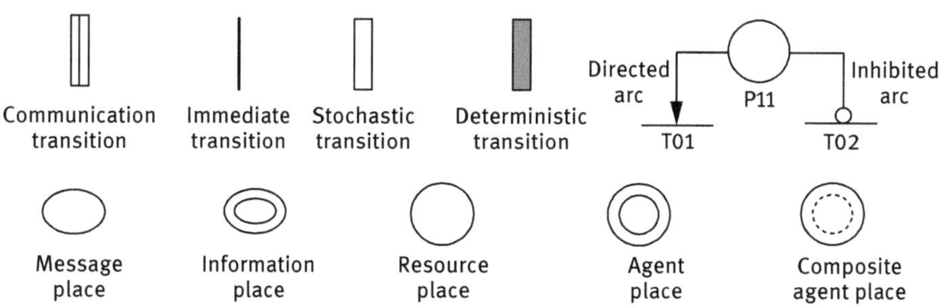

Figure 9.1: Places and transitions of an AOKCTPN.

The icons of places and transitions of the AOKCTON model are shown in Figure 9.1. The knowledge set of places and transitions can be classified into three subsets: basic descriptive knowledge subset, statistical descriptive knowledge subset, and control and decision-making knowledge subset. A detail description of the knowledge classification is shown in Table 9.1.

2. Model operating rule
   The enabling and firing rules of the AOKCTPN model are presented as follows: Transition TC_j is enabled with respect to the color b_{js} if

$$M(AP_i)(a_{ir}) \geq I(AP_i, TC_j)(a_{ir}, b_{js}), \forall AP_i \in APS, a_{ir} \in C(AP_i), b_{js} \in C(TC_j)$$

   When TC_j is fired, the new mark M' becomes:

$$M'(AP_i)(a_{ir}) = M(AP_i)(a_{ir}) + O(AP_i, TC_j)(a_{ir}, b_{js}) - I(AP_i, TC_j)(a_{ir}, b_{js})$$

9.2.2 Modeling Process of an AOKCTPN Model

1. Framework of an AOCKTPN-based AMHS model
   In order to reduce the complexity of the system modeling process and increase the reusability of the system modeling process, agents with the similar structure and function are aggregated to form the layered structure, which includes a basic agent layer, a subsystem layer and a system layer, as shown in Figure 9.2.

2. Basic agent layer modeling
   As the basic elements of an AMHS model, the agents in the basic agent layer include the physical agent and the logical one. Corresponding to the manufacturing resources of AMHSs and SWFSs, the physical agents include a machine agent, a transportation rail agent, a shortcut agent, a turntable agent, and a stocker agent and so on. The logical agents with dispatching and control functions include a lot release agent, an AMHS scheduling agent, a routing agent (included in turntable agent), a zone control agent (included in transportation rail agent), etc.
   (1) Machine agent model
   The machine agent is the core in the basic agent layer, it is used to describe the characteristics of processing machines, wafer lot loading process, and wafer lot unloading process and so on. The AOKCTPN-based machine agent model is developed and shown in Figure 9.3. The detailed reasoning process of the machine agent model is described as follows: the token in the input message place IM_{i1} represents the lot-processing-request message sent by the

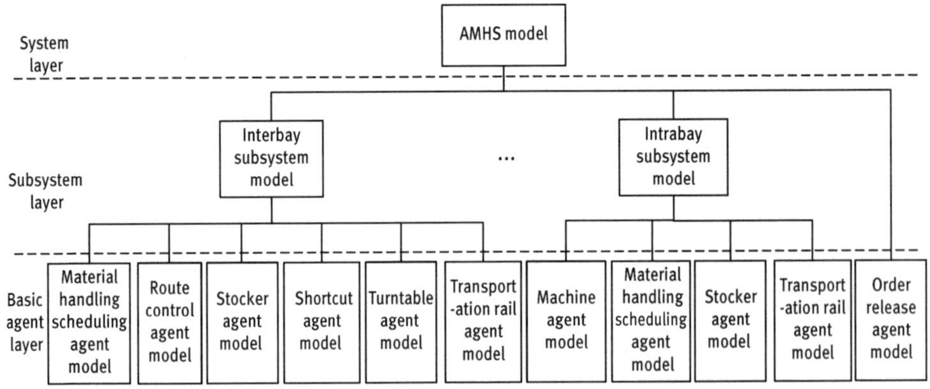

Figure 9.2: Architecture of an AOCKTPN-based scheduling performance evaluation model in AMHSs.

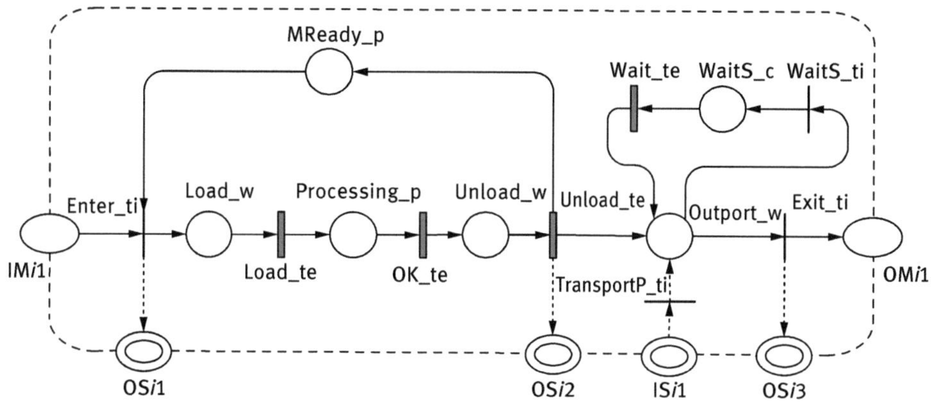

Figure 9.3: An AOKCTPN-based machine agent model.

transportation rail agent. The token in the place *MReady_p* corresponds to machine idleness. When the machine is idle and receives the lot-processing-request message, the immediate transition *Enter_ti* is fired, the lot token enters the place *Load_w*. When the lot is loaded by an idle machine, then the load delay transition *Load_te* is fired, the lot enters the place *Processing_p*. When the machine finished the operation, the delay transition *OK_te* is fired and the lot enters the place *Unload_w*. When the lot is unloaded by a machine, the unload delay transition *Unload_te* is fired, the machine becomes idle and the machine token enters *MReady_p*, and the lot token in place *Unload_w* enters the place *Output_w*. If the place *Output_w* has token and information place IS_{i1} received information token, then the transition *Exit_ti* is fired, and the lot token in place *Output_w* enters the output message place OM_{i1}. Otherwise, the immediate transition *WaitS_ti* is fired, the lot token enters the place *WaitS_c*. After a set unit time t_{stand}, the delay transition *Wait_te* is fired, the lot token re-enters the place *Output_w*. The data types of different color are shown in Table 9.2. The color sets of the machine agent model are shown in Table 9.3. The knowledge set of the machine agent model is described in Table 9.4. In the machine agent model, the information place set $\{OS_{i1}, OS_{i2}, OS_{i3}\}$ is used to output real-time state information of machines, and the information place IS_{i1} is used to input vehicles' state information.

(2) Transportation rail agent model
The AOKCTPN-based transportation rail agent model including a loading-segment transportation rail agent model and an unloading-segment transportation rail agent model is developed and shown in Figure 9.4. The detailed reasoning process of the unloading-segment transportation rail agent model is described as follows: the token in the input message place IM_{i1} represents

Table 9.2: The types of color data.

Data type name	Interpretation	Elements
LI	Lot no.	Natural numbers
P	Product ID	$\{P_1, \ldots, P_j, \ldots, P_m\}$
S	Lot stage	$\{S_1, \ldots, S_j, \ldots, S_n, S_{end}\}$
PT	Product processing time	$\{PT_1, \ldots, PT_j, \ldots, PT_n\}$
KM	Machine no.	$\{KM_1, \ldots, KM_j, \ldots, KM_n\}$
E	Machine state	$\{e_w, e_e\}$
BAY	Material handling subsystem	$\{Bay_1, \ldots, Bay_k, \ldots, Bay_n, Bay_{inter}\}$
PR	Transportation rail no.	$\{PR_1, \ldots, PR_j, \ldots, PR_n\}$
PRT	Moving times of vehicles on transportation rail	$\{PRT_1, \ldots, PRT_j, \ldots, PRT_n\}$
PE	Transportation rail state	$\{p_w, p_e\}$
VT	Vehicle no.	$\{VT_1, \ldots, VT_j, \ldots, VT_m\}$
VE	Vehicle state	$\{v_w, v_e\}$
ST	Stocker no.	$\{ST_1, \ldots, ST_j, \ldots, ST_n\}$
STE	Stocker state	$\{s_w,\ s_e\}$
TT	Turntable no.	$\{TT_1, \ldots, TT_j, \ldots, TT_n\}$
TTE	Turntable state	$\{t_w,\ t_e\}$
ST_{IN}/ST_{OUT}	The time that lot enters/leaves system	$\{0, t_{stand}, \ldots, k\ t_{stand}, \ldots, q\ t_{stand} \ldots\}$

Table 9.3: The color set of an AOKCTPN-based machine agent model.

Name	Function interpretation	Color set
IM_{i1}	Input message of the lot waiting for entering machine	$LI \times \{P_i\} \times \{S_j\} \times KM$
OM_{i1}	Output message of the lot leaving a machine	$LI \times \{P_i\} \times \{S_{j+1}\}$
MReady_p	Machine being idle	$KM \times \{e_e\}$
Enter_ti	Lot enters machine	$LI \times \{P_i\} \times \{S_j\} \times KM \times \{e_e\}$
Load_w	Prepare to load lots	$LI \times \{P_i\} \times \{S_j\} \times KM \times \{e_e\}$
Load_te	Lot be loaded by machines	$LI \times \{P_i\} \times \{S_j\} \times KM \times \{e_w\}$
Processing_p	Machine is processing	$LI \times \{P_i\} \times \{S_j\} \times KM \times \{e_w\} \times \{PT_j\}$
OK_te	Machine finishes process	$LI \times \{P_i\} \times \{S_j\} \times KM \times \{e_w\}$
Unload_w	Prepare to unload lots	$LI \times \{P_i\} \times \{S_j\} \times KM \times \{e_w\}$
Unload_te	Lot be unloaded by machines	$LI \times \{P_i\} \times \{S_j\} \times KM \times \{e_w\}$
Outport_w	Lot being in machine's output-port	$LI \times \{P_i\} \times \{S_{j+1}\} \times KM \times \{e_e\}$
Exit_ti	Lot is preparing to leave a machine	$PR \times \{v_e\} \times LI \times \{P_i\} \times \{S_{j+1}\}$
TransportP_ti	Vehicle has been prepared to carry a lot	$PR \times \{v_e\}$

Table 9.3: (continued)

Name	Function interpretation	Color set
Wait_te	Lot has been waiting for a unit time	$LI \times \{P_i\} \times \{S_{j+1}\}$
WaitS_c	Lot is waiting vehicles	$LI \times \{P_i\} \times \{S_{j+1}\}$
WaitS_ti	Lot starts to wait vehicles	$LI \times \{P_i\} \times \{S_{j+1}\}$
OS_{i1}	Output information "machine starts to process"	$KM \times \{e_w\} \times LI \times \{P_i\} \times \{S_j\}$
OS_{i2}	Output information "machine finishes process"	$KM \times \{e_e\} \times LI \times \{P_i\} \times \{S_j\}$
OS_{i3}	Output information "lot prepares to leave machine"	$LI \times \{P_i\} \times \{S_{j+1}\}$
IS_{i1}	Input information "vehicle is waiting to carry lot"	$PR \times \{v_e\}$

Table 9.4: The knowledge set of an AOKCTPN-based machine agent model.

Name	Knowledge type	Knowledge description
Wait_te	Basic descriptive knowledge of transitions	Transition name: *Wait_te*. Transition type: delay transition. Corresponding agent: Machine agent *PM_ap_i*. Delay time of transition: t_{stand} (t_{stand} is the set unit time of an AMHS system).
	Statistics descriptive knowledge of transitions	The time of transition fired: N^i_m.
Outport_w	Basic descriptive knowledge of places	Place name: *Outport_w*. Place type: resource place. Corresponding agent: machine agent *PM_ap_i*. Place capacity: 1.
	Control and decision-making knowledge of place	Rule 1: If the current place has token and information place IS_{i1} received information "a vehicle is waiting to carry a lot", then transition *Exit_ti* is fired. Rule 2: If the current place has token and information place IS_{i1} didn't receive any information, then transition *WaitS_ti* is fired.

the vehicle-entrance-request message sent by the upstream transportation rail agent. The token in the place *Ready_p* corresponds to unloading-segment transportation rail idleness. When the unloading-segment transportation rail is idle and receives the vehicle-entrance-request message, the immediate

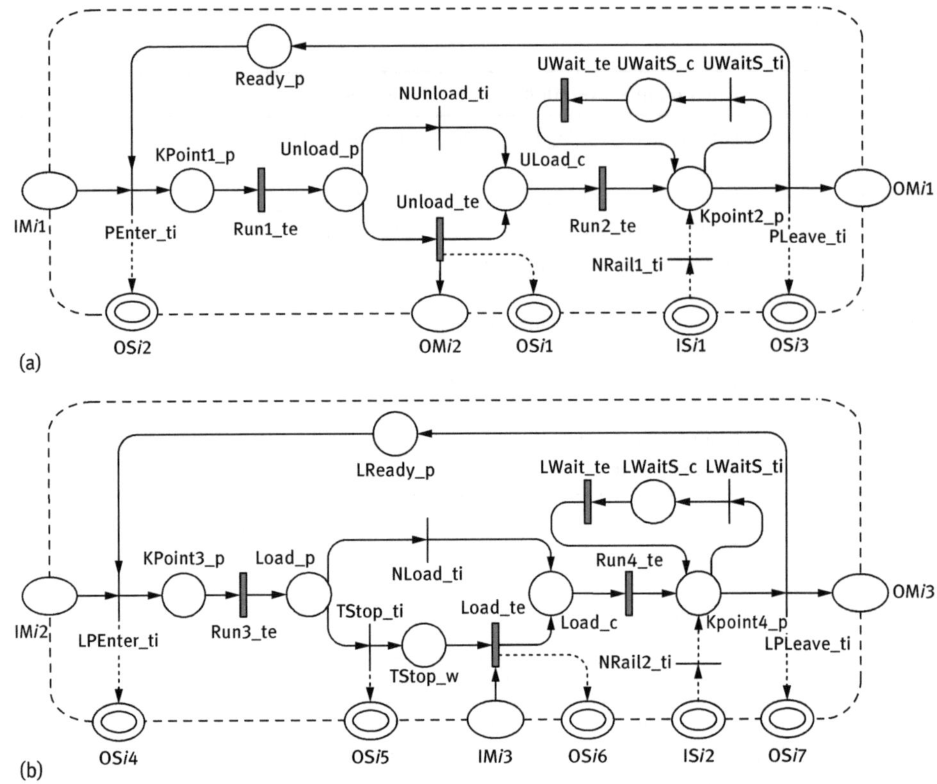

Figure 9.4: An AOKCTPN-based transportation rail agent model: (a) a transportation rail agent model of vehicle unloading positions and (b) a transportation rail agent model of vehicle loading positions.

transition *PEnter_ti* is fired, the vehicle token enters the place *KPoint1_p*, and then the delay transition *Run1_te* is fired. After a period of moving time PRT_j, the vehicle token enters place *Unload_p*. If the color of the vehicle token in place *Unload_p* is $\{v_w\} \times \{KM_i^k \text{ or } ST_i^k\}(KM_i^k \text{ or } ST_i^k$ is the machine or stocker No. of the current transportation rail connected), then the delay transition *Unload_te* is fired. After a specific vehicle unloading time, the vehicle token enters place *Uload_c*, and the lot token enters the output message place OM_{i2}. Otherwise, the immediate transition *NUnload_ti* is fired, and the vehicle token enters place *Uload_c* directly. When *Uload_c* received the vehicle token, the delay transition *Run2_te* is fired. After a specific vehicle moving time, the vehicle arrives the terminal position of the current transportation rail and the vehicle token enters the place *Kpoint2_p*. If the place *Kpoint2_p* has vehicle token and information place IS_{i1} received information token, then the immediate transition *PLeave_ti* is fired, and the vehicle token in place *Kpoint2_p* enters the output message place OM_{i1}. Otherwise, the immediate transition *UWaitS_ti* is fired, the vehicle token enters place *UWaitS_c*. After a set unit time t_{stand}, the delay transition *UWait_te* is fired, the vehicle token re-enters the place *Kpoint2_p*. The reasoning process of the loading-segment transportation rail agent model is similar to that of the unloading-segment transportation rail agent model, due to space limitations, the reasoning process of the loading-segment is not presented in this section.

(3) Shortcut agent model
The AOKCTPN-based shortcut agent model is developed and shown in Figure 9.5. The detailed reasoning process of the shortcut agent model is described as follows: the token in the input message place IM_{i1} represents the vehicle-entrance-request message sent by the upstream turntable agent. The token in the place *SReady_p* corresponds to shortcut idleness. When the shortcut is idle and receives the vehicle-entrance-request message, the immediate transition *SEnter_ti* is fired, the vehicle token enters the place *SPoint1_p*, and then the delay transition *SRun1_te* is fired. After a specific moving time, the vehicle token enters place *Spoint2_p*, If the place *Spoint2_p* has vehicle token and information place IS_{i1} received information token, then the immediate transition *SLeave_ti* is fired, and the vehicle token in place *Spoint2_p* enters the output message place OM_{i1}. Otherwise, the immediate transition *SWaitS_ti* is fired, the vehicle token enters place *SWaitS_c*. After a set unit time t_{stand}, the delay transition *SWait_te* is fired, the vehicle token re-enters the place *Spoint2_p*.

(4) Stocker agent model
The AOKCTPN-based stocker agent model is built up and shown in Figure 9.6. The detailed reasoning process of the stocker agent model is described

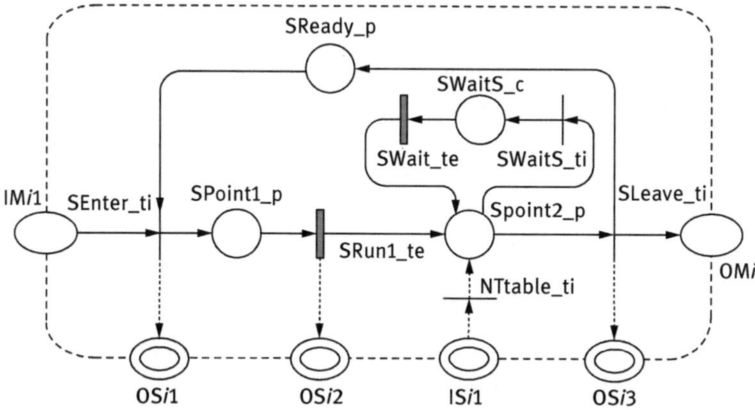

Figure 9.5: An AOKCTPN-based shortcut agent model.

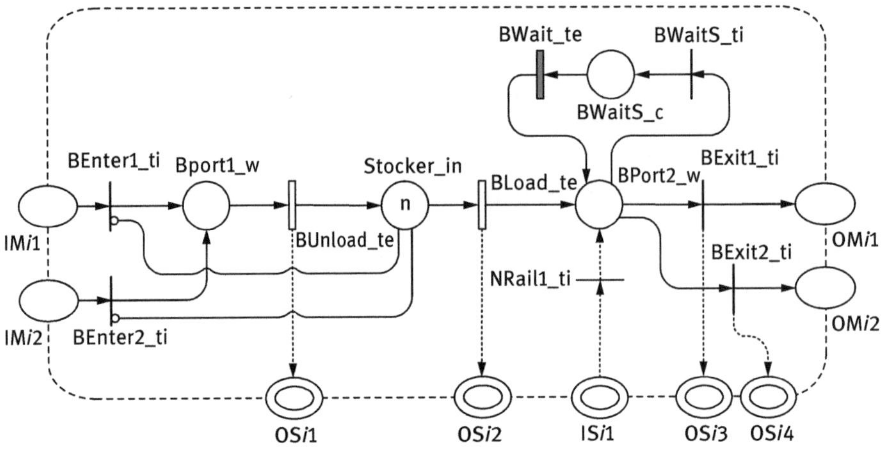

Figure 9.6: An AOKCTPN-based stocker agent model.

as follows: the token in the input message place IM_{i1} and IM_{i2} represents the vehicle-entrance-request message sent by the transportation rail agent of an Intrabay system and an Interbay system, respectively. In the process of restoring wafer lots, when the input message place IM_{i1} or IM_{i2} in the stocker agent receives the wafer lot entrance-request message, the immediate transition *BEnter1_ti* or *BEnter2_ti* is fired, the wafer lot token enters the place *Bport1_w*. After a stochastic storing time (i.e., stochastic transition *BUnload_te* is fired), the wafer lot token enters the place *Stocker_in*. When a quantity of wafer lot token in place *Stocker_in* equals to the capacity of the stocker, the immediate transitions *BEnter1_ti* and *BEnter2_ti* are prohibited to be fired. In the process of pulling out wafer lots, when the place *Stocker_in* has wafer lot tokens, which one wafer lot be taken out firstly is decided based on the knowledge of the place *Stocker_in* (e.g., highest priority rule, longest wait time rule and so on). After the wafer lot token is selected, then stochastic transition *BLoad_te* is fired and the wafer lot token enters place *BPort2_w*. If the place *BPort2_w* has wafer lot token and information place IS_{i1} received information "vehicle in the Intrabay system prepares to carry wafer lot", then the immediate transition *BExit1_ti* is fired, and the wafer lot token in place *BPort2_w* enters the output message place OM_{i1}. Else if the place *BPort2_w* has wafer lot token and information place IS_{i1} received information "vehicle in the Interbay system prepares to carry wafer lot", then the immediate transition *BExit2_ti* is fired, and the wafer lot token enters the output message place OM_{i2}. Otherwise, the immediate transition *BWaitS_ti* is fired, the vehicle token enters the place *BWaitS_c*. After a set unit time t_{stand}, the delay transition *BWait_te* is fired, the vehicle token re-enters the place *BPort2_w*.

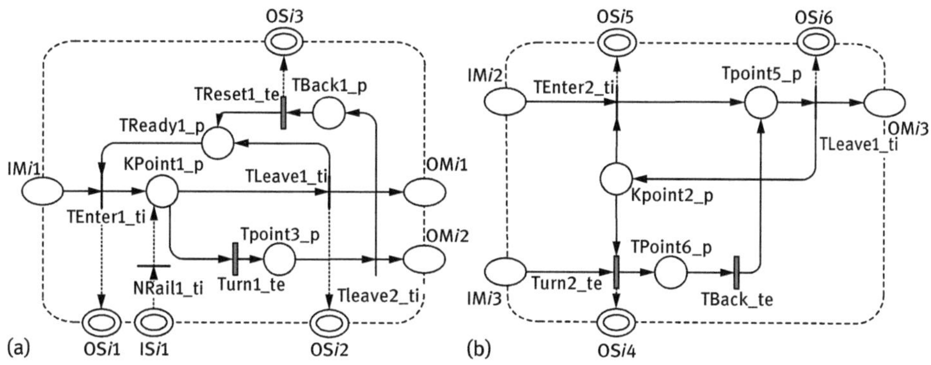

Figure 9.7: An AOKCTPN-based turntable agent model: (a) model with single-input and multi-output interface and (b) model with multi-input and single-output interface.

(5) Turntable agent model
The AOKCTPN-based turntable agent model including a turntable agent model with single-input and a multi-output (SIMO) interface and turntable agent model with multi-input and single-output (MISO) interface is built up and shown in Figure 9.7. The detailed reasoning process of SIMO turntable agent model is described as follows: the token in the input message place IM_{i1} represents the vehicle-entrance-request message sent by the upstream transportation rail agent. The token in the place *TReady1_p* corresponds to turntable idleness. When the turntable is idle and receives the vehicle-entrance-request message, the immediate transition *TEnter1_ti* is fired, and vehicle token enters the place *KPoint1_p*. If the place *KPoint1_p* has vehicle token and information place IS_{i1} received information token, then the delay transition *Turn1_te* and immediate transition *Tleave2_ti* are fired in sequence. The vehicle token in place *KPoint1_p* enters the place OM_{i2}. Otherwise, the immediate transition *TLeave1_ti* is fired and the vehicle token enters the place OM_{i1}, that is, the vehicle enters the downstream transportation rail. Since the reasoning processes of the MISO turntable agent model is similar to that of the SIMO turntable, the detailed reasoning process of the MISO turntable is not introduced in this section.

(6) Order release agent model
Figure 9.8 shows the order release agent model based on an AOKCTPN. The order release process is as follows: the token in Release_s indicates that the system is ready to start release orders. Initially, the token in the place Release_s triggers the instantaneous transition place Release_p and then triggers the instantaneous transition place Exit_ti. The wafer leaves the order release agent model and enters the feeding output processing place

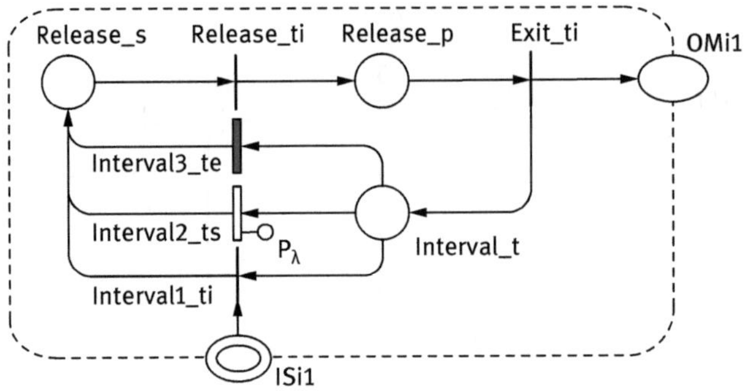

Figure 9.8: An AOKCTPN-based order release agent model.

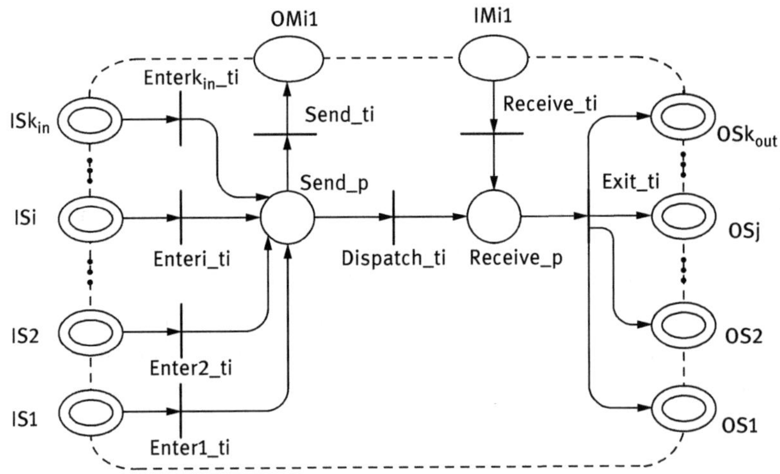

Figure 9.9: An AOKCTPN-based material transportation scheduling agent model.

OMi1. At the same time, the status bit Interval_t is used to trigger the transition time bit of the response according to the material setting information. The random delay transition time Interval2_ts corresponds to the stochastic distribution order release strategy, which determines the delay time interval Interval3_te corresponding to the fixed time interval of the order release strategy.

(7) Material handling scheduling agent model
The material handling scheduling agent model based on AOKCTPN method is shown in Figure 9.9. The material handling scheduling process is as follows: when the transport cart finishes unloading the wafer card or the wafer card arrives at the processing equipment output port for transportation, and the corresponding information place OSi receives the material

handling scheduling request, triggers the instantaneous transition position Enteri_ti and enters the material handling scheduling optimization request status place Send_p. The status bit Send_p triggers the transient transition place Sent_ti to send a material handling scheduling request to the real-time optimization dispatch module. When the real-time optimization scheduling module forms the optimal scheduling scheme, the material handling scheduling agent model receives the optimal scheduling scheme through the transient transition Receive_ti. After the transient transition Exit_ti, the optimal material handling scheduling information is sent out.

3. Subsystem layer modeling
   Based on several basic agent layer models and communication transitions, the subsystem layer model can be constructed. Since AMHSs contain Interbay subsystems and Intrabay subsystems, the subsystem layer model can be subclassified as an Interbay subsystem model and an Intrabay subsystem model.
   (1) Intrabay subsystem model
   The AOKCTPN-based Intrabay subsystem model is developed and shown in Figure 9.10. The detailed reasoning process of the Intrabay subsystem model is described as follows: the token in the input message place IM_{i1} corresponds to the wafer lot requesting for entering the Intrabay subsystem. By firing the communication transition *SP_ti*, the wafer lot token enters the stocker agent place *Stocker_ap*. When the wafer lot in the stocker is preparing to move to the machine corresponding to the current processing step (suppose that the machine agent is *PM_ap_i*), the optimal empty vehicle is designated to carry it based on the Intrabay's intelligent scheduling approach. When the designated empty vehicle token enters the transportation rail agent *VPA_ap_n* + 1, the communication transitions *RK_ti_n* + 1 and *TK_ti_n* + 1 are fired, and the empty vehicle token in *VPA_ap_n* + 1 and the wafer lot token in *Stocker_ap* enter the transportation rail *VPB_ap_n* + 1. The wafer in the stocker is loaded into the empty vehicle. After firing the communication transition *RR_ti_n* + 1 and *CR_ti*, the occupied vehicle token enters the route agent place *Route_ap*. Guided by the running path table of the Intrabay system, occupied vehicle token chooses to enter the downstream transportation rail in sequence. When the occupied vehicle token enters the transportation rail agent *VPA_ap_i*, the communication transitions *TP_ti_i* and *RK_ti_i* are fired concurrently. The wafer lot token in occupied vehicle is unloaded to machine agent *PM_ap_i* and the emptied vehicle token enters the downstream transportation rail agent *VPB_ap_i*. When the wafer lot in machine agent *PM_ap_i* is processed, then the optimal empty vehicle is designated to carry the wafer lot based on the Intrabay's intelligent scheduling approach again. After loading and transporting by designated vehicles, the wafer lot token is moved to the

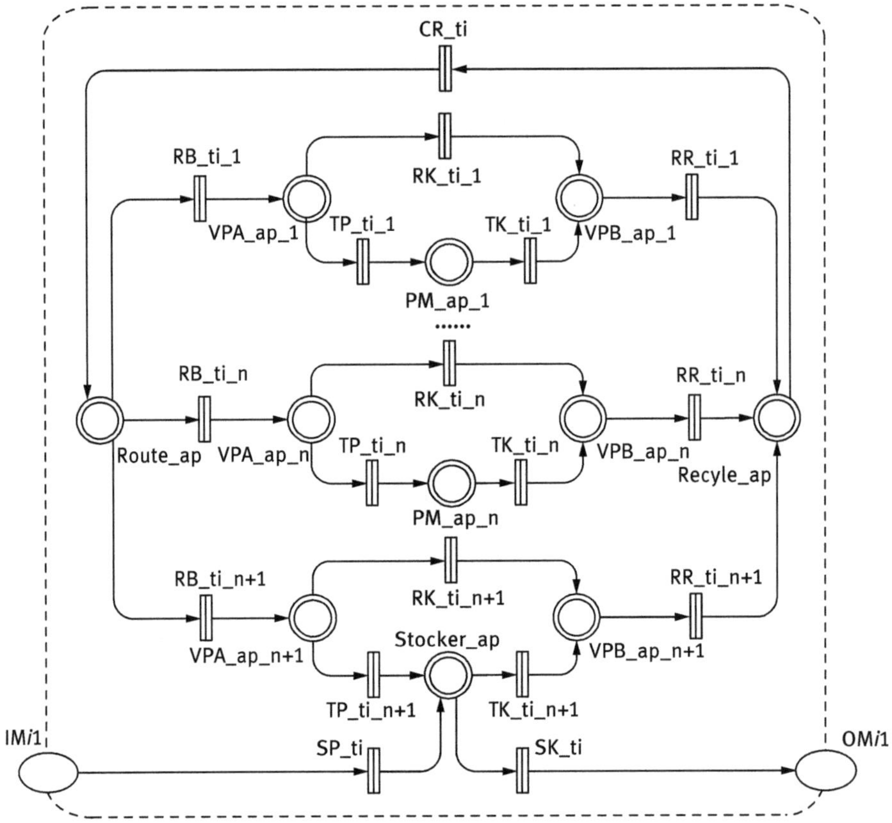

Figure 9.10: The AOKCTPN-based Intrabay subsystem model.

stocker agent *Stocker_ap*. After the wafer lot finished all process in the Intrabay subsystem, the communication transition *SK_ti* is fired and the wafer lot token enters the output message place OM_{i1}, the wafer lot leaves the current Intrabay subsystem.

(2) Interbay subsystem model
The AOKCTPN-based Interbay subsystem model is developed and shown in Figure 9.11. The detailed reasoning process of the Interbay subsystem model is described as follows: the token in the input message place IM_i corresponds to the wafer lot requesting for entering the Interbay subsystem. By firing the communication transition *RP_ti_i*, the wafer lot token enters the stocker agent place *Stocker_ap_i*. When the wafer lot in a stocker requests to move to the destination stocker (suppose that the destination stocker is *Stocker_ap_j*), the optimal empty vehicle is designated to carry it based on the Interbay's intelligent scheduling approach. When the designated empty

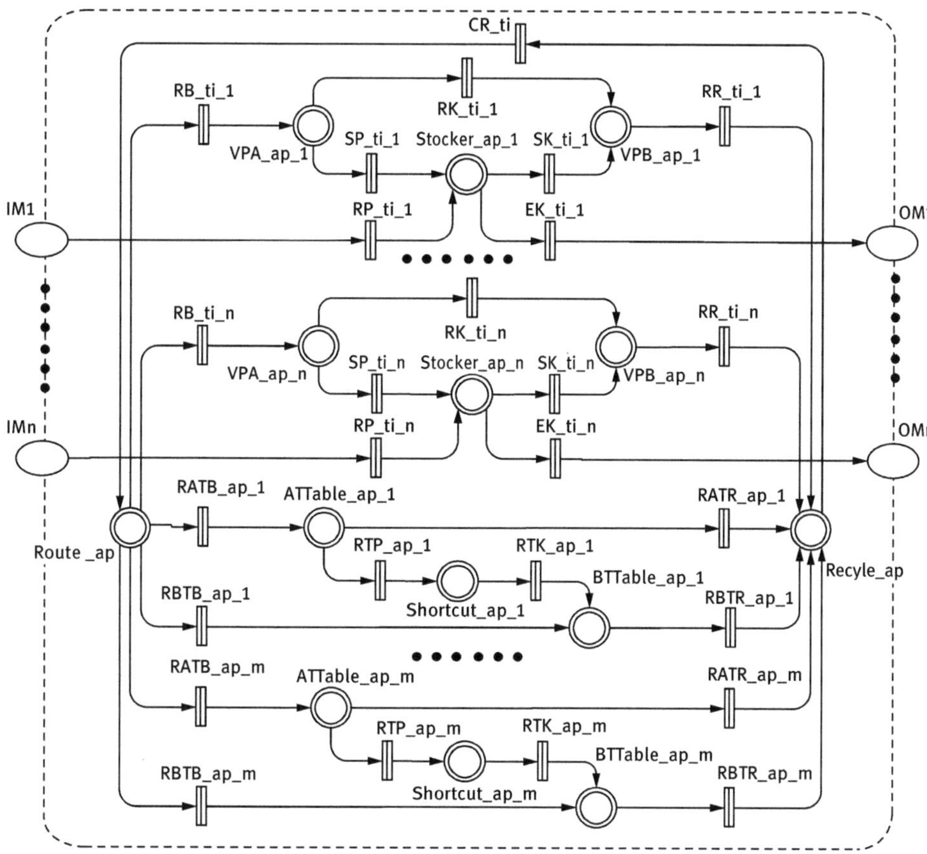

Figure 9.11: The AOKCTPN-based Interbay subsystem model.

vehicle token enters the transportation rail agent *VPA_ap_i*, the communication transitions *RK_ti_i* and *SK_ti_i* are fired, the empty vehicle token in *VPA_ap_i* and the wafer lot token in *Stocker_ap_j* enter the transportation rail place *VPB_ap_i* and the wafer in the stocker is loaded into an empty vehicle. After firing the communication transition *RR_ti_i* and *CR_ti*, the occupied vehicle token enters the route agent place *Route_ap*. Guided by running path table of the Interbay system, the occupied vehicle token chooses to enter the downstream transportation rail or the shortcut until arrives at the transportation rail *VPA_ap_j*. After the occupied vehicle token enters the transportation rail agent *VPA_ap_j*, the communication transitions *SP_ti_j* and *RK_ti_j* are fired. On the one hand, when the communication transition *SP_ti_j* is fired, the wafer lot token is unloaded to the stocker agent *Stocker_ap_j*. After firing the communication transition *EK_ti_j*, the wafer lot enters the output message place OM_j. On the other hand, the

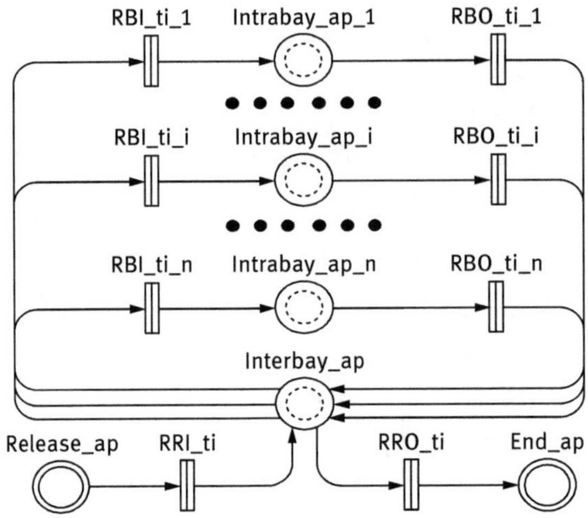

Figure 9.12: The AOKCTPN-based system layer model.

communication transition *RK_ti_j* is fired and the empty vehicle token enters the downstream transportation rail agent *VPB_ap_j*. The empty vehicle runs on the transportation rail continuously.

4. System layer modeling
   The AOKCTPN-based system layer model is developed and shown in Figure 9.12. The detailed reasoning process of the system layer model is described as follows: according to controlling and decision-making knowledge, the release agent place *Release_ap* generates a wafer lot token. After firing the communication transition *RRI_ti*, the wafer lot token enters the composite agent place *Interbay_ap*. Based on the wafer lot's manufacturing process, the corresponding communication transition *RBI_ti_i* is fired and the wafer lot token enters the composite agent place *Intrabay_ap_i*. After finishing the current processing step in the composite agent place *Intrabay_ap_i*, the wafer lot token returns the composite agent place *Interbay_ap*. If the wafer lot has finished all steps, the wafer lot token enters exit agent place *End_ap*. Otherwise, the wafer lot token re-enters the Intrabay subsystem agent for the next processing step. The color sets of the system layer model are shown in Table 9.5. The knowledge sets of the system layer model are described in Table 9.6.

9.2.3 Feasibility Analysis of an AOKCTPN

The feasibility analysis of an AOKCTPN model is used to qualitatively analyze the AOKCTPN model under the guidance of the high-level PN theory to test whether the AOKCTPN model's structure and operation process are correct and feasible. The feasibility of the AOKCTPN model is the prerequisite and basis for the

Table 9.5: The color sets of an AOKCTPN-based system layer model.

Name	Function interpretation	Color set
Release_ap	Wafer lot release	$LI\times\{P_i\}\times\{S_1\}$
RRI_ti	Wafer lot requests to enter an AMHS	$LI\times\{P_i\}\times\{S_1\}$
Interbay_ap	Wafer lot is running in an Interbay subsystem	$LI\times\{P_i\}\times\{S_{j+1}\}\times\{Bay_{\text{inter}}\}$
RBI_ti_i	Wafer lot requests to enter an Intrabay subsystem	$LI\times\{P_i\}\times\{S_j\}\times\{Bay_k\}$
Intrabay_ap_i	Wafer lot is running in an Intrabay subsystem	$LI\times\{P_i\}\times\{S_j\}\times\{Bay_k\}$
RBO_ti_i	Wafer lot requests to enter an Interbay subsystem	$LI\times\{P_i\}\times\{S_{j+1}\}\times\{Bay_{\text{inter}}\}$
RRO_ti	Wafer lot requests to leave an AMHS	$LI\times\{P_i\}\times\{S_{\text{end}}\}$
End_ap	Wafer lot exits an AMHS	$LI\times\{P_i\}\times\{S_{\text{end}}\}$

Table 9.6: The knowledge sets of an AOKCTPN-based system layer model.

Name	Knowledge type	Knowledge description
Release_ap	Basic descriptive knowledge of places	Place name: *Release_ap*. Place type: Agent place.
	Statistics descriptive knowledge of places	The number of wafer lot that leaved place: r_1^i.
	Control and decision-making knowledge of places	Rule1: The token is generated based on the constant interval time. Rule 2: The token is generated based on the stochastic interval time. Rule 3: The token is generated based on the CONWIP (constant work-in-process) rule.

performance evaluation of the scheduling method for AMHSs. The main properties of the feasibility analysis of the model include boundedness and activity, that is, the model is structurally live and bounded. The concrete steps of the feasibility analysis of the AOKCTPN model are presented as follows:

Step 1: Establish the AOKCTPN model of each Intrabay and Interbay subsystem based on the AOKCTPN modeling method.

Step 2: Based on the regularization and simplification of the scheduling/control strategy and knowledge of each basic agent module, the agent bits and the compound proxy bits are expanded to the CTPN model to describe the internal behavior of the agent bit.

Step 3: Considering the complexity and large-scale characteristics of a CTPN model, the CTPN model is simplified by using modularized extended CTPN model under the premise of keeping the model live and bounded in order to reduce the complexity of the model.

Step 4: Establish the reachability map of a simplified CTPN model in order to check the activity and boundedness of the AOKCTPN model.

On the basis of the simplification theory of PN and the simplification method of the timed PN, the simplification rules of the modularized extended CTPN model are presented as follows:

Definition 1: $E = \{P, T, C, I, O, M_0\}$ is defined as the CTPN-based system model, $T = \{TI, TD\}$. $E_A = \left\{P^A, T^A, C^A, I^A, O^A, M_0^A\right\}$ is defined as an agent module based on the CTPN of system model E, $T^A = \{TI^A, TD^A\}$, $E_A \in E$.

Definition 2: $S = (M, F)$ is defined as a state of CTPN model E, where M is the state of model E, F is the set of firing time intervals of the model and $L(E)$ is the transition triggering sequence set. Set t_0 is trigger at the model initial time θ_0, the system state changes from $S_0 = (M_0, F_0)$ to $S_1 = (M_1, F_1)$; at time θ_i the transition t_i triggers the system state to $S_{i+1} = (M_{i+1}, F_{i+1})$. The transition triggering sequence is $\varpi = (t_0, \theta_0)(t_1, \theta_1)...(t_i, \theta_i)$.

Rule 1: For the agent module E_A shown in Figure 9.13, $E_A.P_{in} = \{\mathrm{IP}_1\}$, $E_A.P_{out} = \{\mathrm{OP}_1\}$., $\forall M_0$, $M(P_i^A)(a_{ik}) < I(P_i^A, T_j^A)(a_{ik}, b_{js})$. If:

(1) when IP_1 receives a token, OP_1 receives a token after $\mathrm{T}(0 < T < \infty)$;

(2) when OP_1 receives a token, IP_1 can receive another token.

Then the agent module E_A can be simplified to $E_A{}'$ (Figure 9.13), where (1) $\mathrm{IP}_1{}'^{\bullet} = {}^{\bullet}\mathrm{OP}_1{}' = \{t^{A'}\}$, $t^{A'\bullet} = \{\mathrm{OP}_1{}'\}$, ${}^{\bullet}t^{A'} = \{\mathrm{IP}_1{}'\}$; (2) $C(\mathrm{OP}_1{}') = C(\mathrm{OP}_1)$, $C(\mathrm{IP}_1{}') = C(\mathrm{IP}_1)$, $C(t^{A'}) = C({}^{\bullet}\mathrm{OP}_1) + \{\sum_{i=1}^{ts} td_i^A\}$, ts is the quantity of delay transitions in E_A; (3) $I(\mathrm{IP}_1{}', t^{A'}) = \sum_{P_i \in P^A, T_i \in T^A} I(P_i, T_j)$, $O(\mathrm{OP}_1{}', t^{A'}) = \sum_{P_i \in P^A, T_i \in T^A} O(P_i, T_j)$.

Rule 2: For the agent module E_A shown in Figure 9.14, $E_A.P_{in} = \{\mathrm{IP}_1\}$, $E_A.P_{out} = \{\mathrm{OP}_1, \mathrm{OP}_2\}$.

$$\forall M_0,\ M(P_i^A)(a_{ik}) < I(P_i^A, T_j^A)(a_{ik}, b_{js}).\ \text{If}:$$

(1) when IP_1 receives a token, OP_1 or OP_2 receives a token after T $(0 < T < \infty)$;

(2) when OP_1 or OP_2 receives a token, IP_1 can receive another token.

Then the agent module E_A can be simplified to $E_A{}'$ (Figure 9.14), where (1) $\mathrm{IP}_1{}'^{\bullet} = \{t_1^{A'}, t_2^{A'}\}$, ${}^{\bullet}\mathrm{OP}_1{}' = \{t_1^{A'}\}$, ${}^{\bullet}\mathrm{OP}_2{}' = \{t_2^{A'}\}$, $t_1^{A'\bullet} = \{\mathrm{OP}_1{}'\}$, $t_2^{A'\bullet} = \{\mathrm{OP}_2{}'\}$, ${}^{\bullet}t_1^{A'} = {}^{\bullet}t_2^{A'} =$

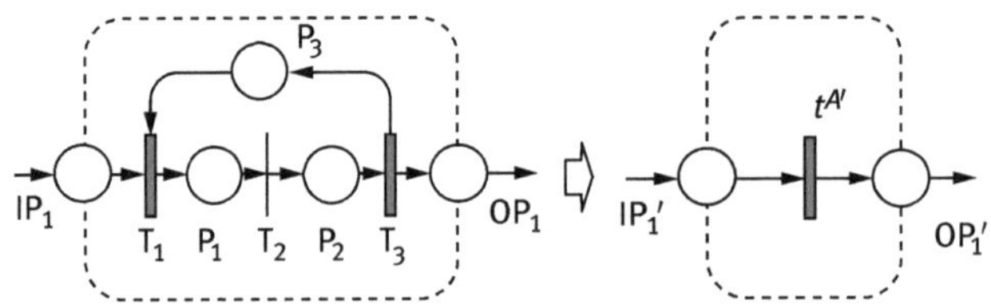

Figure 9.13: Model simplification rule 1.

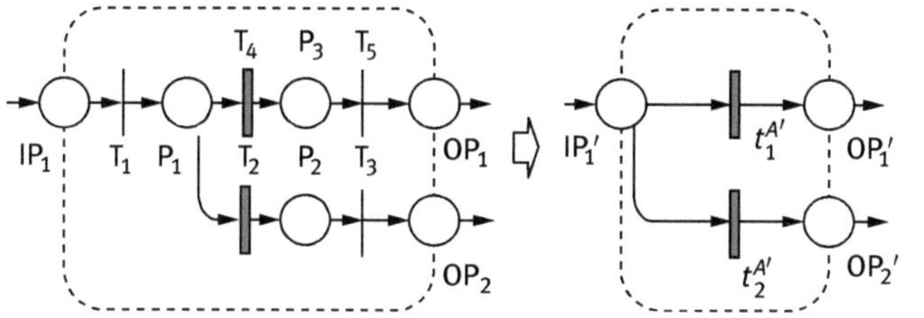

Figure 9.14: Model simplification rule 2.

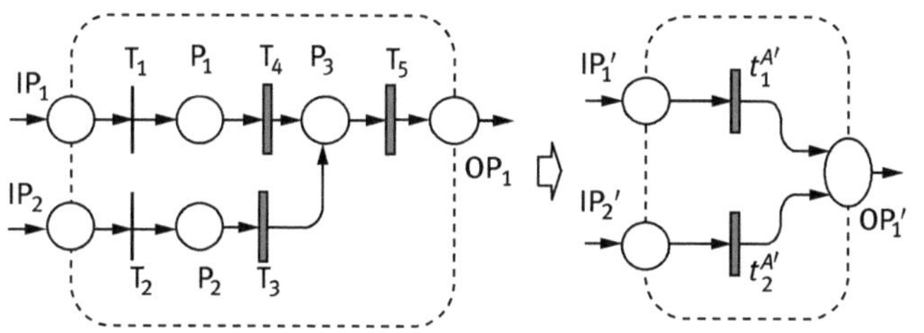

Figure 9.15: Model simplification rule 3.

$\{\mathrm{IP}_1'\}$; (2) $C(\mathrm{OP}_1') = C(\mathrm{OP}_1)$, $C(\mathrm{OP}_2') = C(\mathrm{OP}_2)$, $C(\mathrm{IP}_1') = C(\mathrm{IP}_1)$, $C(t_1^{A'}) = C({}^\bullet\mathrm{OP}_1)+\{\sum_{td_i^A \in \emptyset_1} td_i^A\}$, $C(t_2^{A'}) = C({}^\bullet\mathrm{OP}_2)+\{\sum_{td_j^A \in \emptyset_2} td_j^A\}$, $\emptyset_1$ and $\emptyset_2$ are the sets of locations and transitions sequences in E_A; (3) $I(\mathrm{IP}_1', t^{A'}) = \sum_{P_i, T_j \in \emptyset_1} I(P_i, T_j)$, $O(\mathrm{OP}_1', t^{A'}) = \sum_{P_i, T_j \in S_1} O(P_i, T_j)$, $I(\mathrm{IP}_2', t^{A'}) = \sum_{P_i, T_j \in \emptyset_2} I(P_i, T_j)$, $O(\mathrm{OP}_2', t^{A'}) = \sum_{P_i, T_j \in \emptyset_2} O(P_i, T_j)$.

Rule 3: For the agent module E_A shown in Figure 9.15, $E_A.P_{in} = \{\mathrm{IP}_1, \mathrm{IP}_2\}$,$E_A.P_{out} = \{\mathrm{OP}_1\}$. $\forall M_0$, $M(P_i^A)(a_{ik}) < I(P_i^A, T_j^A)(a_{ik}, b_{js})$. If:

(1) when IP_1 or IP_2 receives a token, OP_1 receives a token after $\mathrm{T}(0 < T < \infty)$;

(2) when OP_1 receives a token, IP_1 or IP_2 can receive another token.

Then the agent module E_A can be simplified to E_A' (Figure 9.15), where (1) ${}^\bullet\mathrm{OP}_1' = \{t_1^{A'}, t_2^{A'}\}$, $\mathrm{IP}_1'^\bullet = \{t_1^{A'}\}$, $\mathrm{IP}_2'^\bullet = \{t_2^{A'}\}$, $t_1^{A'\bullet} = \{\mathrm{OP}_1'\}$, $t_2^{A'\bullet} = \{\mathrm{OP}_1'\}$, ${}^\bullet t_1^{A'} = \{\mathrm{IP}_1'\}$, ${}^\bullet t_2^{A'} = \{\mathrm{IP}_2'\}$; (2) $C(\mathrm{OP}_1') = C(\mathrm{OP}_1)$,$C(\mathrm{IP}_1') = C(\mathrm{IP}_1)$, $C(\mathrm{IP}_2') = C(\mathrm{IP}_2)$,$C(t_1^{A'}) = C({}^\bullet\mathrm{OP}_1)+\{\sum_{td_i^A \in \emptyset_1} td_i^A\}$,$C(t_2^{A'}) = C({}^\bullet\mathrm{OP}_1)+\{\sum_{td_j^A \in \emptyset_2} td_j^A\}$, S_1 and S_2 are the sets of locations and transitions sequences in E_A when $\mathrm{IP}_1 \to \mathrm{OP}_1$ and $\mathrm{IP}_2 \to \mathrm{OP}_1$; (3) $I(\mathrm{IP}_1', t_1^{A'}) = \sum_{P_i, T_j \in \emptyset_1} I(P_i, T_j)$,$I(\mathrm{IP}_2', t_2^{A'}) = \sum_{P_i, T_j \in \emptyset_2} I(P_i, T_j)$,$O(\mathrm{OP}_1', t_1^{A'}) = \sum_{P_i, T_j \in \emptyset_1} O(P_i, T_j)$, $O(\mathrm{OP}_1', t_2^{A'}) = \sum_{P_i, T_j \in \emptyset_2} O(P_i, T_j)$.

Rule 4: For the agent module E_A shown in Figure 9.16, $E_A.P_{in} = \{ \mathrm{IP}_1 \}$, $E_A.P_{out} = \{ \mathrm{OP}_1 \}$. $\forall M_0$, $M(P_i^A)(a_{ik}) < I(P_i^A, T_j^A)(a_{ik}, b_{js})$. If:
(1) when IP_1 receives a token, OP_1 receives a token after $\mathrm{T}(0 < T < \infty)$;
(2) when OP_1 receives a token, IP_1 can receive another token.

Then the agent module E_A can be simplified to E_A'(Figure 9.16), where: (1) $\mathrm{IP}_1'^{\bullet} = {}^{\bullet}\mathrm{OP}_1' = \{ t^{A'} \}$, $t^{A'\bullet} = \{ \mathrm{OP}_1' \}$, ${}^{\bullet}t^{A'} = \{ \mathrm{IP}_1' \}$; (2) $C(\mathrm{OP}_1') = C(\mathrm{OP}_1)$, $C(\mathrm{IP}_1') = C(\mathrm{IP}_1)$, $C(t^{A'}) = C({}^{\bullet}\mathrm{OP}_1) + \left\{ \sum_{i=1}^{ts} w_i \bullet td_i^A \right\}$, ts is the quantity of delay transitions in E_A, w_i is the number of triggers of delay transitions td_i^A in E_A; (3) $I(\mathrm{IP}_1', t^{A'}) = \sum_{P_i \in P^A, T_i \in T^A} I(P_i, T_j)$, $O(\mathrm{OP}_1', t^{A'}) = \sum_{P_i \in P^A, T_i \in T^A} O(P_i, T_j)$.

Rule 5: For the agent module E_A shown in Figure 9.17, $E_A.P_{in} = \{ \mathrm{IP}_1 \}$, $E_A.P_{out} = \{ \mathrm{OP}_1 \}$。 $\forall M_0$, $M(P_i^A)(a_{ik}) < I(P_i^A, T_j^A)(a_{ik}, b_{js})$. If:
(1) when IP_1 receives a token, OP_1 receives a token after $\mathrm{T}(0 < T < \infty)$;
(2) when OP_1 receives a token, IP_1 can receive another token.

Then the agent module E_A can be simplified to E_A' (Figure 9.17), where (1) $\mathrm{IP}_1'^{\bullet} = {}^{\bullet}\mathrm{OP}_1' = \{ t^{A'} \}$, $t^{A'\bullet} = \{ \mathrm{OP}_1' \}$, ${}^{\bullet}t^{A'} = \{ \mathrm{IP}_1' \}$;(2)$C(\mathrm{OP}_1') = C(\mathrm{OP}_1)$, $C(\mathrm{IP}_1') = C(\mathrm{IP}_1)$, $C(t^{A'}) = C({}^{\bullet}\mathrm{OP}_1) + \{ t_{sum}^A \}$, $t_{sum}^A \in [\min \sum_{i=1}^{ts} td_i^A,\ \max \sum_{i=1}^{ts} td_i^A]$, ts is the number of delay transitions in E_A; (3) $I(\mathrm{IP}_1', t^{A'}) = \sum_{P_i \in P^A, T_i \in T^A} I(P_i, T_j)$, $O(\mathrm{OP}_1', t^{A'}) = \sum_{P_i \in P^A, T_i \in T^A} O(P_i, T_j)$。

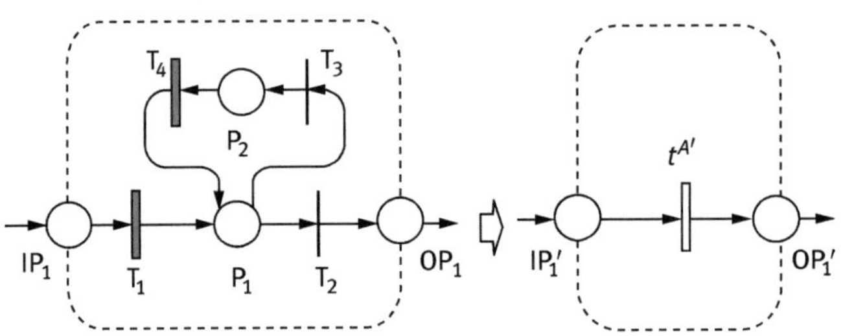

Figure 9.16: Model simplification rule 4.

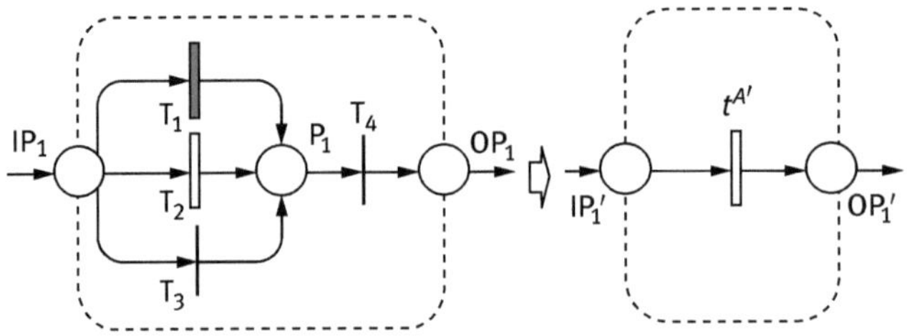

Figure 9.17: Model simplification rule 5.

9.3 Scheduling Performance Evaluation of an AMHS Based on the AOKCTPN

The evaluation indicators of the scheduling performance of an AMHS consist of two aspects: performance indicators of the material handling system and performance indicators of the processing system. The former one includes vehicle's transportation quantity, wafer lot's waiting time, vehicle's transport time, wafer lot's delivery time and vehicle utilization. The latter one includes wafer lot's cycle time, throughput, due date satisfactory rate and WIP quantity. Among them, wafer lot's waiting time, transport time, delivery time, cycle time and due date satisfactory rate can be determined by the statistical analysis of the transition time of an AOKCTPN model, and vehicle's transportation amount, utilization, throughput and WIP quantity can be determined by the trigger times of transient transitions and delayed transitions.

9.3.1 Transition Time

There are two kinds of transitions in an AOKCTPN model: transient transition and delayed transition. The transient transition is triggered immediately when the trigger conditions are satisfied, no time-consuming, such as the status of the wafer lot changes after all processing steps are completed, transitions of communication. The delayed transition takes some time to be triggered after the trigger conditions are satisfied, such as a vehicle loads/unloads a wafer lot. Therefore, an AOKCTPN should confirm all the time parameters of the delayed transition in order to obtain the statistics of the material handling system. For the normal timed PN, the delayed transition can be described by a definite time parameter. For example, in Figure 9.18 (a), transition Tj1 and transition Tj2 denote that wafer lots 1 and 2 are processed on

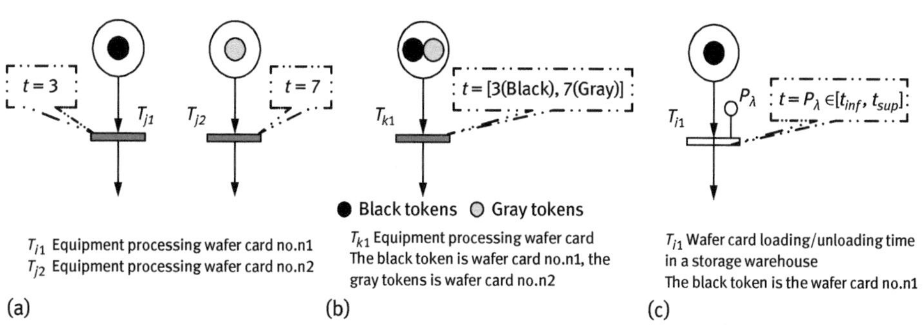

Figure 9.18: Determination of the delayed transition time parameters in an AOCKTPN: (a) definition of the transition time in a classical PN; (b) definition of the delayed transition time in an AOKCTPN; and (c) definition of the random transition time in an AOKCTPN.

machine 1, and the time parameters of Tj1 and Tj2 are the processing times of wafer lots 1 and 2, which are 3 and 7, respectively. For an AOKCTPN model, since it describes the processing procedure by using only a transition, the delay time is different when tokens representing different wafer lots trigger transitions, that is to say, the parameter of delayed transition is not unique. Thus, it is required to determine the appropriate time parameters according to taken colors of each resource in the model. As shown in Figure 9.18(b), transition T_{k1} denotes that machine 1, and wafer lots 1 and 2 are, respectively, represented by black and gray tokens, whose time parameters are defined as [3(black), 7(gray)]. For the existence and influence of various uncertainties in the AMHS, they are described by the random delayed transitions and the sending times of all the delayed transitions are random variables P_λ subject to a certain possibility distribution $P_\lambda \in [t_{\inf}, t_{\sup}]$, as shown in Figure 9.18 (c). For example, the random delayed transition is that wafer lot is loaded and unloaded in the stocker.

9.3.2 Scheduling Performance Evaluation Indicators

The performance indicators of an AMHS and a wafer processing system include vehicle movement M, wafer lot's waiting time WT, WIP (work in process), wafer lot's transport time TT, wafer lot's delivery time DT, vehicle utilization $\eta_{vehicle}$, throughput T, wafer lot's cycle time CT and wafer lot's due date satisfactory rate DS. And the AOKCTPN-based performance indicators are determined by using eqs (9.1–9.9).

$$M = \sum_{i=1}^{Rt} N_{u3}^{i} \tag{9.1}$$

$$WT = \left(\sum_{i=1}^{Mt} N_{m}^{i} t_{\text{stand}} + \sum_{j=1}^{Bt} N_{bw}^{j} t_{\text{stand}} \right) / M \tag{9.2}$$

$$WIP = \sum_{i=1}^{St} (Ns_{in}^{i} - Ns_{out}^{i}) \tag{9.3}$$

$$\begin{aligned} TT = \Bigg\{ & \sum_{i=1}^{Rt} [N_{l1}^{i} t_{Run3\_te} + N_{l3}^{i} t_{Load\_te} + (N_{l1}^{i} + N_{l3}^{i}) t_{Run4\_te} + N_{lw1}^{i} t_{LWait\_te}] \\ & + \sum_{i=1}^{Rt} [N_{u1}^{i} t_{Run1\_te} + N_{u3}^{i} t_{Load\_te} + (N_{u1}^{i} + N_{u3}^{i}) t_{Run2\_te} + N_{uw1}^{i} t_{UWait\_te}] \\ & + \sum_{j=1}^{St} (N_{s1}^{j} (t_{SRun1\_te} + t_{SWait\_te})) + \sum_{j=1}^{St} (N_{At1}^{j} t_{Turn1\_te} + N_{Bt1}^{j} t_{TBack\_te}) \Bigg\} / M \end{aligned} \tag{9.4}$$

$$DT = TT + WT \tag{9.5}$$

$$\eta_{vehicle} = (TT^*M)/\left\{ \sum_{i=1}^{Rt} [N_l^i(t_{Run3\_te} + t_{Run4\_te}) + N_{l3}^i t_{Load\_te} N_{lw}^i t_{LWait\_te}] + \right.$$
$$+ \sum_{i=1}^{Rt} [N_{u1}^i(t_{Run1\_te} + t_{Run2\_te}) + N_{u3}^i t_{Load\_te} + N_{uw}^i t_{UWait\_te}] + \sum_{j=1}^{St} (N_s^j \tag{9.6}$$
$$\left. (t_{SRun1\_te} + t_{SWait\_te})) + \sum_{j=1}^{St} (N_{At}^j t_{Turn1\_te} + N_{Bt}^j t_{TBack\_te}) \right\}$$

$$T = \sum_{i=1}^{St} Ns_{out}^i \tag{9.7}$$

$$CT = \sum_{i=1}^{T} (CT_{lot}^i)/T = \sum_{i=1}^{T} (ST_{OUT}^i - ST_{IN}^i)/T \tag{9.8}$$

$$DS = \sum_{i=1}^{T} sign(CT_{lot}^i)/T,\ sign(CT_{lot}^i) = \begin{cases} 1 & CT_{lot}^i \geq CT_{\text{stand}} \\ 0 & CT_{lot}^i < CT_{\text{stand}} \end{cases} \tag{9.9}$$

Where

N_{u3}^i– Trigger times of transition Load_te of the track agent, which is in the wafer lot load/unload port.

Rt– Quantity of track agents in the wafer lot unload port of an Intrabay/Interbay subsystem;

t_{stand}– Smallest time unit in an AOKCTPN model;

Mt– Quantity of processing machines in an Intrabay/Interbay subsystem;

Bt– Quantity of stockers in an Intrabay/Interbay subsystem;

N_m^i– Trigger times of delayed transition Wait_te of processing agent PM_ap_i;

N_{bw}^j– Trigger times of delayed transition BWait_te of stocker agent Stocker_ap_*j*;

St– Quantity of shortcut agents in an Intrabay/Interbay subsystem;

Ns_{out}^i– Trigger times of transient transition BEnter2_ti of stocker agent Stocker_ap_*i* in an Intrabay/Interbay subsystem;

Ns_{in}^i– Trigger times of transient transition BExit2_ti of stocker Agent Stocker_ap_*i*;

ST_{IN}^i– The global time that wafer lot *i* enters the system;

ST_{OUT}^i– The global time that wafer lot *i* leaves the system;

CT_{stand}– Setting value of wafer lot's average processing time;

CT_{lot}^i– Cycle time of wafer lot *i*.

9.3.3 Scheduling Performance Evaluation Method

By normalizing and weighting indicators: vehicle movement M, wafer lot's waiting time WT, WIP (work in process), wafer lot's transport time TT, wafer lot's delivery time DT, vehicle utilization $\eta_{vehicle}$, throughput T, wafer lot's cycle time CT, and wafer lot's due date satisfactory rate *DS*, the multi-objective function *Obj* of the AOKCTPN-based scheduling evaluation model of an AMHS has the following form:

$$Obj_{multi} = \left[\left(1 - \frac{M}{M_{max} \times t_{set}/t_{stand}}\right) + \frac{WIP}{WIP_{max}} + \frac{TT}{TT_{max}} + \frac{DT}{DT_{max}} + \eta_{vehicle} + \left(1 - \frac{T}{T_{max} \times t_{set}/t_{stand}}\right) + \frac{CT}{CT_{max}} + (1 - DS)\right]/8 \tag{9.10}$$

where

M_{max}– Within t_{stand}, maximum value of vehicle movement;
t_{set}– Predefined time of an AOKCTPN model;
WIP_{max}– Maximum time of WIP quantity in an AMHS;
TT_{max}– Maximum value of wafer lot's transport time;
DT_{max}– Maximum value of wafer lot's delivery time;
T_{max}– Within t_{stand}, maximum value of wafer lot throughput;
CT_{max}– Maximum value of wafer lot's cycle time.

According to eq. (9.10), the AOKCTPN-based scheduling evaluation rule is when the *Obj* becomes smaller, the scheduling method obtains better multi-objective optimization performance.

9.4 Case Study of Scheduling Performance Analysis of an AMHS

A case based on an AMHS of 12-inch semiconductor wafer fabrication line in Shanghai.is investigated to evaluate the effectiveness of the proposed AOKCTPN-based modeling approach. This AMHS consists of 22 Intrabays and one Interbay. The Intrabay and Interbay are connected with stocker. Three kinds of wafer lot product, named A, B and C, are processed in the semiconductor wafer fabrication line [8].

9.4.1 Verify the Feasibility of the AOKCTPN Model

During the feasibility analysis process, the operations of Etch and Cleaning for an Intrabay (including 19 machines and 1 stoker) are investigated as the example. The First Encountered First Served (FEFS) rule is used for the material handling process. Firstly, the Intrabay subsystem model is established on the basis of the AOKCTPN approach. The schedules and knowledge of the basic agents are simplified as the FEFS rule and the CTPN model for each basic agent is shown in Figure 9.19. Then the Intrabay subsystem model can be simplified as a single-machine and single-stoker model (Figure 9.20). At the same time, in order to reduce the quantity of status parameters, suppose that the time delay transition of the CTPN model is

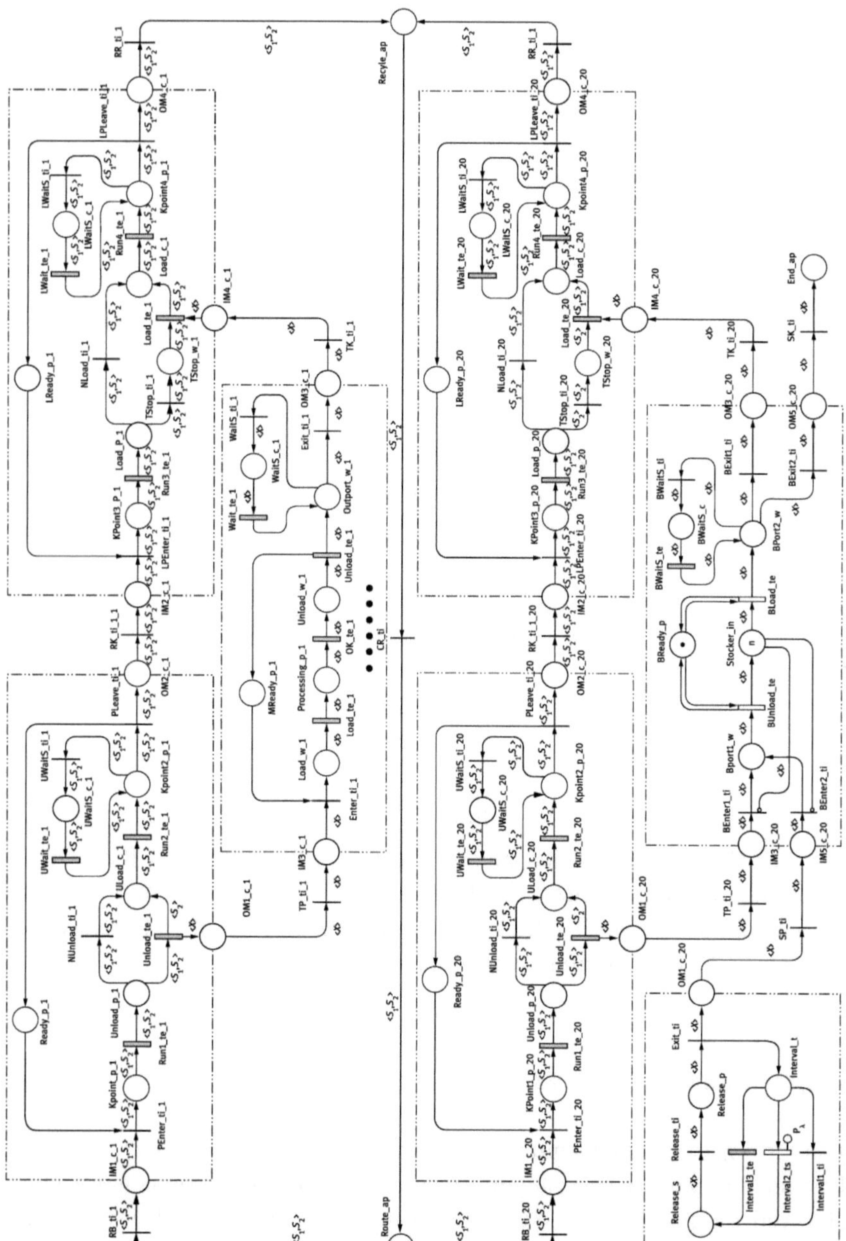

Figure 9.19: The CTPN-based Intrabay subsystem model.

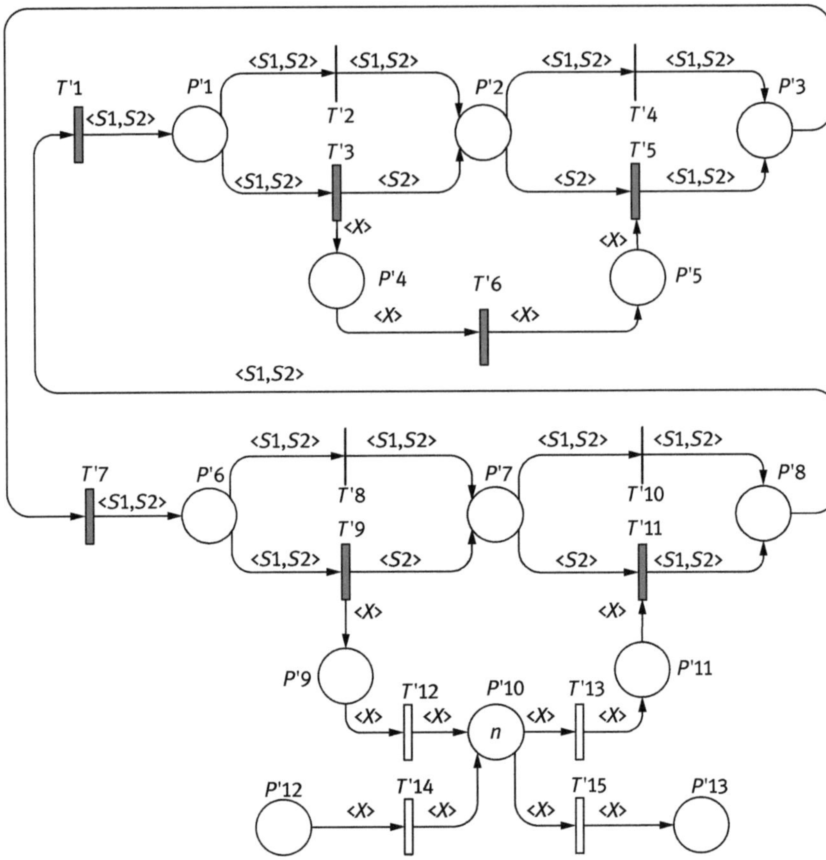

Figure 9.20: The simplified CTPN-based Intrabay subsystem model.

$|C(T_1')|=|C(T_3')|=|C(T_5')|=|C(T_7')|=|C(T_9')|=|C(T_{11}')|=|C(T_{12}')|=|C(T_{13}')|=|C(T_{14}')|=|C(T_{15}')|=|C(T_6')|/2$. Based on the simplified model, the reachability graph of the AOKCTPN model with single AGV is constructed (Figure 9.21). The feasibility of the proposed AOKCTPN modeling approach is thus verified. The verification of an Intrabay model with multiple AGVs and an Interbay model is the same.

9.4.2 Implementation of an AOKCTPN

Based on the object oriented simulation software (i.e., Em-Plant 7.0), the AOKCTPN-based model library of the basic agent layer and the subsystem layer is established in SimTalk. And the real-time scheduling process of an Interbay subsystem and an Intraby subsystem is implemented in C# language, and the information exchange

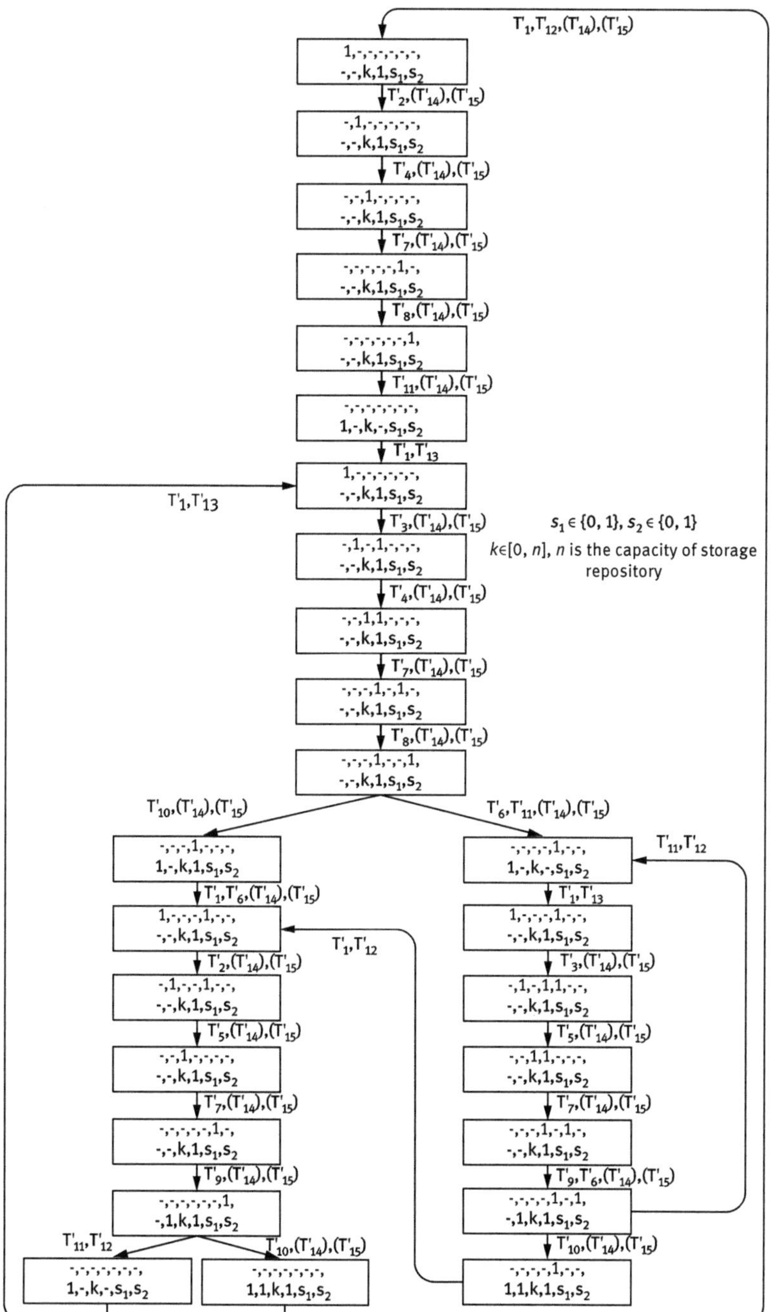

Figure 9.21: The reachable graph for simplified CTPN-based Intrabay subsystem model.

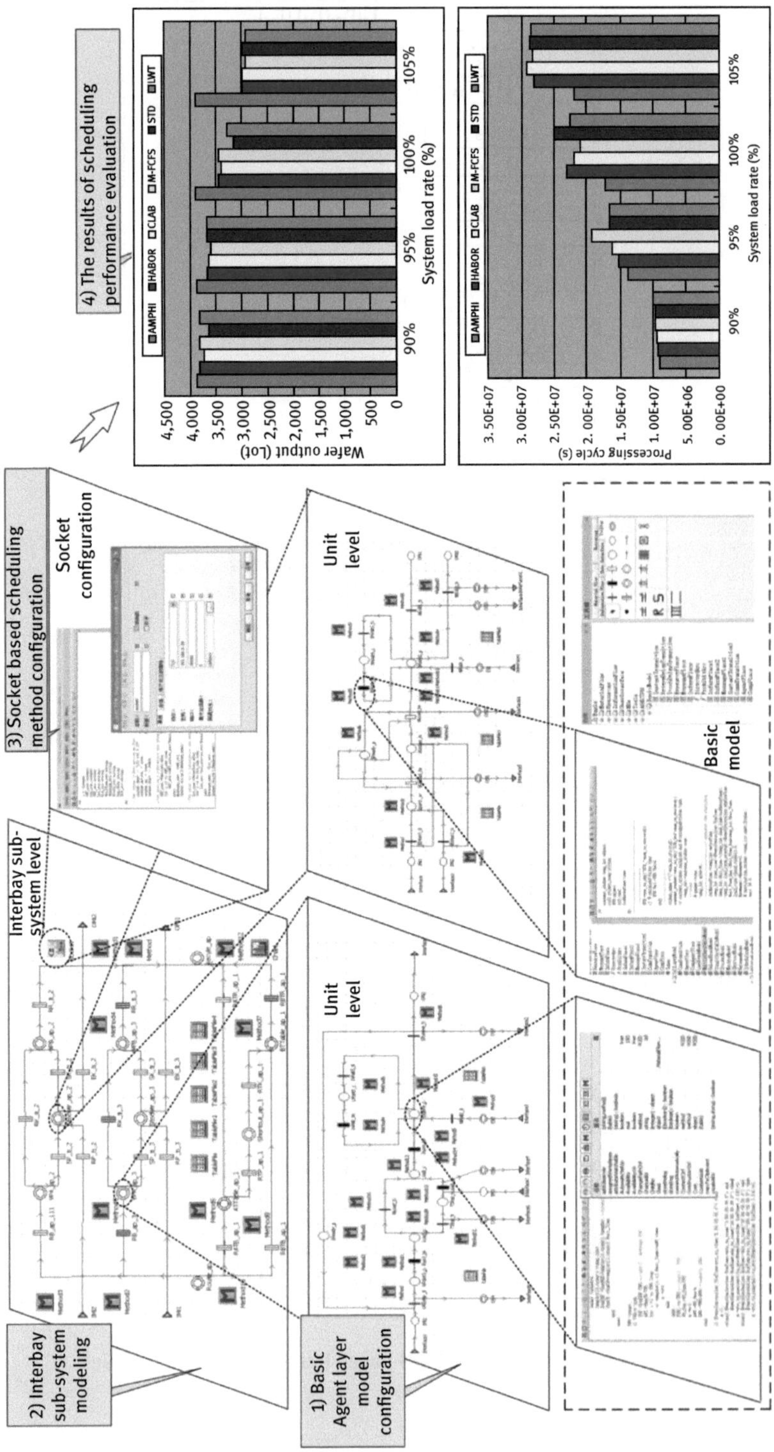

Figure 9.22: AOKCTPN-based scheduling performance evaluation procedures.

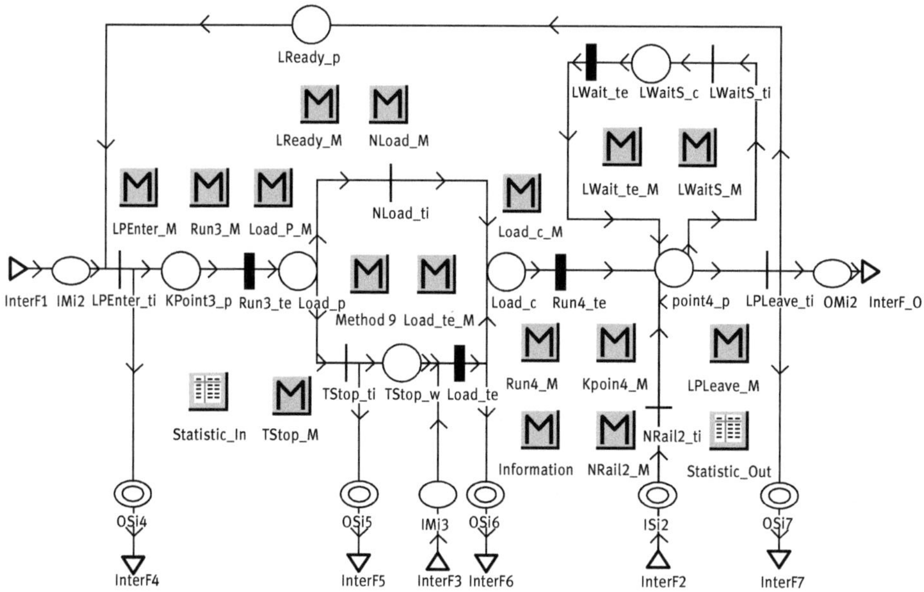

Figure 9.23: The AOKCTPN-based simulation model of transport rails.

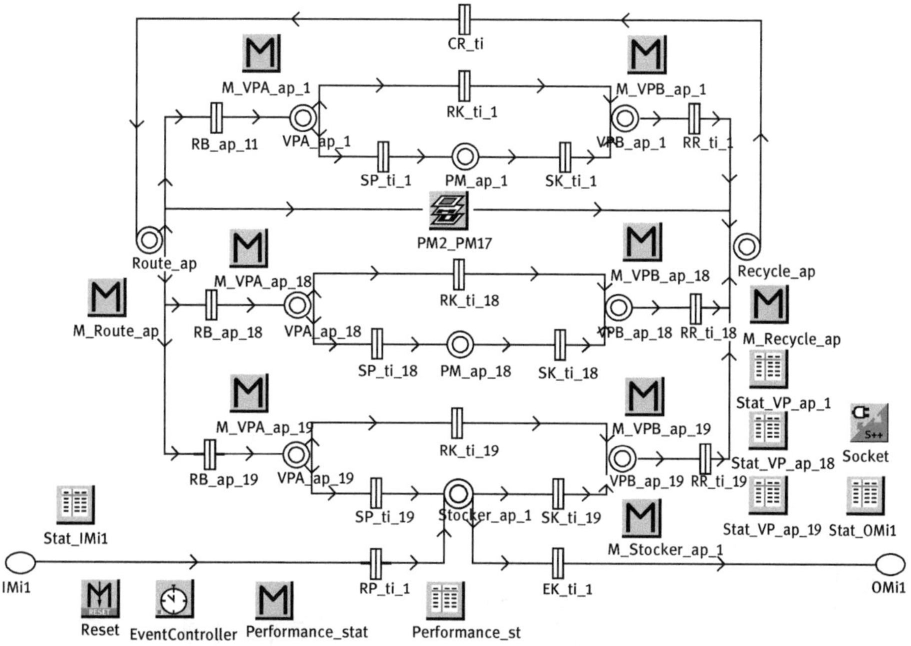

Figure 9.24: The AOKCTPN-based simulation model of an Intrabay subsystem.

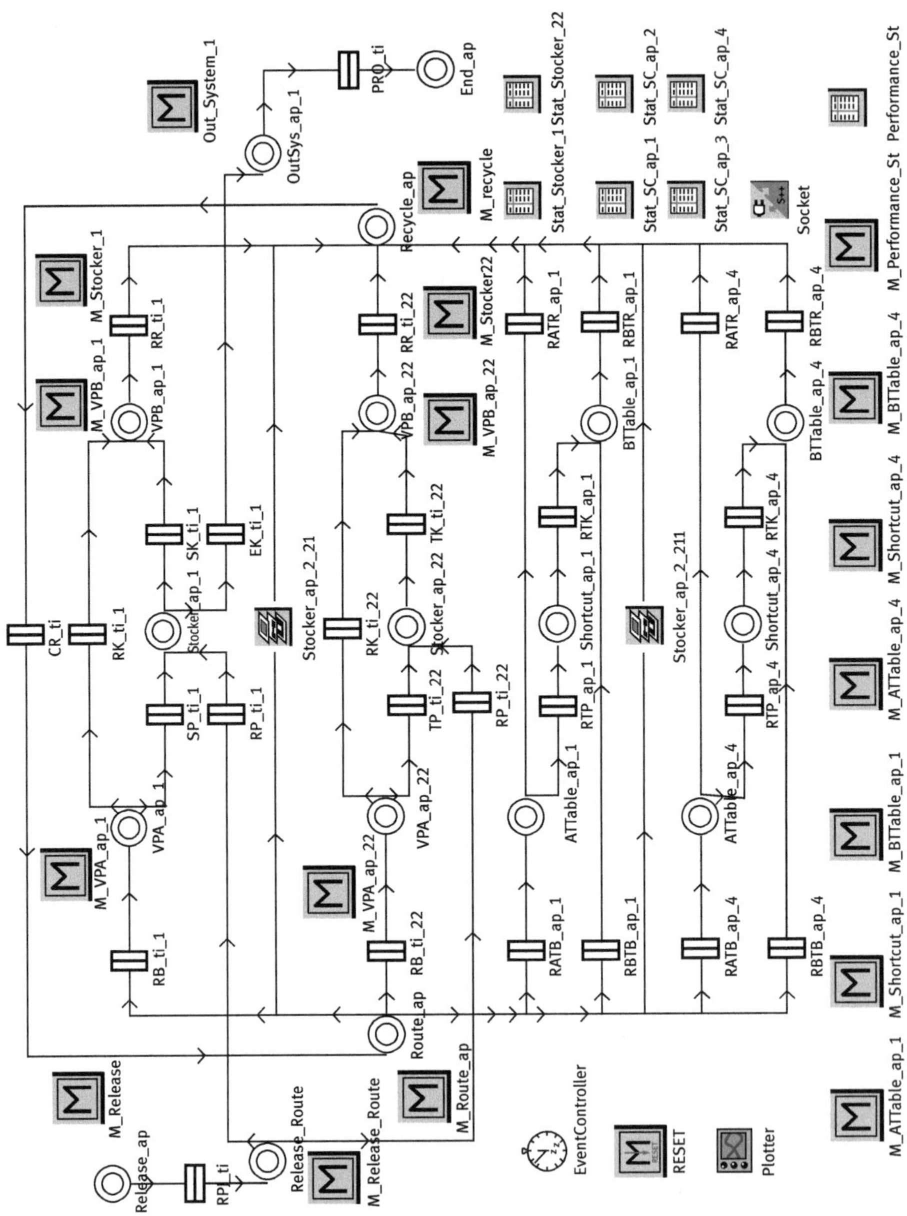

Figure 9.25: The AOKCTPN-based simulation model of an AMHS.

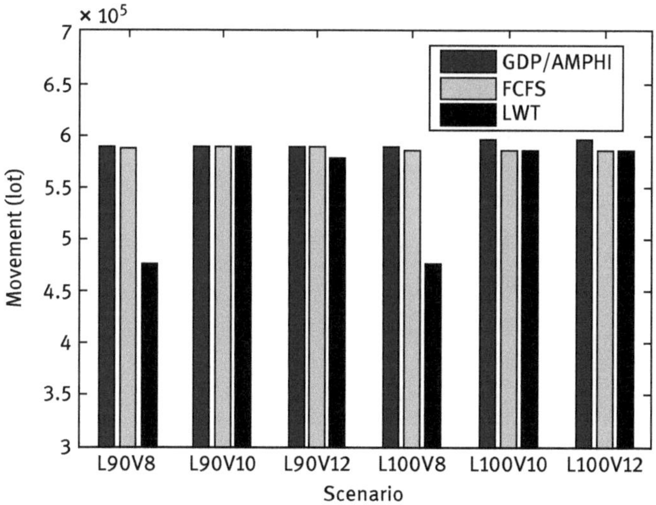

Figure 9.26: Comparison on movement.

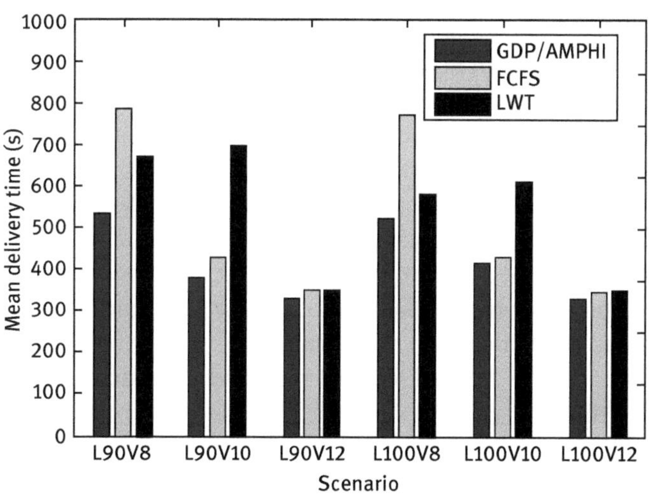

Figure 9.27: Comparison on mean delivery time.

between an Interbay subsystem and an Intraby subsystem and the AOKCTPN-based AMHS model is realized by socket objects.

The performance evaluation procedures of a scheduling method in AMHSs are shown in Figure 9.22 and the details are described as follows:

The simulation model of machines, transportation rails, shortcuts, turntables and stockers are developed based on the AOKCTPN model (e.g., the AOKCTPN-based simulation model of transport rails is indicated in Figure 9.23). On the basis of the

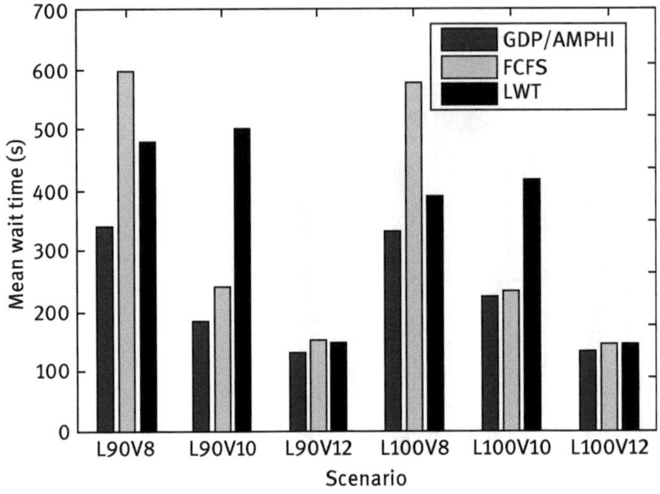

Figure 9.28: Comparison on mean wait time.

AOKCTPN-based simulation model, the simulation model of an Intrabay subsystem and an AMHS are set up, as illustrated in Figures 9.24 and 9.25, respectively.

The intelligent scheduling system adopts the GDP-based dynamic scheduling approach for Intrabay subsystems and the adaptive multi-parameter and hybrid intelligence (AMPHI)-based dynamic scheduling approach for Interbay subsystems, whose performance is evaluated by the proposed distributed simulation platform. The simulation model of AMHSs under these scheduling strategies is examined. The simulation result shows that the GDP/AMPHI-based dynamic scheduling approach exceeds the conventional scheduling rule of both FCFS and LWT in terms of mean movement, delivery time and wait time (as shown in Figures 9.26–9.28). It also demonstrates the effectiveness of the AOKCTPN-based modeling approach and the distributed simulation platform.

9.5 Conclusion

The AOKCTPN-based modeling approach of an AMHS has been proposed in this chapter. This approach improves the ability of describing the information and knowledge of scheduling and controlling processes of an AMHS. By using the proposed scheduling performance evaluation model based on the AOKCTPN, the performances of various scheduling methods could be evaluated effectively and efficiently. With the data of a 12-inch semiconductor wafer fabrication line in Shanghai, the reachability graph of the AOKCTPN model has been constructed and the feasibility of the proposed modeling approach has been verified.

References

[1] Zhang, H., Jiang, Z.B., Guo, C.T., An extended object-oriented Petri nets modeling based simulation platform for real-time scheduling of semiconductor wafer fabrication system. 2006 IEEE International Conference on Systems, Man, and Cybernetics, 2006, 1–6: 3411–3416.

[2] Liu, H.R., Fung, R., Jiang, Z.B., Modeling of semiconductor wafer fabrication systems by extended object-oriented Petri nets. International Journal of Production Research, 2005, 43(3): 471–495.

[3] Liu, H.R., Jiang, Z.B., Fung, R. Modeling of large-scale complex re-entrant manufacturing systems by extended object-oriented Petri nets. International Journal of Advanced Manufacturing Technology, 2005, 27(1-2): 190–204.

[4] Wang, Z.J. and Wu, Q.D., Object-Oriented Hybrid PN Model of Semiconductor Manufacturing Line [C]. Proceedings of the 4th World Congress on Intelligent Control and Automation, 2002, 1–4: 1354–1358.

[5] Zhou, B.H., Pan, Q.Z., Wang, S.J. Modeling of photolithography process in semiconductor wafer fabrication systems using extended hybrid Petri nets. Journal of Central South University of Technology, 2007, 14(3): 393–398.

[6] Zhang, J., Zhai, W.B., Yan, J. Multiagent-based modeling for re-entrant manufacturing system. International Journal of Production Research, 2007, 45(13): 3017–3036.

[7] Zhai, W.B., et al., Research on AOCTPN-based modeling and simulation technology of semiconductor fabrication line. International Journal of Advanced Manufacturing Technology, 2006. 28(7–8): 814–821.

[8] Wu, L.H, Mok, P.Y., Zhang, J. An adaptive multi-parameter based dispatching strategy for single-loop interbay material handling systems. Computers in Industry, 2011, 62(2): 175–186.

Index

https://doi.org/10.1515/9783110487473-010

Bei Fragen zur Produktsicherheit wenden Sie sich bitte an:
If you have any questions regarding product safety,
please contact:

Walter de Gruyter GmbH
Genthiner Straße 13
10785 Berlin
productsafety@degruyterbrill.com